Library and Information Services in Astronomy VIII Meeting (LISA 8)

Astronomy Librarianship in the Era of Big Data and Open Science

EPJ Web of Conferences Volume 186 (2018)

Strasbourg, France
6-9 June 2017

Editors:

S. Lesteven
R. D'Abrusco
B. Kern
B. Dorch

ISBN: 978-1-5108-6806-9

Printed from e-media with permission by:

Curran Associates, Inc.
57 Morehouse Lane
Red Hook, NY 12571

Some format issues inherent in the e-media version may also appear in this print version.

Printed by Curran Associates, Inc. (2018)

For additional information, please contact EDP Sciences – Web of Conferences at the address below.

EDP Sciences – Web of Conferences
17, Avenue du Hoggar
Parc d'Activité de Courtabœuf
BP 112
F-91944 Les Ulis Cedex A
France

Phone: +33 (0) 1 69 18 75 75
Fax: +33 (0) 1 69 28 84 91

contact-edps@webofconferences.org

Additional copies of this publication are available from:

Curran Associates, Inc.
57 Morehouse Lane
Red Hook, NY 12571 USA
Phone: 845-758-0400
Fax: 845-758-2633
Email: curran@proceedings.com
Web: www.proceedings.com

TABLE OF CONTENTS

Participants

A. Accomazzi, Astrophysics Data System (ADS), Smithsonian Astrophysical Observatory, 60 Garden Street, MS 67, Cambridge MA 02138, USA (aaccomazzi@cfa.harvard.edu)

L. N. Akerholt, University of Oslo Library, P.O. Box 1029 Blindern, N-0315 Oslo, Norway (l.n.akerholt@ub.uio.no)

M. Allen, CDS, Observatoire astronomique de Strasbourg, 11, rue de l'Université, 67000 Strasbourg, France (mark.allen@astro.unistra.fr)

N. Anilkumar, Physical Research Library, Navrangpura, Opposite Girls Polytechnic, Ahmedabad 380 009, India (nishtha@prl.res.in)

S. Bennett, P.O. Box 25026, Higher Colleges of Technology Central Services, UAE, Abu Dhabi, United Arab Emirates (sbennett@hct.ac.ae)

T. Boch, CDS, Observatoire astronomique de Strasbourg, 11, rue de l'Université, 67000 Strasbourg, France (thomas.boch@astro.unistra.fr)

D. Bordelon, European Southern Observatory (ESO) Library, Karl-Schwarzschild-Str. 2, 85748 Garching, Germany (dominic.bordelon@eso.org)

C. Bot, CDS, Observatoire astronomique de Strasbourg, 11, rue de l'Université, 67000 Strasbourg, France (caroline.bot@astro.unistra.fr)

R. Botha, Hart RAO, P.O. Box 443, 1740 Krugersdorp, South Africa (roelf@hartrao.ac.za)

D. Bouquin, Harvard-Smithsonian Center for Astrophysics, 60 Garden St. Cambridge, MA 02138, United States (daina.bouquin@cfa.harvard.edu)

M. Brouty, CDS, Observatoire astronomique de Strasbourg, 11, rue de l'Université, 67000 Strasbourg, France (marianne.brouty@astro.unistra.fr)

C. Brunet, CDS, Observatoire astronomique de Strasbourg, 11, rue de l'Université, 67000 Strasbourg, France (catherine.brunet@astro.unistra.fr)

F. Brunetti, INAF — Arcetri Astrophysical Observatory Library, Largo E. Fermi n. 5, 50125 Firenze, Italy (france@arcetri.astro.it)

M. C. Buga, CDS, Observatoire Astronomique de Strasbourg, 11, rue de l'Université 67000 Strasbourg, France (mihaela.buga@astro.unistra.fr)

M. Bunglow Madhava Rao, Raman Research Institute, Sir C V Raman Avenue, Sadashivnagar PO, 560080 Bengaluru, India (meera@rri.res.in)

L. Cambrésy, CDS, Observatoire astronomique de Strasbourg, 11, rue de l'Université, 67000 Strasbourg, France (laurent.cambresy@astro.unistra.fr)

T. Chen, California Institute of Technology, 1200 E California Blvd, 91125 Pasadena, United States (xchen@ipac.caltech.edu)

K. Clube, 40 Whitney Drive, SG1 4BJ Stevenage, United Kingdom (kc@ras.org.uk)

G. L. Coetzer, Hartebeesthoek Radio Astronomy Observatory, Library and Information Services, P.O. Box 443. Krugersdorp 1740, Gauteng, South Africa (glenda@hartrao.ac.za)

R. Cortés, Pontificia Universidad Católica de Chile, Avda. Vicuña Mackenna 4860, 7820436 Santiago, Chile (rcortee@uc.cl)

D. Crabtree, National Research Council Canada, 5071 West Saanich Road, V9E 2E7 Victoria, Canada (Dennis.Crabtree@nrc.ca)

R. D'Abrusco, Harvard–Smithsonian Center for Astrophysics, 60 Garden St. MS 67, Cambridge MA 02138, USA (rdabrusc@cfa.harvard.edu)

C. Dahdouh, INIST - CNRS, 2 allée du Parc de brabois, 54519 Vandoeuvre, France (Claude.DAHDOUH@inist.fr)

J. Damon, ADS, 60 Garden Street, MS-8, Cambridge MA 02138, USA (jdamon@cfa.harvard.edu)

F. David, IOP Publishing, "Temple Circus Temple Way", BS1 6HG Bristol, United Kingdom (davide.fiore@iop.org)

T. A. De Young, Dalene Fischer, South African Astronomical Observatory, P.O. Box 9 Observatory, 7935 Cape Town, South Africa (theresa@saao.ac.za)

T. Delacour, CDS, Observatoire astronomique de Strasbourg, 11, rue de l'Université, 67000 Strasbourg, France (thomas.delacour@astro.unistra.fr)

M. Demleitner, Astronomisches Rechen-Institut, Mönchhofstr. 12-14, 69221 Heidelberg, Germany (msdemlei@ari.uni-heidelberg.de)

D. Depoortere, Av. Vicuna Mackenna 4860 Macul, RM Santiago, Chile (ddepoort@uc.cl)

D. Egret, Laboratoire Univers et Théories (LUTH), Observatoire de Paris, Place Jules Janssen, 92190 Meudon, France (Daniel.Egret@obspm.fr)

A. Eisele, CDS, Observatoire Astronomique de Strasbourg, 11, rue de l'Université, 67000 Strasbourg, France (aline.eisele@astro.unistra.fr)

H. Enke, Leibniz Institute for Astrophysics Potsdam, An der Sternwarte 16, 14482 Potsdam, Germany (henke@aip.de)

L. Fletcher, University of Texas, 1616 Guadalupe, 78701 Austin, United States (l.fletcher@austin.utexas.edu)

K. Frey, Harvard–Smithsonian Center for Astrophysics, John G. Wolbach Library, 60 Garden Street, Cambridge, MA 02138, USA (kfrey@cfa.harvard.edu)

M. Gargano, INAF - Osservatorio Astronomico di Capodimonte, Salita Moiariello 16, 80131 Napoli, Italy (gargano@oacn.inaf.it)

A. Gasperini, INAF - Osservatorio Astrofisico di Arcetri (Library), Largo E. Fermi n. 5, 50125 Firenze, Italy (gasperi@arcetri.astro.it)

F. Genova, CDS, Observatoire astronomique de Strasbourg, 11, rue de l'Université, 67000 Strasbourg, France (francoise.genova@astro.unistra.fr)

G. Jacob, University of Michigan, United States (jkglenn@umich.edu)

M. Gomez, Instituto de Astrofísica de Canarias, C/ Vía Láctea, s/n, La Laguna, 38200 Tenerife, Spain (mog@iac.es)

U. Grothkopf, European Southern Observatory (ESO) Library, Karl-Schwarzschild-Str. 2, 85748 Garching, Germany (uta.grothkopf@eso.org)

S. Guéhenneux, CDS, Observatoire astronomique de Strasbourg, 11, rue de l'Université, 67000 Strasbourg, France (sylvain.guehenneux@astro.unistra.fr)

C. Halter, CDS, Observatoire astronomique de Strasbourg, 11, rue de l'Université, 67000 Strasbourg, France (celine.halter@astro.unistra.fr)

A. Henri, EDP Sciences, 17 avenue du Hoggar, PA de Courtaboeuf, BP 112, 91944 Les Ulis cedex A, France (agnes.henri@edpsciences.org)

A. Holl, Library and Information Centre, Hungarian Academy of Sciences, Arany J. str. 1., H-1051 Budapest, Hungary (holl.andras@konyvtar.mta.hu)

J. Holmquist, Princeton University, Lewis Science Library, Washington Road & Ivy Lane, Princeton University, Princeton, NJ 08544, USA (jane@princeton.edu)

S. Hunt, AURA Inc., 950 N. Cherry Ave., 85719 Tucson, United States (shunt@noao.edu)

C. Ingle, IOP Publishing,"Temple Circus Temple Way", BS1 6HG, Bristol, United Kingdom (chris.ingle@iop.org)

E. Isaksson, University of Helsinki, Helsinki University Library, Kumpula Campus Library, PO Box 64, 00014 University of Helsinki, Finland (Eva.Isaksson@helsinki.fi)

M. E. Jimnez-Fragozo, UNAM, Av. Univ. 3000, Col. UNAM-CU, Deleg. Coyoacán, 04510 Ciudad de México, Mexico (jimenez@astro.unam.mx)

V. Kaestle, CDS, Observatoire Astronomique de Strasbourg, 11 rue de l'Université, 67000 Strasbourg, France (vincent.kaestle@astro.unistra.fr)

P. Kamisato, Keck Observatory, 65-1120 Mamalahoa Hwy, P.O. Box 2279, Kamuela HI 96743, United States (pkamisato@keck.hawaii.edu)

A. Klinkenberg, Elsevier B.V., P.O. Box 839, 1000 AV, Amsterdam, Netherlands (f.klinkenberg@elsevier.com)

M. Kurtz, Harvard–Smithsonian Center for Astrophysics, 60 Garden Street, Cambridge MA 02138, United States (kurtz@cfa.harvard.edu)

G. Landais, CDS, Observatoire Astronomique de Strasbourg, 11, rue de l'Université, 67000 Strasbourg, France (gilles.landais@astro.unistra.fr)

A. Leary, Oxford University Press, Great Clarendon Street, OX2 6DP, Oxford, United Kingdom (adam.leary@oup.com)

S. Lesteven, CDS, Observatoire Astronomique de Strasbourg, 11, rue de l'Université, 67000 Strasbourg, France (soizick.lesteven@astro.unistra.fr)

P. Lopez, SCIENCE-MINER, Les Barbières, 19460 Naves, France (patrice.lopez@science-miner.com)

C. Loup, CDS, Observatoire Astronomique de Strasbourg, 11, rue de l'Université, 67000 Strasbourg, France (cecile.loup@astro.unistra.fr)

M. Louys, CDS, Observatoire Astronomique de Strasbourg, 11, rue de l'Université, 67000 Strasbourg, France (mireille.louys0astro.unistra.fr)

L. Malaver, Pontificia Universidad Católica de Chile, Biblioteca Gauss, 4860 Vicuña Mackenna Ave., Santiago, 8320000, Chile (lmalaver@uc.cl)

F. Martines, INAF — Osservatorio Astronomico di Palermo, Library, Piazza del Parlamento 1, 90134 Palermo, Italy (martines@astropa.inaf.it)

S. Meakins, European Southern Observatory (ESO) Library, Karl-Schwarzschild-Str. 2, 85748 Garching, Germany (smeakins@eso.org)

L. Michaud, Observatoire de Grenoble (OSUG), Domaine universitaire, bâtiment OSUG-D, 122, rue de la piscine, 38400 Saint Martin d'Hères, France (laurence.michaud@univ-grenoble-alpes.fr)

L. Michel, CDS, Observatoire Astronomique de Strasbourg, 11, rue de l'Université, 67000 Strasbourg, France (laurent.michel@astro.unistra.fr)

A. Muench, American Astronomical Society, United States (august.muench@aas.org)

M. Neuville, CDS, Observatoire Astronomique de Strasbourg, 11, rue de l'Université, 67000 Strasbourg, France (magali.neuville@astro.unistra.fr)

J. Novacescu, Space Telescope Science Institute, 3700 San Martin Drive, Baltimore MD 21218, United States (jnovacescu@stsci.edu)

A. Oberto, CDS, Observatoire Astronomique de Strasbourg, 11, rue de l'Université, 67000 Strasbourg, France (lanais.oberto@astro.unistra.fr)

P. Ocvirk, CDS, Observatoire Astronomique de Strasbourg, 11, rue de l'Université, 67000 Strasbourg, France (pierre.ocvirk@astro.unistra.fr)

R. Oderychev, Toiminimi Riikka O, Hiirakonkuja 8 A 5, 00780 Helsinki, Finland (riikka.oderychev@gmail.com)

E. Perret, CDS, Observatoire Astronomique de Strasbourg, 11, rue de l'Université, 67000 Strasbourg, France (emmanuelle.perret@astro.unistra.fr)

T. Pouvreau, CDS, Observatoire Astronomique de Strasbourg, 11, rue de l'Université, 67000 Strasbourg, France (thiphaine.pouvreau@astro.unistra.fr)

S. Prosser, Royal Astronomical Society, Library, Burlington House, Piccadilly, London W1J 0BQ, United Kingdom (sp@ras.org.uk)

V. K. Rai, Inter-University Centre for Astronomy and Astrophysics, Post Bag No., 4, Ganeshkhind, Pune 411 007, India (vkrai@iucaa.ernet.in)

A.K. Robertson, P.O. Box 61073, Honolulu HI 96839, United States (robertso@hawaii.edu)

M. Rode-Paunzen, Dr. Ignaz-Seipel-Platz 2, A-1010 Vienna, Austria (monika.rode-paunzen@oeaw.ac.at)

A. Rots, Harvard–Smithsonian Center for Astrophysics, Chandra X-ray Center, 60 Garden Street - MS 67, Cambridge MA 02138, USA (arots@cfa.harvard.edu)

F. Saconnet, Observatoire astronomique de Paris, 61 avenue de l'observatoire de Paris, 75014 Paris, France (frederic.saconnet@obspm.fr)

A. Schaaff, CDS, Observatoire Astronomique de Strasbourg, 11, rue de l'Université, 67000 Strasbourg, France (andre.schaaff@astro.unistra.fr)

L. Schiavone, INAF — Osservatorio Astrofisico di Torino, Biblioteca, Via Osservatorio 20, Pino Torinese, TO, 10025, Italy (schiavone@oato.inaf.it)

M. Schmitz, California Institute of Technology, 770 S. Wilson Avenue, Pasadena CA 91125, United States (mschmitz@ipac.caltech.edu)

G. Schwartz, American Astronomical Society, United States (greg.schwarz@aas.org)

A. Simoneau, EDP Sciences, 17 avenue du Hoggar, P.A de Courtaboeuf, BP 112, 91944 Les Ulis cedex A, France (anne.simoneau@edpsciences.org)

R. Snow, IOP Publishing,"Temple Circus Temple Way", BS1 6HG, Bristol, United Kingdom, (richard.snow@iop.org)

K. Soldanova, Astronomical Institute, ASCR, Library Fricova 298, 25165 Ondrejov, Czech Republic (katka@asu.cas.cz)

E. Son, CDS, Observatoire Astronomique de Strasbourg, 11, rue de l'Université, 67000 Strasbourg, France (evelyne.son@astro.unistra.fr)

G. Stahlman, School of Information, P.O. Box 210076, 1103 E Second Street 409, Tucson AR 85721, United States (gstahlman@email.arizona.ed)

J. Steffen, American Astronomical Society, 1667 K Street NW, Suite 800, Washington DC 20009, USA (julie.steffen@aas.org)

V. Stoll, Observatoire astronomique de Paris, 61 avenue de l'observatoire de Paris, 75014 Paris, France (veronique.stoll@obspm.fr)

J. Stoop, Elsevier, Radarweg 29, 1043NX Amsterdam, Netherlands (j.stoop@elsevier.com)

R. Svaskova, Astronomical Institute, Czech Academy of Sciences, Library Fricova 298, 25165, Ondrejov, Czech Republic (bibl@asu.cas.cz)

C. Takacs, CNRS LATMOS, 11, boulevard d'Alembert, 78280 Guyancourt, France (cecile.takacs@latmos.ipsl.fr)

M. Thies, Leibniz-Institut für Astrophysik Potsdam, Library, An der Sternwarte 16, 14482 Potsdam, Germany (mthies@aip.de)

D. Thompson, Harvard–Smithsonian Center for Astrophysics, ADS, 60 Garden Street, Cambridge MA 02138, USA (dthompson@cfa.harvard.edu)

A. Tisseau des Escotais, Observatoire astronomique de Paris, 61 avenue de l'observatoire de Paris, 75014 Paris, France (Aleth.Tisseau-des-Escotais@obspm.fr)

R. Van den Heuvel, Radboud Universiteit Nijmegen, Library, Postbus 671, Comeniuslaan 4, 6525 HP Nijmegen, Netherlands (rvandenheuvel@ubn.ru.nl)

C. Van Rooyen, EDP Sciences, 17 avenue du Hoggar, P.A de Courtaboeuf, BP 112, 91944 Les Ulis cedex A, France (charlotte.vanrooyen@edpsciences.org)

P. Vannier, CDS, Observatoire Astronomique de Strasbourg, 11, rue de l'Université, 67000 Strasbourg, France (patricia.vannier@astro.unistra.fr)

M. Veldre, Tartu Observatory, Toravere, Observatooriumi 1, Noo vald, Tartumaa", 61602 Tartumaa, Estonia (veldre@to.ee)

A. Wallis, SPIE, United States (andyw@spie.org)

S. Winkelman, Harvard–Smithsonian Center for Astrophysics, ADS, 60 Garden Street, Cambridge MA 02138, United States (swinkelman@cfa.harvard.edu)

F. Woelfel, CDS, Observatoire Astronomique de Strasbourg, 11, rue de l'Université, 67000 Strasbourg, France (fabienn.woelfel@astro.unistra.fr)

J. Zhang, Purple Mountain Observatory, Chinese Academy of Sciences, 2 West Beijing Rd., Nanjing 210008, The Peoples Republic of China (Jzhang@pmo.ac.cn)

Marion Schmitz and members of the LOC : Registration time

1. Peggi Kamisoto
2. Aline Eisele
3. Patricia Vannier
4. Soizick Lesteven
5. Alberto Accomazzi
6. Tracy Chen
7. Francesca Brunette
8. Francesca Martines
9. Luisa Schiavone
10. Maria Helena Jimenez
11. Vijay Kumar
12. Sharon Hunt
13. Gretchen Stahlman
14. Sherry Winkelman
15. Claude Dahdouh
16. Lydia Fletcher
17. Arnold Rots
18. August Muench
19. Jenny Novacescu
20. Katie Frey
21. Ruud Van Den Heuvel
22. Marcel Thies
23. Michael Kurtz
24. Françoise Genova
25. Fabienne Woelfel
26. Mihaela Buga
27. Magali Neuville
28. Adriaan Klinkenberg
29. José Stoop
30. Jian Zhang
31. Monique Gomez
32. Tiphaine Pouvreau
33. Uta Grothkopf
34. Daniel Egret
35. Celine Halter
36. Caroline Bot
37. Mark Allen
38. Raffaele D'abrusco
39. Andy Wallis
40. Andre Schaaff
41. Thomas Boch
42. Antonella Gasperini
43. Silvia Meakins
44. Marianne Brouty
45. Line Nybakk Akerholt
46. Daina Bouquin
47. Donna Thompson
48. Dennis Crabtree
49. Dominic Bordelon
50. Meera Bunglow Rao
51. Ann K. Robertson
52. Nishtha Anilkumar
53. Harry Enke
54. Denise Depcortere
55. Rodrigo Cortés
56. Lucina Malaver
57. A. Tisseau Des Escotais
58. Marion Schmitz
59. Markus Demleitner
60. Laurence Michaud
61. Samantha Bennett
62. Mauro Gargano
63. Mireille Louys
64. Radka Svašková
65. Anaïs Oberto
66. Katerina Soldanova
67. Greg Schwarz
68. Sylvain Guéhenneux
69. Chris Ingle
70. Eva Isaksson
71. Emmanuelle Perret
72. András Holl
73. Richard Snow
74. Catherine Brunet
75. Theresa De Young
76. Maret Veldre
77. Kim Clube
78. Gilles Landais
79. M. Rode-Paunzen
80. Véronique Stoll
81. Evelyne Son
82. Sian Prosser
83. Glenda Coetzer
84. Frédéric Saconnet
85. Jane Holmquist
86. Roelf Botha
87. James Damon

Members of the SOC: Eva Isaksson, Antonella Gasperini, Raffaele D'Abrusco, Glenda Coetzer, Line Akerholt, Marion Schmitz, Soizick Lesteven, Andràs Holl, Meera Bunglow (Photo: M. Wenger)

Scientific Organizing Committee (SOC)

Eva Isaksson (Finland) (co-chair)
Barbara Kern (USA) (co-chair)
Soizick Lesteven (France) (LOC chair)
Line Nybakk Akerholt (Norway)
Joao Alves (Austria)
Nishta Anilkumar (India)
Daina Bouquin (USA)
Glenda Coetzer (South Africa)
Raffaele D'Abrusco (USA)
Bertil Dorch (Denmark)
Antonella Gasperini (Italy)
Andras Holl (Hungary)
Arna Karick (Australia)
Jill Lagerstrom (USA)
Marion Schmitz (USA)
Lance Utley (USA)

Members of the LOC : Mark Allen, Soizick Lesteven, Gilles Landais, Evelyne Son, Laurent Cambrésy, Fabienne Woelfel, Céline Halter, Caroline Bot (Photo: M. Wenger)

Local Organizing Committee (LOC):

CDS - Centre de Données astronomiques de Strasbourg- France
Soizick Lesteven (CDS), chair
Caroline Bot (CDS)
Laurent Cambrésy (CDS)
Céline Halter (CDS)
Gilles Landais (CDS)
Evelyne Son (CDS)
Fabienne Woelfel (CDS)

Friends of LISA Committe (FOL)

Jenny Novacescu (USA)
Soizick Lesteven (CDS)

We would like to thank the FOL donors:
Corporate donors (ESO, SPIE, EDP Sciences, Strasbourg University)
Individual donations (Uta Grothkopf)

LISA VIII Sponsors

Gold Level

Astronomy & Astrophysics published by EDP Sciences

Silver Level

CDS - Centre de Données astronomique de Strasbourg

Bronze Level

American Astronomical Society
SPIE Digital Library
MNRAS published by Oxford University Press
IOP
ESO
Elsevier
Springer Nature
Strasbourg University - Université de Strasbourg
City of Strasbourg - Eurométropole
La région Grand Est

The Local Organizing Committee would also like to thank
The Congress Cell of the Strasbourg University,
Marc Wenger, our photographer,
Pascal Dubois and Pierre Ocvirk for the visits to the big dome of the Strasbourg Observatory,
Mihaela Buga for running tours every morning
All people of the Strasbourg Observatory.

LISA VIII Events: PAM Award Ceremony

The PAM Division Award is given for a "significant contribution to the literature of physics, mathematics or astronomy or to honor work that demonstrably improves the exchange of information in physics, mathematics or astronomy. The contribution should also significantly benefit libraries or enhance the ability of librarians to provide service." In 2014, the PAM Division Award was presented in honor of the Library and Information Services in Astronomy (LISA) Conference series. From the justification

- LISA conferences are notable for the close cooperation between astronomers and librarians; scientists have always been actively involved as organizers and participants; the LISA meetings have proved an excellent venue for librarians, astronomers and software/data specialists to network and explore trends.
- The published proceedings of the LISA conferences provide a historical overview of the development of electronic research and evolution and sharing of best practices in astronomy libraries
- Another outstanding aspect of the LISA conferences has been the Friends of LISA Committee (FOL). It was established before LISA II to assist astronomy librarians in resource-poor countries to attend the conference. FOL solicits funding from publishers, astronomical institutions, and individual librarians and astronomers. Grants have been awarded to between 18-29 individuals per meeting.

At LISA VII in Naples, Antonella Gasperini and Uta Grothkopf accepted the award on behalf of all organisers, planners, hosts, committee members, speakers, and attendees of LISA conferences. It was decided to make this award a "travelling" award that should be handed over to the current organisers at each new LISA conference. It was assumed that this would be a great way to remember that the PAM Division Award was given to LISA, and would provide an additional incentive to keep the LISA conference up and running. The Award is being passed from one LISA host to another, symbolizing the continued significance of the LISA conferences and the impact it has on astronomy and information services. Long live LISA!

Donna Thompson, Soizick Lesteven, Antonella Gasparini, Uta Grothkopf (Photo: M. Wenger)

LISA VIII Events: 30 years ADS - CDS - NED Collaborations

(Photos: M. Wenger)

Preface

The LISA 8 meeting took place on June 6-9, 2017 and was hosted by Observatoire astronomique de Strasbourg, in Strasbourg, France. The focus of LISA 8 was Astronomy Librarianship in the era of Big Data and Open Science. These topics are both timely and relevant to the entire astronomy community and the immediate future of the field. Over the course of the meeting, presentations were made and posters shared that represented new initiatives, progress in current projects, thoughts and ideas for the future, the creation, development and implementation of new services, support and approaches, and much more. LISA 8 was not only informative — it was engaging. It was an opportunity to learn and to network with professionals and colleagues from around the world.

With 105 participants representing more than 20 countries and almost fifty presentations and an enormous amount of information was shared through presentations and networking. Participants engaged in learning about data curation, the properties of data, data domains, diverse technologies, skills sets needed to work with data, repositories, bibliometrics, database interfaces, past and future projects, LISA and updates to critical resources and tools that we all use. Information regarding cataloging, digitization and preservation projects was shared, and the participants learned the latest news from publishers about research online, supporting authors and what the future may hold. And more! Very importantly, idea of LISA as a community was strengthened. At each LISA meeting, people gather together from all over the world to share ideas, knowledge, expertise and experiences. It takes a dedicated group of people to organize LISA and make the meeting a success.

This conference was an opportunity to celebrate the 30th anniversary of the working relationship between the 3 major data centers that are ADS, CDS and NED with the participation of the three leaders.

The Local Organizing Committee (LOC) was exceptional. They provided a beautiful venue, occasions for networking, and many opportunities to see beautiful Strasbourg. They welcomed participants warmly and enthusiastically. Our thanks to Soizick Lesteven (LOC Chair), Mark Allen (CDS Director), and the entire LOC for their hospitality. The Friends of LISA (FOL) made incredible efforts to seek and secure financial support for the LISA meeting. It is thanks to the FOL and our LISA sponsors and partners that many participants were able to attend the meeting.

The Scientific Organizing Committee (SOC), planned an informative and exciting program that provided participants with a wealth of information from colleagues around the world. The program had both breadth and depth. It stimulated discussions and opportunities for sharing more information even beyond the LISA meeting. Thanks to the SOC for a great program. It is very important to also acknowledge all of the individuals we agreed to Chair the sessions, all of the speakers and poster presenters, and last but not least the participants. Everyone contributed to the success of LISA 8! Finally, we wish to acknowledge the support of our many wonderful friends and partners who sponsored and supported the LISA conference. Finally, we wish to acknowledge the support of our sponsors, scholarly publishers, as well as the general support of the astronomical community, especially AAS and ESO. The formal presentations from LISA 8 are available through these Proceedings, and as such this informative meeting is available to all. We look forward to seeing you at LISA 9!

The Editors, January 2018

Our wonderful tour in the Alsace region

Visit of the "Château du Haut-Koenigsbourg"

(Photos: M. Brouty, M.E. Jiménezi and S. Lesteven)

edp sciences

Web of Conferences

Statement of Peer review

In submitting conference proceedings to *Web of Conferences*, the editors of the proceedings certify to the Publisher that

1. They adhere to its **Policy on Publishing Integrity** in order to safeguard good scientific practice in publishing.
2. All articles have been subjected to peer review administered by the proceedings editors.
3. Reviews have been conducted by expert referees, who have been requested to provide unbiased and constructive comments aimed, whenever possible, at improving the work.
4. Proceedings editors have taken all reasonable steps to ensure the quality of the materials they publish and their decision to accept or reject a paper for publication has been based only on the merits of the work and the relevance to the journal.

Title, date and place of the conference

LIBRARY AND INFORMATION SERVICES IN ASTRONOMY VIII.
"ASTRONOMY LIBRARIANSHIP IN THE ERA OF BIG DATA AND OPEN SCIENCE"
JUNE 6-9, 2017 STRASBOURG, FRANCE

Proceedings editor(s):

Soizick LESTEVEN
Barbara KERN
Raffaele D'ABRUSCO
Bertil DORCH

Date and editor's signature

12/06/2018

Data as a research infrastructure

CDS, the Virtual Observatory, astronomy, and beyond

Françoise Genova[1,*]

[1] *Université de Strasbourg, CNRS, Observatoire astronomique de Strasbourg, UMR 7550, F-67000 Strasbourg, France*

Abstract. The situation of data sharing in astronomy is positioned in the current general context of a political push towards, and rapid development of, scientific data sharing. Data is already one of the major infrastructures of astronomy, thanks to the data and service providers and to the International Virtual Observatory Alliance (IVOA). Other disciplines are moving on in the same direction. International organisations, in particular the Research Data Alliance (RDA), are developing building blocks and bridges to enable scientific data sharing across borders. The liaisons between RDA and astronomy, and RDA activities relevant to the librarian community, are discussed.

1 Introduction

Astronomy has been a pioneer of scientific data sharing. Historical services such as the database of the International Ultraviolet Explorer, a UV satellite which was operational between 1978 and 1996 [1], or SIMBAD [2], operated by the CDS since 1983 succeeding the Bibliographic Star Index (BSI) and Catalogue of Stellar Identifications (CSI), were already accessible remotely before the advent of the web. For many years astronomers have been using remote data services in their daily research work: networked on-line services have been developed since the beginnings of the internet, and further federated in the Virtual Observatory (VO).

Data sharing thus became a routine for astronomers, and an established part of the research landscape for the discipline. But astronomy is not alone: science data sharing is currently a hot topic at the political level and for many funding agencies. Many disciplines are working at setting up data services and a disciplinary interoperability framework. Organisations such as the Research Data Alliance (RDA) work on building blocks and bridges to facilitate data sharing. This paper describes several aspects of this fast evolving landscape, the political push towards Open Data (Section 2), the path towards and scientific impact of open data, taking astronomy as example (Section 3), examples of other disciplines for which data already is or is becoming a research infrastructure (Section 4), and the role of the Research Data Alliance (Section 5). The conclusion (Section 6) summarizes the fast evolution of the context and how astronomy should fit in, and the evolving role of librarians in data curation.

* e-mail: francoise.genova@astro.unistra.fr ORCID iD 0000-0002-6318-5028

2 The Open Data context

Scientific data sharing has been a hot topic in the recent period, including on the political agenda. The G8 Science Ministers made strong statements on the subject in June 2013[1], including:

- "To the greatest extent and with the fewest constraints possible publicly funded scientific research data should be open"
- "Open scientific data should be easily discoverable, accessible, assessable, intelligible, useable, and wherever possible interoperable to specific quality standards"

They made additional statements in October 2015 [2], following a report of the Group of Senior Officials on Global Research Infrastructures (GSO), stating in particular that "further progress on sharing and managing scientific data and information should be achieved, especially by continuing engagement with community based activities such as the Research Data Alliance." Their Open Science statement of May 2016 [3] proposes to establish a working group on open science, and to promote international coordination and collaboration to develop the necessary elements.

Among the additional recent advances relevant to Open Data, one can cite the development of the FAIR Guiding Principles for scientific data management [3]. FAIR means Findable, Accessible, Interoperable, Reusable, and although this naming is recent and makes a buzz, the concepts have for a very long time been guiding principles for astronomy, where they are implemented and fully in use as discussed in Section 3.

It is important to understand that scientific data sharing is not only a political subject and not only a way to enable the private sector to take the full benefit of scientific data, which can be the dominant objective of some of the Open Science promoters. A good description of the expected impact was given by the G8 Ministers of Science in their June 2013 statement: "We are committed to openness in scientific research data to speed up the progress of scientific discovery, create innovation, ensure that the results of scientific research are as widely available as practical, enable transparency in science and engage the public in the scientific process."

Astronomy was a pioneer and can be taken as a case study of Open Science. In astronomy data is open, and moreover it is used by the scientific community in its daily research work. The way astronomers work now demonstrates that scientific data sharing first of all enables a change in paradigm in the way science is done.

The layman view of astronomical infrastructures is, of course, our space and ground-based observatories. Data sharing is a powerful way to optimize the scientific return of these large infrastructures, as demonstrated by the well-known example of the bibliometric impact of the HST archive: more than half of the papers published each year from HST observations are using archival data[4]. In addition, data sharing is indeed at the core of astronomy scientific needs, allowing multi-wavelength/multi-messenger/multi-technique astronomy, the study of time variability and of phenomena at different scales, the comparison of theoretical models with observations, etc. The fact that the community succeeded in setting up world-wide sharing of FAIR data, makes the distributed astronomical data one of the essential infrastructures of the discipline, as recognized for instance in the European Roadmap of astronomical infrastructures developed by ASTRONET[5] [4]. As explained in Section 3, the observatory archives are not the only essential building blocks of this data infrastructure. For instance, the added-value services developed by the CDS, which are only one element of the landscape, receive more than 800,000 queries/day in average.

---

[1] https://www.gov.uk/government/news/g8-science-ministers-statement
[2] http://www.g8.utoronto.ca/science/2015-berlin.html
[3] http://www.g8.utoronto.ca/science/2016-tsukuba.html
[4] https://archive.stsci.edu/hst/bibliography/pubstat.html
[5] ASTRONET gathers the astronomy funding agencies operating in Europe

3 The astronomical data infrastructure

3.1 The basic elements of the astronomical data infrastructure

The early definition and long term maintenance of a common data format, FITS [5], which includes both the data and relevant metadata, has certainly been one of the key elements to enable data sharing in astronomy. The discipline strong tradition of international collaboration, in particular to build telescopes and instruments, has also been a fertile ground for setting up the disciplinary data sharing framework beyond borders, together with the fact that observational data is in general open. Observations often become public after a proprietary period, which reserves data obtained from competitive Calls for Proposals to the original proposers for a limited period of time, an important factor to make data sharing acceptable by the community. Finally, the fact that the development of observatory archives and data services is driven by community needs has been essential to ensure widespread adoption of remote data usage as a routine of astronomers' scientific work.

The astronomical data infrastructure has very diverse components, among which of course is the observations: the archives from the ground and space based telescopes and the large sky surveys, which contain homogeneous observations of up to billions of objects. It also includes data from publications, value-added databases which gather homogenized information in particular from publications or observations, such as SIMBAD, VizieR [6] and the Aladin image database [7] at CDS, NED [8], and the ADS [9] for bibliographic data, and more and more modelling results.

SIMBAD, for instance, contains information gathered from catalogues and publications, including all the names of a given object. It is used together with NED and VizieR, by observatory archives to transform object names provided by their users to their own user interface into coordinates, which are their main access key. On the other hand, the references in which the object was cited provided by SIMBAD are used, together with NED, by the ADS to provide the references citing an object. An additional added value is the capacity to sort the references attached to an object by relevance [10]. It is important to note the huge amount of work behind the scenes, and the key role of the CDS "documentalists" [11] [12].

3.2 Networking and Interoperability - The Astronomical Virtual Observatory

This short description of SIMBAD already demonstrates that astronomical on-line services are networked. This networking begun as soon as the on-line resources began to be developed, in the early stages of the internet in 1993-1994. It was implemented through collaboration between the providers of added-value services, journals and observation archives, taking advantage of the capacity of the web to link web pages. It is still available and used, and is well illustrated for instance by the multiple links to distant resources available in the ADS.

Around 2000, the idea to go beyond this already powerful networking of resources by providing seamless access to on-line data emerged as the Virtual Observatory concept. The Virtual Observatory is a framework of standards and tools which allows users to discover, access and use data. The interoperability standards are defined by the International Virtual Observatory Alliance [6].

The IVOA was created in 2002 as an alliance of national VO initiatives (it also includes the European VO initiative Euro-VO and the international organization ESA, (the European Space Agency). Its standardisation procedures are inspired from the World-Wide-Web Consortium (W3C). When possible, it uses generic elements, such as OAI-PMH (the reference for digital libraries), for its registry of resources [13], and W3C SKOS/RDF for its vocabularies [14]. The VO standards deal with different

[6]http://ivoa.net

aspects of the FAIR principles, in particular Findability, Accessibility and Interoperability. For instance, the IVOA registry of resources includes disciplinary extensions which allow users to discover resources of interest for them beyond the Dublin-Core-type criteria used in the library world.

Like astronomy (compared for instance to particle physics), the Virtual Observatory has no central point; it stages a multipolar world in a global endeavour. It is an open and inclusive model, a thin interoperability layer on top of distributed and heterogeneous data holdings. Anyone can build a VO-enabled tool or register a data service. There are currently more than 100 "authorities" with a registered service, including large organisations as well as small research teams. The VO usage by data providers has been evolving beyond enabling interoperable access to their data, and more and more of them now embed VO building blocks in their archives and services.

3.3 Current status of the Virtual Observatory

The VO is currently operational and used. Its successful development and implementation has three pillars, as recognized from the early phase of the European VO coordination [15]: support to science users in their uptake of the VO, support to data providers in their uptake of the VO, and the technological work to develop and update the VO standards and tools. The VO has several kinds of users: scientists using VO-enabled data services and tools, data providers implementing the VO framework, and developers of VO-enabled tools. The VO is, most of the time, invisible from its science users - the price to pay when one really aims to provide seamless access. One key for success amongst science users has been the high quality of data and services available in the VO. This is thanks to the work of the data and service providers. Another one is seamless access to data through interoperable tools such as Aladin or Topcat, which are well suited to science needs thanks to the work of the VO developers. The VO was adopted by the data and service providers because it gives more visibility to their data by making them discoverable, and also because it provides access tools and building blocks for their systems: there is no need to reinvent the wheel, the VO developers already worked and pooled resources and expertise to propose relevant and handy solutions.

The current priorities of the IVOA are linked to the needs of the future large astronomical projects: multi-dimensional data, with a first milestone reached in May 2017 with a complete initial set of relevant standards; and time domain. The current European VO project is closely aligned with these priorities. The European Commission has been funding a series of projects since 2001 to support the coordination of VO activities in Europe [16]. The current one is a Work Package of the *Astronomy ESFRI*[7] *and Research Infrastructure Cluster* ASTERICS[8], called *Data Access, Discovery and Interoperability* (DADI), which is foreseen for 4.5 million euros over 4 years starting in May 2015 (the whole budget of ASTERICS is 15 million euros). DADI aims at "making the ESFRI and pathfinder project data available for discovery and usage by the whole astronomical community, interoperable in a homogeneous international framework (the Virtual Observatory), and accessible by a set of common tools." DADI and ASTERICS involve both astronomy and astroparticle physics telescopes, which is particularly important with the rapid development of multi-messenger astronomy. DADI includes the Euro-VO partners, i.e. VO initiatives from France, Germany, Italy, Spain and UK, and representatives of the large projects Cherenkov Telescope Array (CTA), the European Gravitational Observatory (EGO)/Einstein Telescope (ET), the Cubic Kilometre Neutrino Telescope (KM3Net) and the Square Kilometre array (SKA) and their pathfinders. The European Southern Observatory (ESO), which is in charge of the Extremely Large Telescope (ELT), is associated to the project, which is also working in close collaboration with ESA.

[7]ESFRI http://www.esfri.eu/ is the European Strategic Forum for Research Infrastructures, which develops a roadmap for pan-European Research Infrastructures

[8]https://www.asterics2020.eu/ , 2015-2019

3.4 Big and smaller data in astronomy

Another trendy word in the data world is "Big Data." Astronomy can be taken as an example of "Big Data" science, because the ground and space-based telescopes and the large surveys produce large data volumes. But the discipline also takes care of its "smaller data", the so-called Long Tail of data [17], in particular the research results which are linked to publications. This started in 1993, with the agreement between the CDS and *Astronomy & Astrophysics* that the "long" tables in the papers would be published on line in an ftp server by the CDS [18]. The agreement was later extended to other journals. These data come with a standard description, which has been the basis to ingest them in the VizieR system when it was implemented. VizieR can store any kind of astronomical quantity which can be found in a publication. One can note here that variety is also one of the characteristics of Big Data. The standard description and the UCD controlled vocabulary for astronomical quantities [19] provide a homogeneous frame for the very heterogeneous database content. In August 2017 VizieR contains more than 16,000 "catalogues", which are actually data sets which can include tabular as well as non-tabular data such as images, spectra, time series, etc. attached to a publication, or object lists and other data from large surveys [20].

VizieR approach to the Long Tail of Data is thus to preserve data validated by a publication, and to make it fully discoverable and interoperable in the Virtual Observatory, usable and citable through ADS (and soon through a DOI). "VizieR Photometry Viewer" [21] is an example of value-added service made possible by the homogeneous data description and the rich information stored about the data [22]: it extracts all the spectral points available in VizieR in a given direction on the sky, and displays them on a single plot taking into account the photometric system used in the original observation. Each point is linked to the original reference.

In astronomy there is thus not really "big" and "little" data, but rather Useful, Validated and Documented data which are fully FAIR.

4 Data as an infrastructure: examples from different disciplines

The notion and management of Research Infrastructures can be different from country to country. This is a domain where European coordination made a difference and enabled the development of a regional strategy, which includes global infrastructures such as SKA. The European landscape of Research Infrastructures is structured by the ESFRI Roadmap, which covers all scientific disciplines and is regularly updated. The ESFRI Research Infrastructures are infrastructures of pan-European interest defined as "facilities, resources or services of a unique nature identified by European research communities to conduct top-level research activities in all fields". The first Roadmap was published in 2006, the last update [23] in 2016, and the next version is foreseen for 2018. More and more European countries also develop a National Roadmap, taking into account the ESFRI Roadmap as well as their national priorities.

All research infrastructures have aspects linked to data, and ESFRI Roadmap candidates have to answer questions about their data management and policy, and their computing and network needs. This is also the case for instance for the French National Roadmap candidates. Some of the infrastructures on the Roadmaps are fully "virtual", for instance the CDS is in the French Roadmap.

The way different disciplines develop their data infrastructure is strongly dependent from the disciplinary culture and organisation. It is interesting to observe that the Humanities have been present in the ESFRI Roadmap from the beginning, with two projects, DARIAH and CLARIN, which are distributed data infrastructures: DARIAH[9] supports digital research in the arts and humanities in gen-

[9]http://www.dariah.eu/

eral, CLARIN[10] is focused on digital language resources. In France, the national Humanities research infrastructure Huma-Num[11] provides added-value services in support of the data life cycle and in addition manages disciplinary consortia set up for 4 years (with possible renewal) to organise communities around data sharing topics relevant to their needs. This is very useful for disciplines which are more loosely organised than astronomy. It also coordinates French participation in DARIAH and CLARIN. The national organisation in Earth Sciences is different: there are 4 disciplinary "Poles": Solid Earth, Ocean, Atmosphere, Continental Surfaces and Interfaces, in which all the relevant research organisations, including the French Space Agency CNES, participate. "Inter-Poles" Technical discussions and collaboration on topics of common interest are organised. An overarching structure is being built, which is a candidate for the national Roadmap. Each Pole has strong links with the ESFRI and international projects in its domains.

The way different disciplines organise themselves to tackle the development of their disciplinary interoperability framework was discussed during a session [24] of the SciDataCon 2016 Conference[12], which was held in Denver (USA) on 2016, September 11th – 13th. The panel included representatives from humanities, linguistics, astronomy, earth sciences, material sciences and crystallography. Many commonalities were found:

- the development of the data infrastructure must be science driven;
- defining the disciplinary part of the interoperability standards is mandatory but difficult;
- it is important not only to share data, but also tools, user interfaces, etc;
- sociological aspects are more challenging than the technical ones.

Once again, incentives to data sharing appeared to be a key issue for all. Governance is more diverse, linked, as explained earlier, to the discipline organisation and history. It was also found that many sociological and technological aspects can be shared, and it was suggested that the RDA is a good vehicle for that, with roles also for CODATA[13] and WDS[14] of the International Council for Science ICSU.

As explained in [24], international organizations such as GEO, CODATA and the RDA can be used to help disciplines which are not strongly organised at the international level to discuss their data framework. International discussion forums can be established as CODATA Task Groups and/or RDA Interest and Working Groups (Section 5). In addition, the WDS "community of excellence" is a host for individual data providers as regular members, but also for disciplinary networks once they are constituted, and it organizes interaction between its members. The IVOA is a Network member of the WDS, and several astronomical data centres, including the CDS, are regular members.

5 The Research Data Alliance

The Research Data Alliance[15] is a community-driven international organisation created in 2013 by the Australian government, the European Commission and the USA NSF, with the mission to build the social and technical bridges to enable open sharing of data. In August 2017, the RDA has more than 5900 members from 129 countries, 66% from academia and research, 14% from public administration, 11% from enterprise and industries. It hosts focused Working Groups and Interest Groups

[10]https://www.clarin.eu/
[11]http://www.huma-num.fr/about-us
[12]http://www.scidatacon.org/2016/
[13]Committee on Data for Science and Technology, http://www.codata.org/
[14]World Data System, http://www.icsu-wds.org/
[15]https://www.rd-alliance.org/

tackling many different aspects of data interoperability challenges, in themes such as domain science, community needs, data referencing and sharing, data stewardship and services, or base infrastructure. Any member can propose a Group, and in August 2017 there are 88 active Groups, 30 Working Groups which prepare implementable deliverables in 18 months, and 58 Interest Groups which organise discussion and exchange of good practices on a specific topic and can also produce outputs. The activities of the Working Groups and Interest Groups, including their email exchanges and the documents they produce, are documented on the RDA web site, as well as the recommendations and outputs, and adoption cases. RDA members have a large variety of profiles, which makes it a remarkable international, neutral forum gathering very diverse expertise. They include data providers, researchers, project and program managers, and publishers. Librarians make more than 10% of membership.

There is not Astronomy Group in the RDA, because it would duplicate the IVOA, but astronomy is present, and the experience gained in building the IVOA and the astronomical data infrastructure is used in the RDA context. For instance, the author of this paper is currently co-chair of the RDA Technical Advisory Board. It is also interesting to note that the RDA Group working on an International Materials Resource Registry will reuse the concepts of the IVOA Registry of Resources. The IVOA Registry of Resources is already customized to their needs by planetary sciences and the Virtual Atomic and Molecular Data Centre VAMDC. The RDA allows wider dissemination. Data experts from the astronomical community participate in several other RDA Groups, for instance those dealing with Data Provenance. Astronomy is also involved in the *Disciplinary Collaboration Framework Interest Group* recently proposed as an inter-disciplinary forum for discussion, which aims at becoming the voice of scientific disciplines in the RDA.

Many RDA activities are of interest for the LISA community. In particular, one of the RDA recommendations is a catalogue of requirements [25] and a set of procedures [26] for core audit and certification of data repositories. It was prepared by the *Repository Audit and Certification DSA–WDS Partnership Working Group*, which gathered members from the Data Seal of Approval DSA[16] and the World Data System. These two organisations had been managing two different basic certification processes for data repositories, which were merged and improved by the Working Group. This is a very important clarification and simplification of the landscape, at a time when it is more and more important for data providers to demonstrate that there are trustworthy through certification. The CDS for instance had successively and successfully applied for WDS and then DSA certification [22] because it was not easy to choose one. Now it is looking for renewal of its certification in the unified system.

Another remarkable aspect of the RDA is the importance given to data publishing and persistent identifiers. For instance, a Working Group made widely adopted recommendations for dynamic data citation (when data evolves in time keep track of the queries in a "query store"). Several Working Groups involving publishers and data providers, managed in collaboration between the RDA and the WDS, have also been tackling different aspects of data publishing. This includes for instance Publishing Data Workflows and Bibliometrics, and currently the *Scholarly Link Exchange Working Group* (Scholix) which aims to enable a comprehensive global view of the links between scholarly literature and data.

The exchange of good practices and lessons learnt in the *Interest Group on the Long Tail of Research Data* and the one on Domain Repositories can also be of interest for the LISA community. Finally, the *Libraries for Research Data Interest Group* produced a very successful output, *23 Things: Libraries for Research Data* [27], which provides an overview of practical, free, online resources and

[16]http://www.datasealofapproval.org

tools that one can begin using today to incorporate research data management into one's practice of librarianship. The document is available in 11 languages[17].

6 Conclusions

These are exciting times for scientific data sharing! Astronomy has been at the forefront and, thanks to the efforts of many organisations and individuals, data is now one of the disciplinary research infrastructures. This includes many different distributed on-line resources, the IVOA being a rare example of an operational, global interoperability framework. Even if many astronomers do not realize it, the Virtual Observatory enables scientists to routinely Find, Access and Interoperate data and tools.

The context of scientific data sharing is evolving very fast, and many disciplines are moving on. It is important for the astronomy community to be involved in initiatives such as the RDA, to ensure that our requirements are taken into account in the emerging data sharing building blocks. Trust is an essential component of the landscape. It mainly relies on the quality of data and of the accompanying documentation, which requires lots of underlying work at the disciplinary level by people who know the data. The astronomy data services have been building trustworthiness with their user community for a long time (45 years for the CDS). In the new context, it can be worth applying for an external certification. Even if the process is not completed up to submission of an application for review the catalogue of criteria is a very good basis for in-house assessment of the repository processes.

The evolving role of librarians has been an important topic of the LISA conferences for a while. Specialized librarians, the so-called "documentalists", have been one of the essential profiles at CDS since the beginning, one of its three pillars with researchers and IT engineers [28]. There are several levels of data curation, including disciplinary repositories, non-specialist repositories such as Zenodo, Figshare or the institutional ones set up by many Universities or research organisations. Going beyond Dublin-Core-type data description requires disciplinary knowledge, and discipline-savvy librarians have a key role to play in the new research landscape, now and in the future.

Acknowledgements

The author acknowledges support from the Research Data Alliance - Europe 3 and ASTERICS Projects, funded by the European Commission (projects 65194 and 65477 respectively).

References

[1] W. Wamsteker, R.E.M. Griffin, Astrophysics and Space Science **228**, 385 (1995)

[2] M. Wenger, F. Ochsenbein, D. Egret, P. Dubois, F. Bonnarel, S. Borde, F. Genova, G. Jasniewicz, S. Laloë, S. Lesteven et al., Astronomy and Astrophysics Supplement **143**, 9 (2000)

[3] M.D. Wilkinson, M. Dumontier, I.J. Aalbersberg, G. Appleton, M. Axton, A. Braak, N. Bomberg, J.W. Boiten, L. Bonino da Silva Santos, P.E. Bourne et al., Sci. Data **3**, 160018 (2016)

[4] M.F. Bode, M.J. Cruz, F.J. Molster, *The ASTRONET Infrastructure Roadmap: A Strategic Plan for European Astronomy* (2008), ISBN 978-3-923524-63-1

[5] D.C. Wells, E.W. Greisen, R.H. Harten, Astronomy and Astrophysics Supplement **44**, 363 (1981)

[17] https://www.rd-alliance.org/group/libraries-research-data-ig/outcomes/23-things-libraries-research-data-supporting-output

[6] F. Ochsenbein, P. Bauer, J. Marcout, Astronomy and Astrophysics Supplement **143**, 23 (2000)

[7] M. Buga, P. Fernique, C. Bot, M.G. Allen, F. Bonnarel, M. Brouty, *Bibliography, catalogs, pixel data: management of heterogeneous Big Data at CDS by the documentalists*, in *These Proceedings* (2018)

[8] M. Schmitz, *Evolution of NASA/IPAC Extragalactic Database (NED) into a data mining discovery engine*, in *These Proceedings* (2018)

[9] A. Accomazzi, the ADS Team, *New ADS functionality for the curator*, in *These Proceedings* (2018)

[10] A. Oberto, S. Lesteven, S. Derriere, C. Bonnin, M. Buga, M. Brouty, C. Bruneau, C. Brunet, A. Eisele, F. Genova et al., *A New Bibliographical Feature for SIMBAD: Highlighting the Most Relevant Papers for One Astronomical Object*, in *Open Science at the Frontiers of Librarianship*, edited by A. Holl, S. Lesteven, D. Dietrich, A. Gasperini (2015), Vol. 492 of *Astronomical Society of the Pacific Conference Series*, p. 284

[11] M. Buga, C. Bot, M. Brouty, C. Bruneau, C. Brunet, L. Cambresy, A. Eisele, F. Genova, S. Lesteven, C. Loup et al., *How Documentalists Update SIMBAD*, in *Open Science at the Frontiers of Librarianship*, edited by A. Holl, S. Lesteven, D. Dietrich, A. Gasperini (2015), Vol. 492 of *Astronomical Society of the Pacific Conference Series*, p. 47

[12] C. Brunet, A. Oberto, M. Brouty, M. Buga, C. Loup, E. Perret, B. Vollmer, F. Woelfel, *COSIM: the necessary evolution of a cross-identification tool along with data evolution*, in *These Proceedings* (2018)

[13] R. Hanisch, IVOA Resource Registry Working Group, NVO Metadata Working Group, *Resource Metadata for the Virtual Observatory Version 1.12*, IVOA Recommendation 02 March 2007

[14] A.J.G. Gray, N. Gray, F.V. Hessman, A. Preite Martinez, S. Derriere, T. Linde, R. Seaman, B. Thomas, *Vocabularies in the Virtual Observatory Version 1.19*, IVOA Recommendation 07 October 2009, 1110.0520

[15] P. Padovani, *The AVO to EURO-VO Transition*, in *Astronomical Data Analysis Software and Systems XV*, edited by C. Gabriel, C. Arviset, D. Ponz, S. Enrique (2006), Vol. 351 of *Astronomical Society of the Pacific Conference Series*, p. 771

[16] F. Genova, M.G. Allen, C. Arviset, A. Lawrence, F. Pasian, E. Solano, J. Wambsganss, Astronomy and Computing **11**, 181 (2015)

[17] C.L. Borgman, *Big Data, Little Data, No Data - Scholarship in the networked world* (The MIT Press, 2015), ISBN 9780262028561

[18] F. Ochsenbein, J. Lequeux, Vistas in Astronomy **39**, 227 (1995)

[19] A. Preite Martinez, S. Derriere, N. Delmotte, N. Gray, R. Mann, J. McDowell, T. Mc Glynn, F. Ochsenbein, P. Osuna, G. Rixon et al., *The UCD1+ controlled vocabulary Version 1.23*, IVOA Recommendation 02 April 2007

[20] P. Ocvirk, G. Landais, *Associated data in VizieR*, in *These Proceedings* (2018)

[21] M.G. Allen, F. Ochsenbein, S. Derriere, T. Boch, P. Fernique, G. Landais, *Extracting Photometry Measurements from VizieR Catalogs*, in *Astronomical Data Analysis Software and Systems XXIII*, edited by N. Manset, P. Forshay (2014), Vol. 485 of *Astronomical Society of the Pacific Conference Series*, p. 219

[22] G. Landais, F. Genova, J.Y. Hangouet, the VizieR Team, *VizieR catalogue system certified by the Data Seal of Approval*, in *These Proceedings* (2018)

[23] ESFRI, *Strategy Report on Research Infrastructures - Roadmap 2016* (2016), ISBN 978-0-9574402-4-1

[24] F. Genova, C. Arviset, B.M. Almas, L. Bartolo, D. Broeder, E. Law, B. McMahon, Data Science Journal **16**, 16 (2018)

[25] L. Rickards, M. Vardigan, M. Diepenbroek, I. Dillo, F. Genova, H. L'hours, W. Hugo, V. Khodiyar, G. Li, J.B. Minster et al., *DSA–WDS Partnership Working Group Catalogue of Common Requirements*, RDA (2016)

[26] L. Rickards, M. Vardigan, M. Diepenbroek, I. Dillo, F. Genova, H. L'hours, W. Hugo, V. Khodiyar, G. Li, J.B. Minster et al., *DSA–WDS Partnership Working Group Catalogue of Common Procedures for Certification*, RDA (2016)

[27] M. Witt, RDA Libraries for Research Data Interest Group, *23 Things: Libraries for Research Data*, RDA (2017)

[28] E. Perret, T. Boch, F. Bonnarel, C. Bot, M. Buga, M. Brouty, C. Bruneau, C. Brunet, L. Cambrésy, S. Derrière et al., *Working Together at CDS: The Symbiosis Between Astronomers, Documentalists, and IT Specialists*, in *Open Science at the Frontiers of Librarianship*, edited by A. Holl, S. Lesteven, D. Dietrich, A. Gasperini (2015), Vol. 492 of *Astronomical Society of the Pacific Conference Series*, p. 13

 https://doi.org/10.1051/epjconf/201818602001

Bibliography, catalogs, pixel data: Management of heterogeneous Big Data at CDS by the documentalists

Buga M.[1,*], *Fernique* P.[1,**], *Bot* C.[1], *Allen* M. G.[1], *Bonnarel* F.[1], and *Brouty* M.[1]

[1] *Université de Strasbourg, CNRS, Observatoire astronomique de Strasbourg, UMR 7550, F-67000 Strasbourg, France*

Abstract. High speed Internet and the evolution of data storage space in terms of cost-effectiveness has changed the way data are managed today. Large amounts of heterogeneous data can now be visualized easily and rapidly using interactive applications such as "Google Maps". In this respect, the Hierarchical Progressive Survey (HiPS) method has been developed by the Centre de Données astronomiques de Strasbourg (CDS) since 2009. HiPS uses the hierarchical sky tessellation called HEALPix to describe and organize images, data cubes or source catalogs. These HiPS can be accessed and visualized using applications such as Aladin.

We show that structuring the data using HiPS enables easy and quick access to large and complex sets of astronomical data. As with bibliographic and catalog data, full documentation and comprehensive metadata are absolutely required for pertinent usage of these data. Hence the role of documentalists in the process of producing HiPS is essential. We present the interaction between documentalists and other specialists who are all part of the CDS team and support this process. More precisely, we describe the tools used by the documentalists to generate HiPS or to update the Virtual Observatory standardized descriptive information (the "metadata"). We also present the challenges faced by the documentalists processing such heterogeneous data on the scales of megabytes up to petabytes.

On one hand, documentalists at CDS manage small size textual or numerical data for one or few astronomical objects. On the other hand, they process large data sets such as big catalogs containing heterogeneous data like spectra, images or data cubes, for millions of astronomical objects. Finally, by participating in the development of an interactive visualization of images or three-dimensional data cubes using the HiPS method, documentalists contribute to a long-term management of complex, large astronomical data.

1 Introduction

1.1 The context of Big Data

Strasbourg astronomical Data Center (CDS[1]) collects and distributes worldwide astronomical data and related information. Let me go back to the popular sense of the word "astronomical": extremely

* e-mail: Mihaela.Buga@astro.unistra.fr ORCID: 0000-0002-0845-7705
** e-mail: Pierre.Fernique@astro.unistra.fr
[1] http://cds.unistra.fr

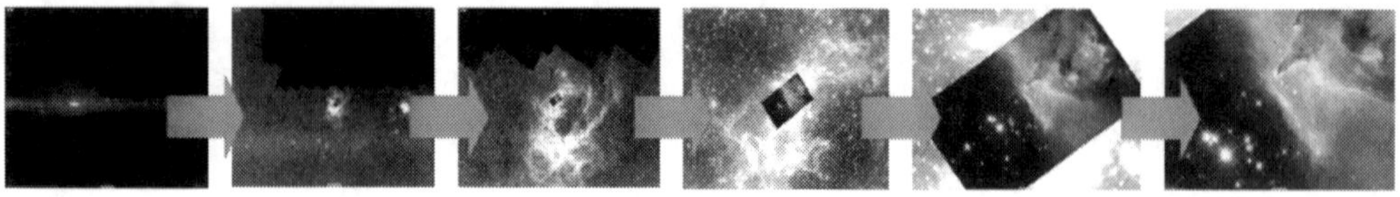

Figure 1. The more you zoom in on a particular area the more details show up

large, enormous. Hence, managing astronomical data has always been a matter of Big Data. But how big? Present or future missions bring astronomy into the era of petabyte surveys, with large volumes of high quality images and catalog data. The quantity of data is continuously growing. To point this out, we take as example the future LSST project[2]. Due to the high speed with which the telescope will map the sky, but also to the depth it can see, LSST will produce about 15 TB of data per night, with a total amount of collected data of 60 PB. Processing this data will produce a 15 PB catalog database. But already today the Gaia[3] satellite, launched in 2013, is observing one billion stars several times in order to get their brightness, proper motions, distances and positions over time. CDS documentalists, astronomers and an IT specialists team made the Gaia Data Release 1 (DR1) available for the scientific community[4]; it can be accessed in various ways using the Gaia@CDS[5] page: either by using the catalog service VizieR Web interface, Cone Search or TAP tools or by cross-identifying the Gaia sources with other catalogs using the X-Match service. Additionally, today, Gaia identifiers are present for almost 1.8 million of SIMBAD[6] sources.

Another way of accessing interactively large data is via Aladin[7]: pixel or catalog data structured using the Hierarchical Progressive Surveys (HiPS) [1] method can be displayed by panning or zooming just like in google map, but looking at the sky and using sky survey data. HiPS is an IVOA[8] standard (since May 2017) for packaging, storing, querying and describing astronomical data. It has been developed at CDS since 2009 and it has enjoyed a big success: various data centers such as IRAP[9], IAS[10], SSC-XMM[11], CADC[12], JAXA[13] or ESAC[14] use this method to provide hundreds of terabytes of data representing hundreds of HiPS. Fig. 1 shows an Aladin view of a large data set structured in HiPS, accessed at finer and finer resolution by zooming on a wide area until reaching a detailed view of the survey. Restructuring images into HiPS involves a resampling of the original pixels and a preservation of the associated meta data, as described in section 2 and 3.

This paper describes how documentalists structure original pixel data into HiPS. Section 2 outlines the mechanism behind the HiPS method. Section 3 shows the main steps documentalists follow to organize data using the HiPS method. In section 4 we discuss challenges faced by the documentalists and we provide our conclusions.

[2]https://www.lsst.org/about

[3]http://icc.ub.edu/research/key_projects/gaia

[4]check Boch T. October 2016 IVOA "Interop" Meeting presentation for more details about the Virtual Observatory access to Gaia DR1: http://wiki.ivoa.net/internal/IVOA/InteropOct2016Apps/Apps2-TBoch-VO-access-GaiaDR1.pdf

[5]http://cdsweb.u-strasbg.fr/gaia

[6]http://simbad.u-strasbg.fr/simbad/

[7]http://aladin.u-strasbg.fr/aladin.gml

[8]http://www.ivoa.net/

[9]https://ov-gso.irap.omp.eu/doku.php

[10]https://idoc.ias.u-psud.fr/ipod

[11]http://xmmssc.irap.omp.eu/

[12]http://www.cadc-ccda.hia-iha.nrc-cnrc.gc.ca/en/about.html

[13]http://global.jaxa.jp//

[14]http://www.esa.int/About_Us/ESAC

1.2 Documentalists at CDS

Documentalists, or in other words, 'information scientists', manage content coming from electronic publications of scientific literature, dedicated web sites, internal or external servers. Additionally, sometimes documentalists receive data directly from the authors. Next step after getting the data is processing it using various internal tools developed by the IT specialists, such as DJIN [3], COSIM[4] or HiPSgen[15]. The astronomers' scientific expertise complements documentalists' knowledge in order to provide pertinent and useful information to the scientific community. Namely, documentalists update the following CDS services:

- *the SIMBAD database* [6]: with cross-identifications, basic data, measurements and bibliography for individual astronomical objects outside the solar system;
- *the VizieR catalog service* [7]: with astronomical sources , published tables, observation logs, spectra, light curves, polarization data, models , statistical analyses, compilations, etc grouped in a collection of astronomical catalogs with associated data [5];
- *the interactive sky atlas Aladin* [2]: with images and data cubes structured in HiPS, accessed and visualized interactively[16].

Therefore, in addition to the burst of the data volume in astronomy, documentalists at CDS also face data complexity. Indeed, CDS hosts numerical or text string data, related to a single SIMBAD object or organized in VizieR tables and catalogs; associated data such as spectra, time series, etc; images or data cubes in original format or structured into IVOA standards such as HiPS and MOC (Multi-Order coverage Map)[17]. Since 2016, documentalists joined the IT specialists and astronomers team handling big pixel data. Documentalists structure images and data cubes into HiPS, allowing Aladin or other specialized sky browsing tools developed by data centers such as JUDO[18], ESASky[19], MIZAR[20] to give a progressive view of surveys covering a part of, or the entire sky, at various spatial scales. Data reorganization into HiPS uses a hierarchical multi-resolution division of the sky called HEALPix, that we will present next.

2 HiPS way of structuring astronomical data

2.1 HEALPix tessellation technique

Tessellation is used by architects for brickwork, by artists for decoration and bees use this technique to build their honeycombs. In astronomy we use tessellation to map data on the sky. Let us go into a little more detail. Different hierarchical multi resolution divisions of the sky are used in astronomy: WWT[21] uses a triangular partition of the sphere and Google[22] a cylindrical one (second and third upper panel in Fig. 2). The CDS Hierarchical Progressive Survey method uses a curvilinear partitioning called HEALPix (Hierarchical Equal Area isoLatitude Pixelization)[8] (the lower panel in Fig. 2). The HEALPix way of dividing the sphere has a wide usage in astronomy and it has been chosen by CDS as a balance between performance and quality. HEALPix has a base resolution used to divide the sphere into 12 equal cells, called the order 0 HEALPix map (the most left frame from the lower panel

[15]http://aladin.u-strasbg.fr/hips/HipsIn10Steps.gml

[16]Big sources catalogs can also be structured into HiPS for a progressive view in Aladin (subsets of the catalog are displayed in function of the angular resolution), but this is done by other members of the team, not by documentalists

[17]http://ivoa.net/documents/MOC/

[18]https://darts.isas.jaxa.jp/astro/judo2/about.html

[19]http://sky.esa.int/

[20]http://sitools2.github.io/MIZAR/

[21]http://www.worldwidetelescope.org/

[22]https://www.google.fr/intl/fr/sky/

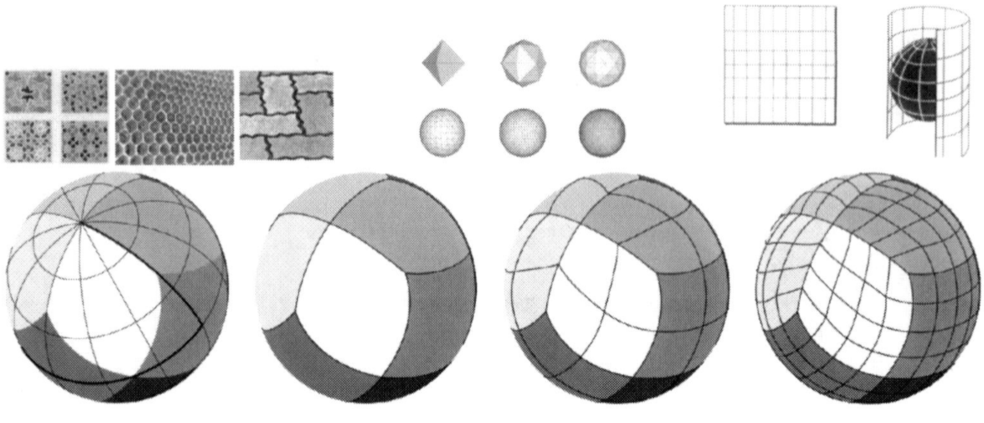

Figure 2. Tessellation usage. The first most left upper panel shows examples of tessellation in art, nature and architecture. The lower panel shows the type of tessellation called HEALPix. The second and third upper panels show other types of hierarchical multi-resolution tessellation of the sky used in astronomy, namely triangular and cylindrical division of the sphere.

in Fig. 2). Increasing resolution is reached by sub-dividing each cell into 4 equal cells, recursively, until reaching the maximum survey resolution (order 1 HEALPix map . . . max order HEALPix map). The lower panel in Fig. 2 shows the hierarchical structure of HEALPix : each pixel is divided into four similar pixels at each successive order.

2.2 HiPS structure: maps, tiles and pixels

Documentalists use the HiPS generator tool called HiPSgen to structure hierarchically large volumes of images or data cubes into HiPS. The associated astrometrical calibration has to be provided in the header of the FITS[23] image file or as an ascii separate file. A valid WCS[24] header is required in order to specify the coordinates on the sky for a given pixel location in the original image.

The result of structuring this astronomical data into HiPS, using the HEALPix tessellation of the sky, is a set of HEALPix maps over a range of successive orders: the higher the order, the better the resolution of the image. The tiles define the basic unit of storage for the HIPS. For instance, the projection of the sphere from Fig. 3 with various shades of gray indicating the successive orders of the HEALPix maps, shows three HIPS tiles of different orders: the bigger the HEALPix map order, the smaller the tile, the higher the resolution of the data representation. Therefore, HIPS enables a multi-resolution data representation, allowing a visualization of the original data in a progressive way, from the full sky to individual image pixels.

To sum up, a HiPS representation of original data is a set of tiles and the associated metadata, stored all together as a directory tree, like in Fig. 4: NorderNN directories correspond to various shades of gray from the Fig. 3; NpixNNN .png or .fits tiles correspond to "smaller" or "bigger" tiles as those from the Fig. 3; other files such as properties or index.html store the metadata. A HiPS can be

[23]Flexible Image Transport System; VO standard format for astronomical images and tables
[24]http://www.atnf.csiro.au/people/mcalabre/WCS/

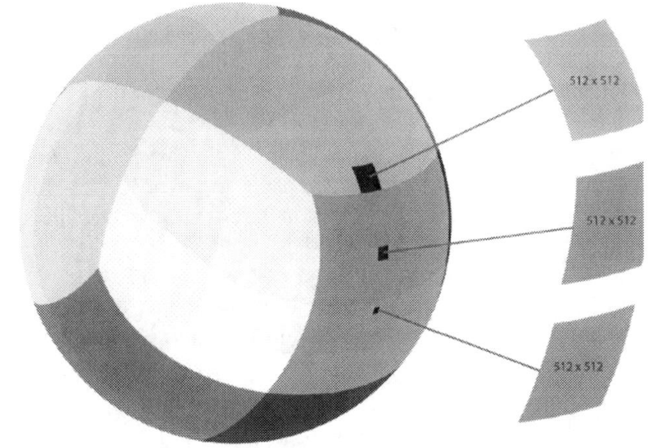

3rd order HiPS tile, 7°.33 side length, 512x512 pixels of 51".5

4th ———"——— 3°.66 ———"——— ——— 25".8

5th ———"——— 1°.83 ———"——— ——— 12".9

Figure 3. HIPS tiles and pixels representation. The relationship between the tiles and the HEALPix pixels are discussed in detailed in Fernique et al. 2015 [1]. Each tile (in black) has a constant number of pixels over all of the orders, here 512x512 pixels/tile.

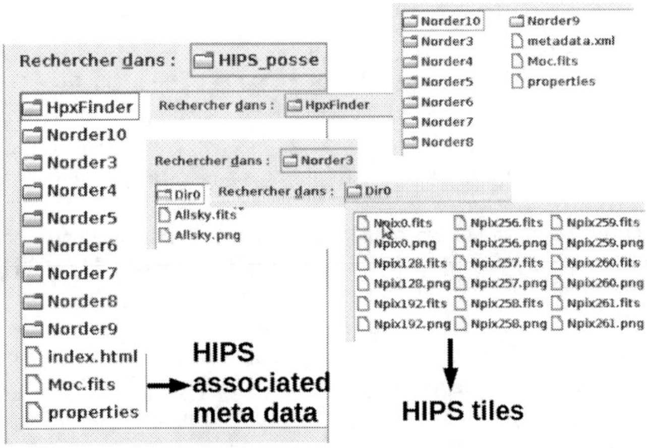

Figure 4. HiPS representation of the original images from the DSS POSS survey: tiles and associated metadata. The directories OrderNN (NN=3..10) contains all the tiles associated to the HEALPix maps.

used locally by Aladin or other dedicated clients by using its root directory name. To enable on-line[25] use of HiPS, the directories and files are simply copied in a HTTP server.

[25]Here is a list of HiPS surveys published by different HiPS servers http://aladin/hips/list

3 HiPS creation process and metadata update

First, together with the astronomers and IT specialists, we establish a priority list and check the data availability, during meetings or by email.

Second, linux commands are used to get the data from internal or external server or from other file hosting services such as Dropbox.

Then the data are prepared. In other words, there are cases when bad margins have to be removed from the original data such as for the photographic plates digitized using the MAMA[26] facility in Paris. In other cases, multi-filter data has to be structured in folders in order to create one HiPS per color, as for the DSS[27] survey.

Finally, the HiPSgen tool parameters are handled and the HiPS tiles and metadata generation process starts. The HiPS generation tool accepts a long list of parameters; usually, values are set for 10 to 15 of them. The only required parameter is the source image directory. Moreover, other optional parameters are used to handle:

- *the original images*: sky background, margins or metadata;
- *the HiPS tiles*: .fits extension tiles are created by default, but .png and .jpeg tiles can also be generated for faster access;
- *HiPS metadata*: the survey associated MOC following the IVOA MOC standard, information about the associated original data to be displayed for the final user in Aladin or other advanced HiPS clients.

The HiPS generation process can take from few minutes to several days, depending on the size of the data. Finally, the associated metadata are updated, the result is checked and together with the team, we validate the final HiPS. Here are two examples of associated metadata documentalists updates:

- *the HiPS generic metadata* (properties file from Fig. 4): All this generic metadata can be visualized either in a short HTML presentation associated to the HiPS (index.html file from Fig. 4) or in Aladin. Besides adding a short description of the survey, information about the progenitor data associated with the tiles, bibliography, etc[28], calculations and conversions using VO standards metadata conventions are performed, such as for the survey wavelength unit or epoch representation;
- *the original images metadata*: The information about the original images is included into the JSON format tiles (stored in the HpxFinder directory from Fig. 4) during the HiPS generation. The metadata.xml file (see Fig. 4) is a VOTable[29] template. Its content is updated in order to allow the final user to visualize information associated with the original images, such as the one present in the lower part of Fig. 5. In this panel one can see an Aladin all sky view of the sphere. The small squares invite us to select the original images used to build the HiPS and to access the associated meta information displayed in the lower part of the Aladin view. Any changes in the choice of the original images displayed information can be done by simply modifying the metadata.xml file.

4 Challenges for documentalists

The diversification of the astronomical data types and size brings changes in the way documentalists at CDS deal with databases content. Next, we will discuss how documentalists handle issues related

[26] Machine Automatique a Mesurer pour l'Astronomie

[27] http://www.cadc-ccda.hia-iha.nrc-cnrc.gc.ca/fr/dss/

[28] A list with the mandatory and optional keywords describing the associated meta information can be found in the IVOA Proposed Recommendation http://www.ivoa.net/documents/HiPS/20170406/PR-HIPS-1.0-20170406.pdf

[29] http://www.ivoa.net/documents/VOTable/20091130/REC-VOTable-1.2.html

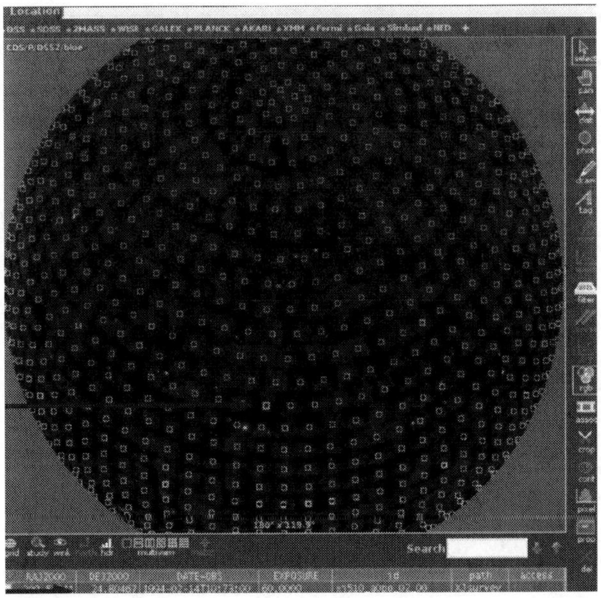

Figure 5. Aladin all sky view of DSS blue images structured into HiPS

to big volume data, heterogeneity, incomplete or incorrect associated metadata or big astronomical survey complexity.

The time required by the HiPS process can be very long, depending on the size of the original images, not to mention the time needed to transfer the final result to the server. During this time, the documentalist either switches to SIMBAD updating with the bibliography, cross-identifications, basic data and measurements or he/she updates the metadata of other existing HiPS. Working with large pixel data implies switching between servers, disks or partition, checking the currently running processes, handling new tools and linux console commands (wget, rsync, get, grep, gunzip, HiPSgen, etc). Multitasking skills of documentalists become an important requirement in today's Big Data era.

An astronomical survey may have images with randomly distributed bad pixels, different sky backgrounds, resolutions and filters. Given this input data heterogeneity, tests using a small number of images need to be performed before launching the calculation of the whole image survey. Furthermore, applying the same parameters to heterogeneous data may be a big challenge. Sometimes the HiPS generation process has to be restarted partially or entirely due to the irregularities created by the rich diversity of images or due to the lack of images for a certain location on the sky. The more surveys we handle, the more we learn about the heterogeneity problems which may arise, the faster we handle properly the structuring of images and data cubes into HiPS.

An insufficiency or inaccuracy of the mandatory keywords from the original data FITS format files (ex: input data is a HEALPix map which do not respect the HEALPix conventions) may slow the HiPS creation process. Consequently, contact is established with the data providers in order to get files with complete and correct metadata headers. Updating the resulting HiPS associated information is necessary for further usage. This may imply discussions with the members of the team concerning new formulas for converting various measurements into IVOA standard units. Although in most of the cases full metadata is provided for the new HiPS, aspects like the limits of the visualization tools in loading very large images may prevent the display of the HiPS original data.

Handling pixel data coming from various astronomical surveys implies communicating directly or by email with astronomers and IT specialists. HiPS validation, a better understanding of the input data, the choice of the pertinent metadata or information to be displayed in advanced HiPS clients, are just a few issues demanding that astronomers, IT specialists and documentalists have a close collaboration.

5 Conclusions

We presented the way documentalists use the HiPS IVOA standard to structure larger and larger volumes of astronomical data such as images and data cubes. This hierarchical, multi-resolution way of structuring data allows a progressive view of astronomical surveys, by zooming and panning on specific areas of the sky. Documentalists at CDS produce HiPS and update the associated metadata. This allows important information about the original data to be stored in the same directory as the HiPS itself and to be easily displayed by the HiPS client/browser tools.

Therefore, documentalists at CDS handle bibliographies and catalogs but also pixel data such as images and data cubes. The data are structured in HiPS tiles with associated metadata and can be visualized progressively, just like in google maps. Documentalists adapt to the era of large surveys and contribute to an easy and fast visualization and reuse of Big Data in astronomy.

References

[1] P. Fernique, M.G. Allen, T. Boch et al., A&A **578**, A114 (2015)

[2] F. Bonnarel, P. Fernique, O. Bienaymé et al., A&AS **143**, 33 (2000)

[3] S. Lesteven, C. Bonnin, S. Derrière et al. 2010, *DJIN: Detection in Journals of Identifiers and Names* edited by Eva Isaksson, Jill Lagerstrom, András Holl, and Nirupama Bawdekar (2010), Vol. 433 of Astronomical Society of the Pacific Conference Series, p. 317–32

[4] C. Brunet, M.G. Allen, M. Brouty et al., *COSIM: the necessary evolution of a cross-identification tool along with data evolution*, These Proceedings (2018)

[5] P. Ocvirk, G. Landais, L. Michel et al., *Associated data: indexation, discovery, challenges and roles*, These Proceedings (2018)

[6] M. Wenger, F. Ochsenbein, D. Egret et al., A&AS **143**, 9 (2000)

[7] F. Ochsenbein, P. Bauer, J. Marcout, A&AS **143**, 23 (2000)

[8] K.M. Górski, E. Hivon, A.J. Banday et al., ApJ **622**, 759 (2005)

EPJ Web of Conferences **186**, 02002 (2018) https://doi.org/10.1051/epjconf/201818602002
LISA VIII

Associated data: Indexation, discovery, challenges and roles

Pierre Ocvirk[1,★], *Gilles* Landais[1], *Laurent* Michel[1], *Heddy* Arab[1], *Sylvain* Guehenneux[1], *Thomas* Boch[1], *Marianne* Brouty[1], *Emmanuelle* Perret[1], *François-Xavier* Pineau[1], *Tiphaine* Pouvreau[1], and *Patricia* Vannier[1]

[1]*Observatoire astronomique de Strasbourg, Université de Strasbourg, 11 rue de l'Université, 67000 Strasbourg*

Abstract. Astronomers are nowadays required by their funding agencies to make the data obtained through public-financed means (ground and space observatories and labs) available to the public and the community at large. This is a fundamental step in enabling the open science paradigm the astronomical community is striving for. In other words, tabular data (catalogs) arriving to CDS for ingestion into its databases, in particular VizieR, is more and more frequently accompanied by the reduced observed dataset (spectra, images, data cubes, time series). While the benefits of making this associated data available are obvious, the task is very challenging: in this context "big data" takes the meaning of "extremely heterogeneous data", with a diversity of formats and practices among astronomers, even within the FITS standard. Providing librarians with efficient tools to index this data and generate the relevant metadata is therefore paramount.

1 Vizier

VizieR is the reference database for tabular data from astronomical catalogues and tables published in scientific papers. It contains reference catalogs such as Gaia and SDSS, and also provides access to more advanced data products, such as the CoRoT time series [1]. Its workflow and usage is described in [2]. Tables are stored, along with homogeneous "metadata". This metadata consists of a range of information describing the data, such as the magnitude system used, the coordinate system and its epoch, the units of the columns, the type of data (magnitudes, sizes, positions...), stored as Unified Content Descriptors. This set of metadata can be queried, allowing users to discover catalogs relevant for their research. Another strength of this approach is that VizieR allows for instance to perform cone queries in all of the >14 000 catalogs stored in one simple click. Vizier is fully VO-compatible and communicates with a number of important astronomical resources and software such as the CDS SIMBAD database, Aladin and TOPCAT. These features have made VizieR a very popular resource among astronomers, and as a result, there have been 300 000 queries/day during the last year. Thousands of astronomers have decided that VizieR is the right place to make their data available to the community, and institutional data providers such as ESO and ESA have also trusted the service to distribute their surveys.

Significant developments to improve VizieR pipeline for non-tabular data attached to papers were pursued, based on the ObsCore VO standard [3]. Also an important agreement was made with the

★e-mail: pierre.ocvirk@astro.unistra.fr ORCID: 0000-0002-8488-504X

AAS to use ObsCore as a common description framework. The VizieR associated data service has been been developed using the Saada data publishing tool [4] in collaboration with Laurent Michel (High Energy Team), and has been available in test mode in 2016. Among the datasets already implemented are the CoRoT data (150,000 time series), and the LAMOST DR1 spectra. These are queriable with an interactive interface, and a feature of the service is that these are also made automatically available to the VO using the relevant IVOA standards.

2 Associated data: the challenge of automatic indexation

2.1 Metadata and ObsCore data model

By "associated data" we mean here, non-tabular data, such as spectra, images, data cubes obtained by integral field spectroscopy or mm/submm arrays such as ALMA, associated to the catalog published in Vizier. Our main goal is to make this data discoverable, for instance using a positional query of the type: "select all optical spectra within 1 arcmin of the center of NGC300", through a dedicated service. The positional information is therefore primordial, as is often the case in astronomy. Therefore this positional information must be used for our indexation, along with other important properties of the observation. We have chosen the VO ObsCore data model to store the associated data's metadata, as it offered the best compromise between completeness and simplicity, while covering the most important aspects of the metadata we wanted to capture (name, position, wavelength range, time).

2.2 Heterogeneous FITS headers

At the moment we consider as associated data only reduced data, in the form of FITS files, because of the presence of directly accessible metadata in the FITS headers. In these FITS headers, we try to find, in order of priority, the pointing of the observation, the time of observation, the electromagnetic wavelength range, and the instrument in use. Although FITS headers can be quite verbose, they are extremely heterogeneous. Indeed, telescope design and software, instruments data reduction pipelines, guidelines and tasks in MIDAS or IRAF and custom user-written routines have an impact on the content of the FITS headers and keywords. Therefore, the latter have evolved significantly since the introduction of the FITS standard. For instance, among the common keywords for the target name, we find : "TARGET", "OBJECT", "NAME", and all possible abbreviations and combinations of these basic words, linked with possibly hyphens or underscores, "OBJ", "OBJ_NAME", "TARG-NAME", "OBJNAME", etc ... The same holds for the rest of the metadata, i.e. coordinates in space and time, and although lists of commonly used or recommended keywords exist (such as HEASARC's), there is no sign that astronomers nor engineers will restrict their creativity to stay within the existing dictionary of keywords. Therefore, we try to guess the mapping from the FITS keywords to the ObsCore data model as well as possible, and the mapping algorithm will for instance try to find the name within a curated collection of keywords. The main engine for this mapping is built in Saada, a publishing tool for catalogs developed by L. Michel at CDS. Because of the huge diversity of FITS formats, the automatic mapping from FITS keywords to ObsCore data model is often incomplete and must be completed by hand, i.e. requiring the expertise of a documentalist or an astronomer. This is, at the moment, the main factor preventing the full deployment of the associated data pipeline. CDS does not currently (as of August 2017) the manpower to properly index all of the incoming associated data, although work is being done to make the mapping software more robust, so as to limit human intervention in the process as much as possible.

2.3 Authors to the rescue

Since the most accurate knowledge of the data resides with the authors themselves, the VizieR catalog submission webpage (http://cdsarc.u-strasbg.fr/vizier.submit/) has been significantly overhauled to allow users to deposit and index their associated data along with the catalogs. Once the associated data is uploaded, the automatic mapping engine provides a first guess, which the author can check, edit and correct if need be, as shown on the left of Fig. 1. Since this feature is pretty recent, we cannot tell yet how successful it will be at properly indexing the data and removing some of the indexation workload from the shoulders of the CDS staff. Only time will tell.

3 The Vizier associated data service

Once the mapping of FITS keywords onto the ObsCore data model has been successfully performed, the associated data is ingested into a dedicated database which can then be queried.

The new service has been released to provide easy and interoperable access to data associated with journal publications. These associated data (images, spectra, time series) have always been part of VizieR in addition to the catalogue collection, but they are now much easier to find and use. The Saada database engine is used in collaboration with L. Michel (OAS XMM-SSC). The data can be queried by an interactive interface and the service automatically publishes the data via VO protocols. Important data sets already in the system include 170,000 CoRoT time series and 2.4 million LAMOST DR1 spectra. The service can be accessed at http://cdsarc.u-strasbg.fr/assocdata/, leading to the page shown in Fig. 2.

The benefits of a dedicated service for the associated data are three-fold:

- Better indexation and therefore discoverability and reusability thanks to homogeneous metadata using the ObsCore data model.
- The metadata can be queried for discovery using VO protocols: SIA [5], SSA [6], ObsTAP.
- Consequently the service can be queried by VO tools: plat, CASSIS, TOPCAT, Aladin,as shown in Fig. 3.

4 Conclusion

While the service is very new, we plan to advertise it widely among the astronomical community and hope its use will become standard practice, just as the publication of catalogs in VizieR has become. Work will continue to improve the robustness of the automatic mapping of FITS keywords to the ObsCore data model.

References

[1] S. Chaintreuil, A. Bellucci, F. Baudin, et al., *II.5 Where to find the CoRoT data?* (2016), p. 109
[2] F. Ochsenbein, P. Bauer, J. Marcout, A&AS **143**, 23 (2000)
[3] D. Tody, A. Micol, D. Durand, et al., *Observation Data Model Core Components, its Implementation in the Table Access Protocol Version 1.0*, IVOA Recommendation 28 October 2011 (2011), 1111.1758
[4] L. Michel, P. Bantzhaff, C. Frère, et al., *A New Web Interface for Saada*, in *Astronomical Data Analysis Software and Systems XXI*, edited by P. Ballester, D. Egret, N. Lorente (2012), Vol. 461 of *Astronomical Society of the Pacific Conference Series*, p. 415

[5] P. Dowler, D. Tody, F. Bonnarel, ArXiv e-prints (2016), 1601.00519
[6] D. Tody, M. Dolensky, J. McDowell, et al., *Simple Spectral Access Protocol Version 1.1*, IVOA Recommendation 10 February 2012 (2012), 1203.5725

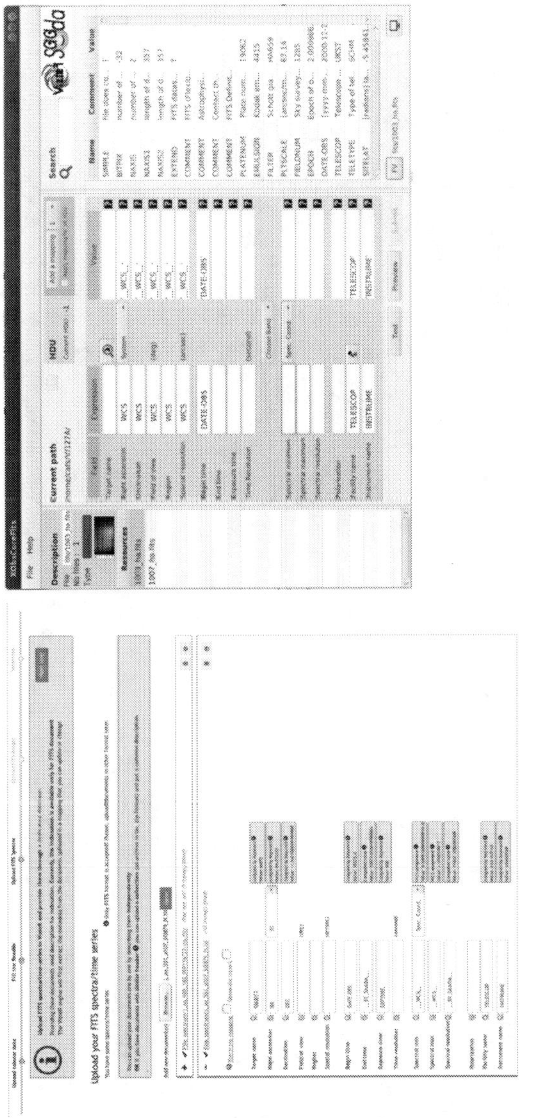

Figure 1. *Bottom:* supervised mapping of the FITS metadata to ObsCore using the interface built in the overhauled VizieR catalog submission web page. *Top:* main panel of the CDS-side mapping software for the documentalists.

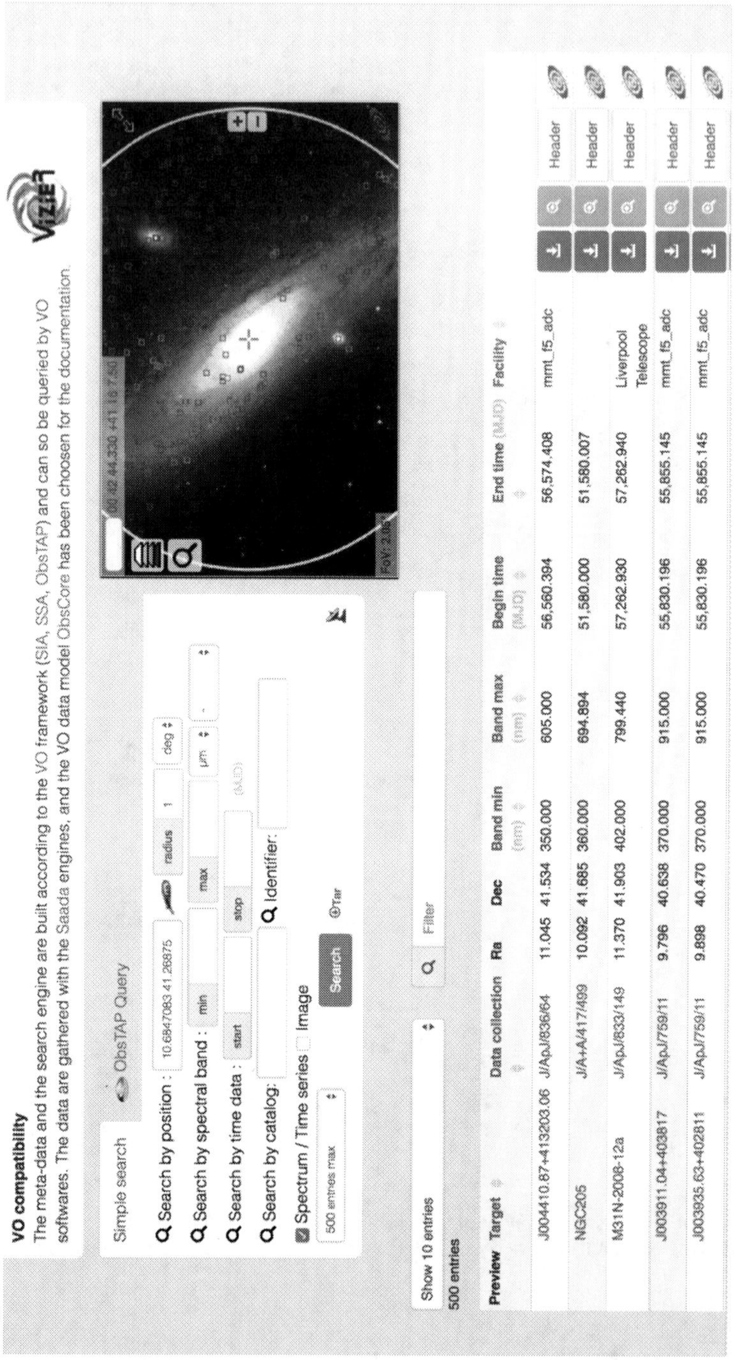

Figure 2. Homepage of the dedicated associated data service (http://cdsarc.u-strasbg.fr/assocdata/).

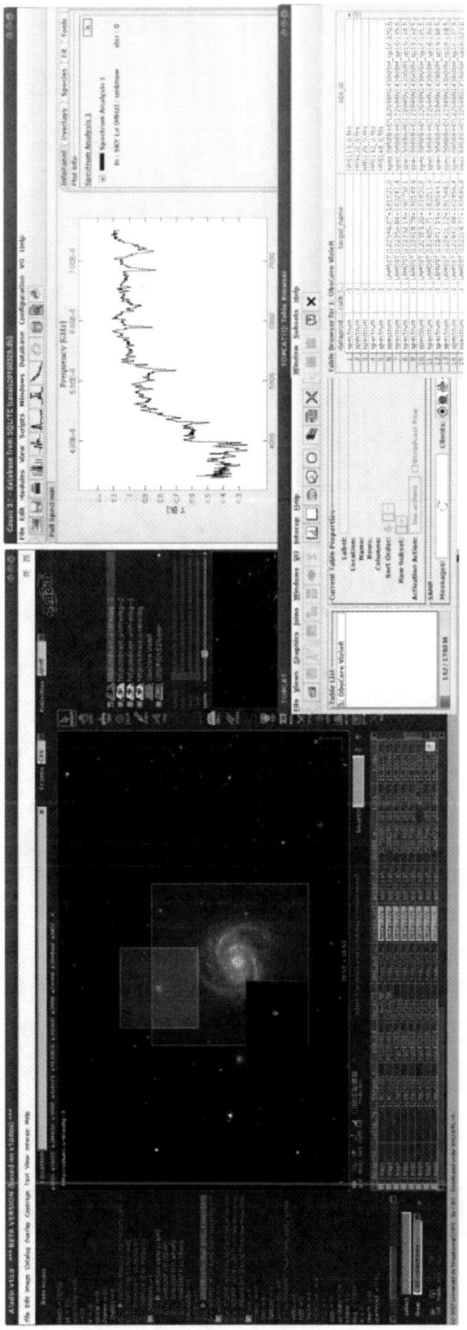

Figure 3. Examples of VO tools accessing the associated data service (Aladin, CASSIS, TOPCAT).

VizieR catalogue system certified by the Data Seal of Approval

Gilles Landais[1,*], *Françoise* Genova[1,**], *Jean-Yves* Hangouët[1,***], and VizieR Team[1]

[1]*Université de Strasbourg, CNRS, Observatoire astronomique de Strasbourg, UMR 7550, F-67000 Strasbourg, France*

Abstract. Astronomy is a fertile environment with collaborations leading to the elaboration of disciplinary standards. Deploying standards in data centers can be beneficial for long term preservation and it puts the discipline in the open data era.
VizieR [1] is the CDS service dedicated to catalogues. It gathers data and tables, but also associated data like spectra or images coming from published papers or collaborations with various agencies. VizieR gives access to data through services to the astronomical community and preserves data for the long term. Several years ago, it became clear that it was worth applying for an external evaluation to certify that VizieR is a trustable data repository.
The Data Seal of Approval (DSA) [2] is granted to repositories that are committed to archiving and providing access to scholarly research in a sustainable way. CDS applied to the DSA in particular for the VizieR catalogue service. The VizieR information system is based on accurate data documentation provided by specialists. The data life cycle from ingestion to access and preservation is presented, with reference to the Open Archive Information System (OAIS [3]).

1 Open Data in VizieR

1.1 Role of the data center

The CDS VizieR catalogue service, created in 1995, provides the astronomical community with free access to data. Ingestion workflows have been in action since the beginning. VizieR uses a standardized description of the resources which is provided by specialized documentalists, in some cases with an initial description provided by the author or the journal.

The CDS is clearly a major actor in Open Data in astronomy and beyond. Open data is currently a "hot" concept [4], demanded by authorities for publicly funded science data.

The first mission of an Open Data Center consists of preserving the data and giving free access. In return users have to cite the data origin (author, article) with a persistent identifier. VizieR provides a bibcode (the bibliography identifier used by the astronomical community [5]) and internal identifiers which are both persistent and enable citations and access. Furthermore, a Virtual Observatory

*e-mail: gilles.landais@astro.unistra.fr ORCID: 0000-0003-4868-5873
**orcid: 0000-0002-6318-5028
***orcid: 0000-0001-7338-8075

identifier (ivoid) is assigned for tables, enabling data reuse in a larger context.

The second mission consists of providing data that is useful and usable for research. Usability is well described by the FAIR (Findable, Accessible, Interoperable, Reusable) [6] principles, and leads data centers to implement vocabularies, formats, protocols, identifiers, etc.

The data quality depends on the assigned meta-data. In VizieR we distinguish "basic" meta-data and "rich" meta-data.

- Basic meta-data is the minimal description required for understanding the data.
  e.g. for a table : column description, units, etc.
- Rich meta-data enables data to be reused by software.
  In astronomy, the IVOA (International Virtual Observatory Alliance) elaborates advanced standards in particular for registries, data models, protocols and vocabularies which enable FAIR interoperability. Providing data in the Virtual Observatory framework should be compulsory today for an astronomical data center.

1.2 The Open data challenges

VizieR, as is the case for every data center, is between the data producers and the consumers which each have their own demands.

The producers for VizieR are the agencies which run space and ground-based telescopes, teams which produce large surveys, and the academic journals, with huge data diversity and increasing volume, especially for large surveys catalogues with tables over a billion of records (e.g. Gaia mission).

The consumers are the astronomers, research pipelines or software which require data and meta-data quality that we have to ensure. Quality has a cost. For example, it takes resources to find the information necessary to produce the rich meta-data of the Virtual Observatory standards.

Responding to these demands is a challenge, which becomes possible through interaction and cooperation with others. Firstly, cooperation with data producers is needed to improve the ingestion pipeline and build the initial knowledge about the data. Secondly, CDS long term success demonstrates that pooling different skills is beneficial: at CDS, astronomers, documentalists and engineers work together.

These collaborations locally and with data producers are needed to jump on the big data bandwagon, which is moving faster and faster.

VizieR is a reference service for catalogue dissemination recognized by its funding bodies, its partners and the global astronomical community. The next step was to get certification that VizieR is a trustable repository through an external evaluation. The CDS had already been certified as a member of the ICSU World Data System (WDS - http://www.icsu-wds.org/ [7]) community of trusted data services for Global Science. We decided in 2014 to submit an application to the Data Seal of Approval for VizieR, which fulfils preservation tasks in addition to data dissemination.

2 The Data Seal of Approval

2.1 The criteria

The Data Seal of Approval (DSA) is dedicated to scholarly research data repositories. It is a trust-based certification, simpler to get than the ISO one. It does not include audit visits but is based on a self-assessment of 16 criteria, which is then examined by the DSA Board which grants the Seal. The label is provided for 3 years before needing renewal.

The DSA worked with the WDS under the auspices of the Research Data Alliance to align their certification criteria and processes, resulting in an updated catalogue of common criteria in 2016, which better underlines the importance of data reuse. The two organisations established a common Certification Board in 2017 a common Certification Board.

The DSA criteria cover different aspects of the information system:

- Organization : these criteria include the mission of the data center, the continuity of access strategy, etc.
- Data management : there are several criteria to describe the workflows from input to output including identifiers and meta-data assignment, integrity checking and availability of scientific expertise.
- Technology and security is composed of 3 criteria on data preservation aspects, architecture, monitoring and preservation plans.

2.2 DSA evaluation

The DSA proposes an auto-evaluation with a compliance scale composed of 5 levels depending on the implementation status for each criterion. Among the levels, one is "Not applicable." For instance, in VizieR the criteria concerning confidentiality and ethics is not applicable because VizieR data are not subject to disclosure risks.

The DSA board (now the common DSA-WDS one) examines the responses and the documentation provided by the data center. Peer evaluation is performed by one member of DSA and one from WDS. The reviewers write a report to the Board which decides whether to award the Seal.

During the VizieR evaluation, which was under the sole DSA Board responsibility at that time, we got comments from the DSA reviewers on our initial application. It was helpful assistance to improve the documentation. Finally, we submitted a more complete response with linked documentation provided on our website. The description of the CDS processes in the OAIS framework described in the next section was among the documents provided for review.

Writing the documents submitted for review was a team effort at CDS. The initial work was mainly conducted by the technical staff: software engineers and people in charge of the network and system infrastructure. The documents were completed and reviewed by documentalists and the CDS governance.

For the renewal this year, the documentalists contributed to the writing of the documents, which resulted in a more precise description of the workflows.

The DSA review was also a guideline. It helped us to improve our Information System by providing a good list of relevant questions. It resulted in some improvements. As an example, checksums are now computed for each table ingested in the pipeline and verified during the data life cycle.

2.3 Describing the Information System with OAIS

DSA proposes to use OAIS (Open Archive Information System) to describe the data center processes. OAIS is registered as an ISO norm (14721:2012); it defines a reference model and an editing guideline to describe an Information System (including the system, architecture, staff).

OAIS is dedicated to open archives. Data and meta-data are inseparable in the OAIS concepts, they are gathered into information packages and evolve together during the whole data life cycle.

In the next sections we will describe some of the OAIS concepts:

- the Information System described in six OAIS entities (section 3.1)
- the three types of information packages of the OAIS norm (section 4)

3 VizieR described in the OAIS norm

3.1 The OAIS entities

The OAIS norm divides the Information System into six entities:

- "Entry", "Archival storage" and "Access" are the entities related to the workflows which manage data and meta-data through the different steps of the data life cycle from ingestion to output. "Archival Storage" also includes preservation and mirroring aspects.

- The processes are conducted by the "Data management" entity which includes rich meta-data assignment and submission validation.

- The workflows are supervised by the "Preservation Planning" entity which assures the migration and monitoring strategy, redundancies and technology.

- Finally, the "Administration" entity administrates the whole system. It includes the governance of the data center, the relations with producers and users, etc. The Administration also validates the technology proposed by the Preservation planning entity, etc.

The six entities are completed by "Management" which is outside of the Information System. At CDS, we have the Scientific Council composed by an international committee that evaluates the CDS scientific impact. They audit the CDS each year and write a report which is a guideline for the next year.

Figure 1 shows the VizieR information System described in the OAIS-like norm.

VizieR benefits from for than 20 years of experience, complemented by the even longer term expertise of the CDS in astronomical data management and dissemination which has been built up since its creation in 1972. In particular VizieR benefits from well established workflows for data description and of CDS expertise in the implementation of the Virtual Observatory standards. Both were key elements of VizieR in the description used to obtain certification.

3.2 VizieR Workflows according to the CDS data curation

The data provenance as well as the content (including meta-data) are defined by the Administration entity of the CDS, which is composed of the astronomers and the governing body.

Depending of the origin, the data are submitted by authors or documentalists into the "Entry" entity. Documentalists are the building block of the data ingestion. They offer assistance to authors for their data submission and then check and complete the description or fully write the description. Finally, the astronomers validate the documentalist's work and allow access to the public database.

Data curation includes homogenization of the formats and assignment of basic and rich meta-data. There is also a data quality validation for each document ingested.

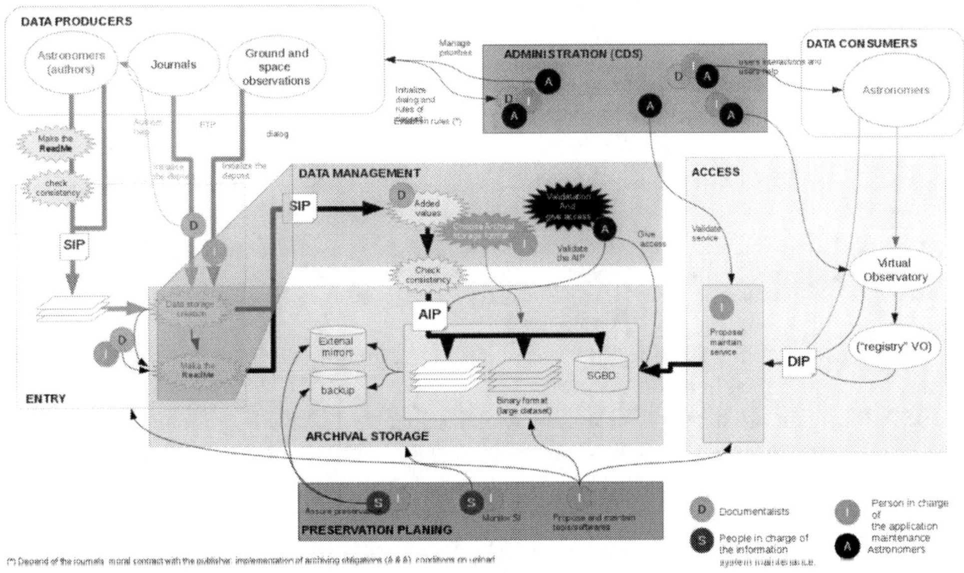

Figure 1. The VizieR Information System described with OAIS

4 The VizieR Information Package life cycle in OAIS norm

4.1 Input Data

The input data, or the "Submit Information Package" (SIP) in the OAIS norm, is the data submitted by authors or generated by documentalists.

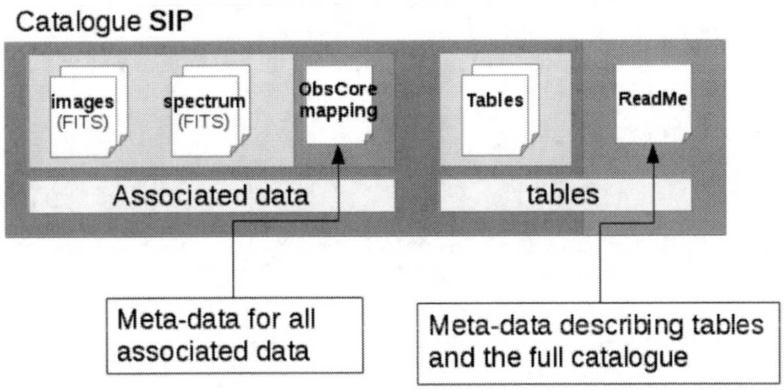

Figure 2. The VizieR Submission Information Package (SIP)

In VizieR, the meta-data of the SIP are gathered in the ReadMe file. The ReadMe file describes the catalogue and the tables with meta-data including the authors, the abstract, references links, columns descriptions, etc.

Associated data [8] such as images or spectra, often in FITS format, are described with the ObsCore Data Model [9] of the Virtual Observatory. The ObsCore meta-data of the FITS documents is separated from the ReadMe and provided as key-value mapping.

The tables are submitted in the CDS ascii format with aligned columns. The CDS provides tools for authors to migrate CSV, FITS or VOTable into the CDS ascii format. This format is adapted for long term preservation; it is independent of the technology and is described with a byte-by-byte section in the ReadMe file. All the columns of the tables are described with their unit, data type and a description text.

4.2 Archival Storage

Data are preserved by the "Archival Storage" entity. Data and meta-data are the Archival Information Package (AIP) described in the OAIS norm.

The AIP contains all data and meta-data available locally. In VizieR, it includes the basic and rich meta-data with the identifiers, but also some private data useful to rebuild the system, like the history of the ingestion processes, the email correspondence with authors, etc.

The AIP are the result of curation by the CDS documentalists. This process is the core of VizieR, and is required to provide useful data and added-value.

Among the meta-data, a configuration file completes the ReadMe file. It includes:

- the United Content Description (UCD [10]) for each column which are used by VO enabled software;
- links to other VizieR catalogues or external links;
- target name resolution to add positional computed columns;
- plot visualization for time series;
- photometric system and filters for photometry columns.

Finding the information is not trivial, for example looking for the photometric meta-data requires particular attention. Providing this kind of information cannot be automatized and the role of specialized documentalists is crucial.

4.3 Data Provision

The data provided by VizieR are the Diffusion Information Package (DIP) described in the OAIS norm.

In OAIS, the DIP includes the information package and the dissemination protocols. The Virtual Observatory standards including data model, formats and protocols constitute a framework which is a good example of what can be a DIP.
As an example, the VO registries [11] index the astronomical resources in a common list of resources independent of the data center. All resources have an identifier (ivoid) and are linked with a description. The registry technology is based on the OAI-PMH protocol, a standard used for instance in the digital library context or in EUDAT B2FIND registry (https://www.eudat.eu/services/b2find), which allows federation with other resources.

5 Conclusion

VizieR DSA certification was facilitated by more than 20 years of experience with well established workflows which are already in place for data management and description. The implementation of the Virtual Observatory standards was the second pillar that enabled us to document the CDS response to the criteria. These standards match well with the dissemination requirements of the OAIS norm.

The documentation in VizieR is provided by specialized documentalists. Their work is crucial to the high quality of the service and to gathering the information needed to build and rebuild the system.

The full certification process required about 3 person month. It was team work to which all the team members contributed. While there was no immediate update of the system there was better documentation, which was necessary to obtain certification. This demonstrated the high quality of the processes built over time, and the detailed assessment of each criterion led to technical improvements. Some of them are still in development today. One of the most interesting output was the end-to-end description of the processes, which led to a better understanding by all of the whole system, and of their role in it.

We would also like to express our gratitude to the VizieR team for their work: the documentalists P.Vannier, M.Brouty, E.Perret, S.Guehenneux, T.Pouvreau, the astronomers P.Ocvirk and C.Bot and the engineers T.Boch and F.X.Pineau

Also, many thanks to F.Ochsenbein who built VizieR.

References

[1] F. Ochsenbein, P. Bauer, J. Marcout, A&AS **143**, 23 (2000)

[2] I. Dillo and L. de Leeuw, https://www.datasealofapproval.org

[3] CCSDS, *Reference model for an Open Archival Information System* (OAIS), Magenta Book. (Issue 1) (2002)

[4] F. Genova, *Data as an infrastructure: CDS, the Virtual Observatory, astronomy, and beyond*, These Proceedings (2018)

[5] M. Schmitz, G. Helou, C. Lague et al., *NED and SIMBAD conventions for bibliographic reference coding* in *Information & On-line Data in Astronomy*, edited by D. Egret and M.A. Albrecht. Astrophysics and Space Science Library, Vol. 203. (1995) p. 259-270, ISBN: 978-94-010-4178-2

[6] M.D. Wilkinson, M. Dumontier, I.J. Aalbersberg et al., Scientific Data 3, art. num. 160018 (2016)i DOI: https://doi.org/10.1038/sdata.2016.18

[7] I. Gärtner-Roer, S. Harrison, S. Sorvari, Geophysical Research Abstracts **19**, EGU2017-4917, 2017, EGU General Assembly (2017)

[8] P. Ocvirk, G. Landais, L. Michel et al., *Associated data: indexation, discovery, challenges and roles*, These Proceedings (2018)

[9] D. Tody, A. Micol, D. Durand et al., *Observation Data Model Core Components, its Implementation in the Table Access Protocol Version 1.0* (2011)

[10] S. Derrière, A. Preite Martinez, R. Williams et al., *An IVOA Standard for Unified Content Descriptors Version 1.10*, ADS bibcode: 2005ivoa.spec.0819D (2005)

[11] M. Demleitner, P. Harrison, M. Molinaro et al., *IVOA Registry Relational Schema Version 1.0*, ADS bibcode: 2014ivoa.spec.1208D (2014)

EPJ Web of Conferences **186**, 02004 (2018) https://doi.org/10.1051/epjconf/201818602004
LISA VIII

COSIM: The necessary evolution of a cross-identification tool along with data evolution

Catherine Brunet[1,*], *Mark* Allen[1], *Marianne* Brouty[1], *Mihaela* Buga[1], *Cécile* Loup[1], *Anaïs* Oberto[1], *Emmanuelle* Perret[1], *Bernd* Vollmer[1], and *Fabienne* Woelfel[1]

[1] *Université de Strasbourg, CNRS, Observatoire astronomique de Strasbourg, UMR 7550, F-67000 Strasbourg, France*

Abstract. SIMBAD is a bibliographic added-value database on astronomical objects, where the data on individual objects are cross-identified as far as possible. The data comes exclusively from what has been published by the scientific community. To treat large tables, the work is done semi-automatically with the help of a customized software. Since 2014, we are using a new one, called COSIM (Comparison of Objects for SIMBAD). It meets the new requirements which is a consequence of the evolution of the available astronomical data. It has increased in number, accuracy and diversity. On the basis of the data presented in a published table, COSIM searches for objects that are already known in SIMBAD, by name or by coordinates. A combination of scores based on the available and comparable parameters, like the main object type, coordinates, velocity and magnitudes, suggests whether the candidate is good for cross-identification or not. As soon as the result of the search is clear, indicating that there is either no matching candidate or only one good candidate, COSIM creates the commands necessary for updating the SIMBAD database. The documentalists can act on the method of calculation of each score, according to the nature of the objects in the table. Thus, with COSIM the documentalists manage to obtain a good cross-identification level with a minimum risk of omitted or false cross-identifications in a relatively short time compared to the treated data number.

1 The SIMBAD database

The SIMBAD database [1] is a service offered by the Strasbourg astronomical Data Center (CDS) [2]. It was created in 1979 and has been maintained since. It is a bibliographic added-value database on astronomical objects outside the solar system, searchable via the Internet.

SIMBAD provides basic data, identifiers, bibliographic references and observational measurements on astronomical objects. All data comes from the scientific literature [6]. SIMBAD now contains about 9 million objects and 24 million identifiers for 330,000 references. This research tool is complementary to VizieR, another database maintained at the CDS, containing long lists of objects and large surveys.

* e-mail: catherine.brunet@astro.unistra.f ORCID: 0000-0001-9392-8817
[1] http://simbad.u-strasbg.fr/simbad/
[2] http://cdsweb.u-strasbg.fr/

SIMBAD is constantly maintained and updated by a multi-skilled team, including astronomers, documentalists (also called information scientists) and computer scientists [1, 2].

2 The documentalists' work

One of the characteristics of SIMBAD is that the information on an astronomical object is gathered together. That means there is a difficulty: the cross-identification. As the authors do not always designate their objects with a standardized name, identifying them is a challenge that the documentalists who update SIMBAD constantly face. For each object, we have to answer the question: which object does this name designate? Is it already in SIMBAD or is it a new one?

For each object name found in the article (in the text, in the abstract, in a figure, in a table...), SIMBAD is searched to see if it is already known. Two searching modes are used:

- first by identifier, if necessary with the help of the Dictionary of Nomenclature [3] made at the CDS [1];
- if the identifier is not found, a search is made around the coordinates (when they are given).

At the beginning of SIMBAD, the number of volumes was reasonable and the documentalists could afford to read the whole article, looking for astronomical objects in the text, the figures and the tables. We thus updated SIMBAD manually till the 90's [3, 4].

Then a software was developed and from then on, long lists of objects have been treated in a semi-automatic manner. From that moment, the documentalist team has been specialized into two branches: one for identifying objects in text, and one for dealing with the long lists [2].

Each of the two team branches has adapted itself to the evolution of the available astronomical data that have increased in number, accuracy and diversity:

- For the processing of texts, a software was created in 2008, *DJIN* [1, 5], in order to assist the detection of object names. The branch became the *DJIN team.*
- For the long lists, a new version of the first software had become necessary. It has been deployed since 2014 and has been called *COSIM (Comparaison d'Objets pour SIMBAD)*. The branch became the *COSIM team.*

3 The operational COSIM software

The aim of COSIM is to help the documentalists to enter long lists of astronomical objects into SIMBAD. It is a software written in Java language. Starting from a dedicated format table, it will search SIMBAD for good cross-identifications, through a direct link to the database. At the end of the process it will write update commands.

3.1 The place of COSIM in the whole procedure

It is first important to emphasize that before and in parallel with the automatic work, there is a lot of human work, too. That's why we generally speak about semi-automatic work. We present in figure 1 the complete COSIM team procedure. First, in a weekly meeting between astronomers and documentalists, a list of priorities is discussed and it is decided which data are likely to be put into SIMBAD. As a result, the documentalist gets a small notice (step 1), that allows her to extract the useful data for COSIM. It is in this phase generally that the Dictionary of Nomenclature is consulted,

[3]http://cds.u-strasbg.fr/cgi-bin/Dic-Simbad

to re-write identifiers if necessary. A file is given to COSIM (step 2). COSIM consults SIMBAD to search for cross-identifications and returns an output file (step 3). This file is analyzed by the documentalist. This is an important part, which is sometimes long and difficult. Not only does the documentalist use whatever tool or database she has at her disposal, such as SIMBAD, VizieR, Aladin and others, but she also consults her colleagues (other documentalists, astronomers, sometimes the computer scientist) or authors. Sometimes some preliminary manual updates are necessary. The documentalist adjusts the parameters (step 4), until COSIM writes update commands for all objects of the list. The update in SIMBAD can be then automatically done (step 5).

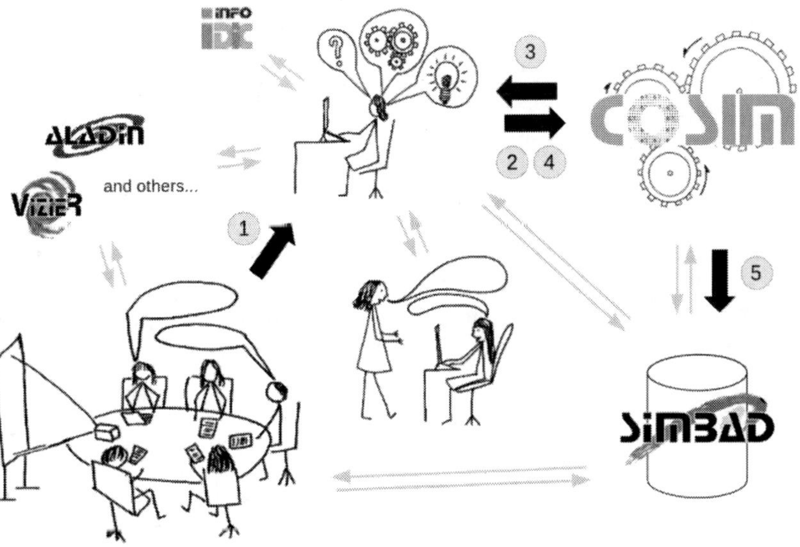

Figure 1. The COSIM team procedure. *Drawings are by Anna Preite Martinez.*

3.2 The COSIM software overview

COSIM searches SIMBAD for cross-identifications by identifier and/or around the coordinates. Then, each candidate found is compared to the entering object. To do this, a score is attributed to each comparable parameter, such as the object type, the coordinates, the velocity, the magnitude. Then all those scores are combined in order to evaluate the candidate (is it good for cross-identification or not?). According to the number of good candidates found, a decision is taken: to create a new object, or to update an existing one. Or, no automatic decision can be taken and the case is to be analyzed by the documentalist.

3.3 The object search in SIMBAD

COSIM searches SIMBAD for the object first by identifier and if not found, it searches around the coordinates. The search radius is automatically adapted to the accuracy of the coordinates in the table. But it can be modified. For example, for a table of galaxy clusters, one will search a larger area.

On the contrary, for deep field observations, one will search a smaller area. One can also decide to force the search around the coordinates, even if the object has been found by an identifier. In certain circumstances this can be useful, for example in order to look for mergers in SIMBAD.

As a result, one has a list of SIMBAD objects, either found by identifier, or that lie in the search area, or both.

3.4 The comparison

3.4.1 The scores attribution and calculation

Each candidate is attributed some scores: one for the object type (OT), one for the coordinates (COO), one for the radial velocity or redshift if any (V), and one for the magnitudes if any (M). In some very particular cases, one or two more scores are introduced: one which tests the presence or absence of an acronym among identifiers (ACRO), and one which tests the presence or absence of a reference among the bibliography (B).

The OT score is basically the result of a compatibility table between object types. For example, if the candidate is a star cluster whereas the entering object is a star, the score returned is -1; if the candidate found is a young stellar object candidate whereas the entering object is a young stellar object, the score returned is +1. One has the possibility to modify the results of the compatibility table, at will. For example, one can decide to consider a star and a galaxy as compatible, because the objects are point-like objects in a deep field and no human check has been done.

The COO, V and M scores are all three the result of a formula, that takes into account the uncertainty of both sides, the σ, Δ being the difference between entry and SIMBAD.

$$1.5 - 0.5\frac{\Delta}{\sqrt{\sigma^2_{\mathrm{SIMBAD}} + \sigma^2_{\mathrm{TABLE}}}}$$

The σ is generally given by the authors; if not, we estimate it, according to the instrument used and the wavelength, but also according to the object type. Indeed, for extended objects, we will artificially increase the coordinates sigma.

For each score there are two limit values (see figure 2): a *minimum*, under which the score is *low*, a priori for bad candidates, and a *maximum*, over which the score is *high*, a priori for good candidates. Between both, the score is said to be *medium*. The role of the documentalist is here to adapt those two values, in order to minimize the medium zone. Of course the best situation is when one can let coincide the minimum and the maximum, so that the score is either low or high.

The possibility of determining minima and maxima for each parameter separately is a new thing offered by COSIM. Thanks to this one can, for example, accept large differences in velocity but only for a very low distance.

3.4.2 The combination of scores

Once each comparable parameter has led to a qualified score, COSIM will combine the scores in order to give a status to the candidate.

All the calculated and not ignored scores are taken into account for the candidate evaluation. The OT and COO scores being considered as essential ones, as soon as one or the other is low, the candidate is labeled "bad" (see figure 3, c). It is not the case for the V and M scores: if one of the two is low whereas all the others are high, the candidate remains labeled "undefined" (see figure 3, d). In order to label a candidate "good", all of its scores must be high if it has been found with a search

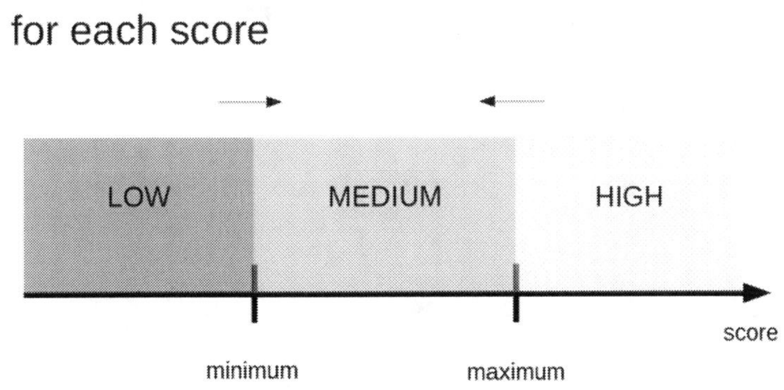

Figure 2. The limit values for each score

around the coordinates (see figure 3, a); or, all of its scores must be at least medium if it has been found by identifier (see figure 3, b). Actually, we are more confident when the candidate is found by identifier, because one can think that the author has already done the cross-identification work.

In order to reduce the number of cases where the software cannot automatically label the candidates either "good" or "bad", the documentalist has at her disposal some options. One can decide to ignore one score (because we estimate there are measure errors, or we estimate that the parameter is not a good comparison criteria for this sort of object). All candidates with this low score can be then nevertheless labeled "good" (see figure 4, a). One can on the contrary decide to attach a great importance to one score, as for the OT and COO scores. This time, all candidates with this low score will be labelled "bad" (see figure 4, b). In some particular cases, where the cross-identifications are very delicate, one can decide to attach even more importance to one score, by asking to cross-identify only if the corresponding parameter is present (see figure 4, c).

3.5 The update decision

According to the number of good and bad candidates found and when an automatic decision can be taken, the software writes update commands. Here is the exhaustive list where the automatic decision can be taken:

- No candidate has been found, neither by identifier nor around the coordinates. COSIM writes update commands to create a new object in SIMBAD.
- No candidate has been found by identifier, and all candidates found around the coordinates have been labeled "bad". COSIM again writes update commands to create a new object in SIMBAD.
- One or several candidates have been found, either by identifier or around the coordinates, but only one has been labeled "good". The others, found around the coordinates, have been labeled "undefined" or "bad", it does not matter. COSIM writes update commands to update the SIMBAD existing object.

At the end of the analysis process, all objects entering the table should lead to one of two decisions (new object created or existing object updated). But before reaching this ideal situation, a lot of various, problematic cases can be encountered.

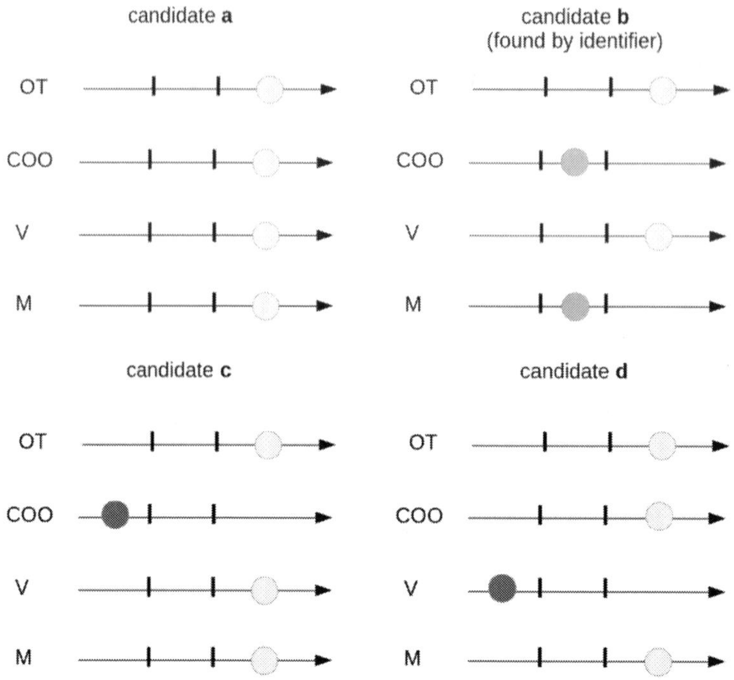

Figure 3. Examples of candidate evaluation by default: candidates which are labeled "good" (a, b); candidate which is labeled "bad" (c); candidate which is labeled "undefined" (d).

Basically, numerous combinations exist where the automatic decision is impossible. Nevertheless, thanks to the good comparison criteria used by COSIM and thanks to the physical character of the data, they do not happen so often and the situation is generally manageable. We have to analyze zero to ten percent of the entering objects manually, knowing that the size of the tables we treat is generally between several hundred to several thousand objects. If too many objects would need an individual analysis, we prefer to miss some cross-identifications and create more new objects.

Among the cases where no automatic decision can be taken, a few ones are notable:

- the "Possible Merger": two (or more) candidates have been labeled "good". The software cannot choose and has not been programmed to do the update on several objects. Actually, even if the "possible mergers" put a break on the automatic update decision, they are interesting, because they allow us to "tidy" SIMBAD. Thanks to the great possibility of COSIM to refine the comparison parameters, we can really rely on COSIM for detecting those cases, and only a small portion are wrong. SIMBAD has been improved by a lot of mergers since the use of COSIM.
- the "Conflict": the author has given several identifiers for one object, and they correspond to different objects in SIMBAD. Here again, the "conflict" puts a break on the software process: COSIM can't choose. The documentalist will analyze the situation, helped by the status given by COSIM (it can be at the same time a "possible merger") and do modifications, either in the input file, or directly in SIMBAD.

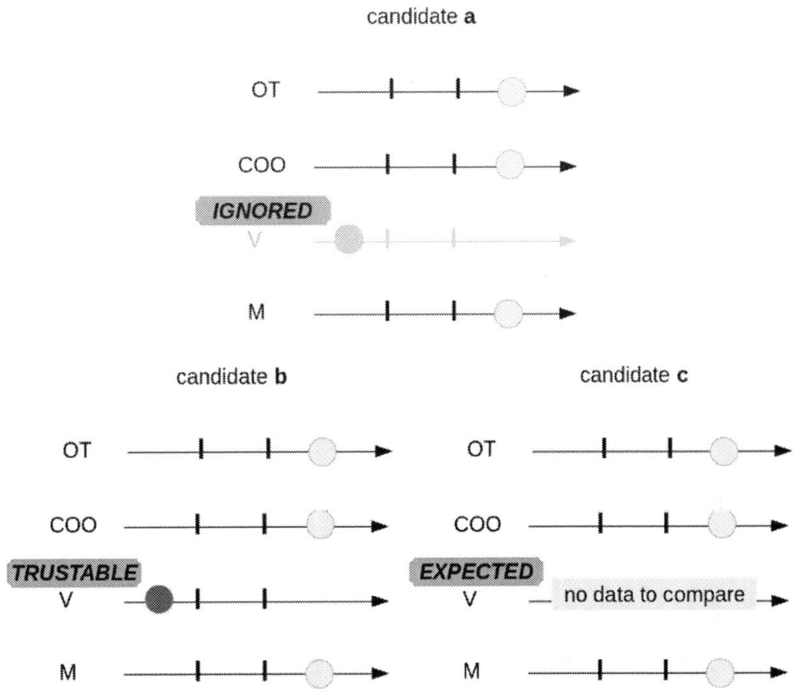

Figure 4. Examples of candidate evaluation with options: candidate which is labeled "good" (a); candidates which are labeled "bad" (b, c).

- the "Already Connected": the same SIMBAD object is labeled "good" for a second entry in the same table. As we suppose that all entries of the table correspond to different objects without repetition (unless it is clearly written), COSIM is not authorized to update twice the same SIMBAD object.

In all of these three cases and others, the documentalist will analyze the situation and give the appropriate answer so that COSIM will be able to write update commands in the next step:

- either in adapting the criteria of comparison in COSIM, if not already done (modifying the search radius, the sigma in the formulas, the limit values of each score, or adding options);
- or in doing some individual modifications, either in the input file, or directly in SIMBAD.

The latest intervention being time-consuming, the first will be as far as possible preferred.

3.6 Statistics and sorted lists

With a list of several hundred or several thousand objects, it would be impossible and counterproductive to check whether the automatic decision is correct for each object. So to help us in the analysis of the output file, COSIM provides statistics and sorted lists. In one glimpse, we have the number of "Possible Mergers", the number of "Conflicts", the number of "Already Connected", and the number of different final decisions: NEW, UPDATE on a candidate found by identifier, UPDATE on a candidate found around the coordinates, UNDEFINED cases. And overall, we have information that really matters in the process: a list of the nearest candidates which have been rejected, and a list of the furthest candidates accepted for cross-identification, both lists being ordered by object type. This allows us to check if the level of cross-identification is correct.

4 Conclusion

The COSIM software is an essential part of the SIMBAD operations, with the main advantages being that:

- all the parameters in the comparison are adjustable
- statistics and sorted lists give a precious help
- it is faster than the precedent software

As a result we can rely on these automatic parts of the process, allowing us to use precious time on aspects that need human expertise. Also, by using COSIM we can treat more objects and have confidence in the quality of the cross-identifications. COSIM is thus an important update of CDS procedures that helps us to address the increasing volume of published data. It is also important to understand that while COSIM is a useful tool, the global process must be driven by real people with expertise in astronomy and documentation, which remains essential.

References

[1] M. Brouty, F. Woelfel, C. Bruneau et al., *Information Scientists: Between Editors and Data Centers* in *21st Century Astronomy Librarianship, From New Ideas to Action*, edited by Eva Isaksson, Jill Lagerstrom, András Holl, and Nirupama Bawdekar (2010), Vol. 433 of Astronomical Society of the Pacific Conference Series, p. 195—200

[2] M. Buga, C. Bot, M. Brouty et al., *How Documentalists Update SIMBAD* in *Open Science at the Frontiers of Librarianship*, edited by András Holl, Soizick Lesteven, Dianne Dietrich, and Antonella Gasperini (2015), Vol. 492 of Astronomical Society of the Pacific Conference Series, p. 47–50

[3] S. Laloë, A. Beyneix, S. Borde et al., Bulletin d'Information du CDS **43**, 57 (1993)

[4] S. Laloë, Vistas in Astron. **39**, 179 (1995)

[5] S. Lesteven, C. Bonnin, S. Derrière et al. 2010, *DJIN: Detection in Journals of Identifiers and Names* in *21st Century Astronomy Librarianship, From New Ideas to Action*, edited by Eva Isaksson, Jill Lagerstrom, András Holl, and Nirupama Bawdekar (2010), Vol. 433 of Astronomical Society of the Pacific Conference Series, p. 317–323

[6] M. Wenger, F. Ochsenbein, D. Egret et al., A&AS **143**, 9 (2000)

EPJ Web of Conferences **186**, 03001 (2018) https://doi.org/10.1051/epjconf/201818603001
LISA VIII

Evolution of the NASA/IPAC Extragalactic Database (NED) into a Data Mining Discovery Engine. II. Current Contents and Future Plans

Marion Schmitz[1,2,*]

[1] *California Institute of Technology*
[2] *IPAC, MS 100-22, Pasadena, CA 91125*

Abstract. Recent advances have been made in evolving the NASA/IPAC Extragalactic Database (NED) into a data mining discovery engine. Infrastructure changes and data integration techniques are enabling more than a 10-fold expansion. NED will soon contain over a billion objects with their fundamental attributes (such as names, positions, redshifts, fluxes, and diameters) determined via cross-identifications among the largest sky surveys and over 100,000 smaller but scientifically important catalogs and journal articles. In addition, enhancements to the user interface, including new APIs, VO protocols, and queries involving derived physical quantities, will provide new pathways for multi-wavelength studies of large extragalactic samples.

1 Introduction

The NASA/IPAC Extragalactic Database (NED[1]) is an information system provided for the astronomical community that facilitates and accelerates multi-wavelength research on objects beyond our Milky Way that would otherwise be impossible or impractical to accomplish. The NED team is continuously integrating data from the literature, NASA mission archives, and large sky surveys to produce and serve a comprehensive census of the observed universe. Currently, the database consists of information synthesized from over 22 NASA missions, large surveys such as GALEX and SDSS, and information gleaned from over 103,000 journal articles, catalogs, and astronomical telegrams. Content includes object names, coordinates, redshifts, redshift-independent distances, fluxes, sizes, classifications, and attributes, along with derived quantities such as cross-identifications (XIDs), redshift-based distances, metric sizes, spectral energy distributions (SEDs), foreground Galactic extinction estimates, luminosities, velocity corrections, and cosmological corrections. A repository for images and spectra contributed by authors of journal articles ("data behind the plots") is also supported to simplify reproducibility of published results and making new discoveries in combination with other data in NED.

*e-mail: mschmitz@ipac.caltech.edu ORCID: 0000-0002-2055-7549
[1]http://ned.ipac.caltech.edu/

2 NED Basics

At the time of this writing (July 2017) the NED contents included 253 million objects (with names and positions) obtained from 108 thousand distinct references. Data for these objects includes photometry (2.3 billion data points), redshifts (over 11 million), redshift-independent distances for 27,000 objects, 2.5 million images, and 600,000 spectra. The Level5 Knowledgebase contains 893 articles of interest for extragalactic astronomy and cosmology.

The NED interface allows the database to be queried by name, near name or position (cone search), by reference, and by author. Galaxy samples can be constructed by parameter constraints on redshift, sky area, object types, survey names, and flux density (magnitudes), or by filtering galaxy classifications and attributes. In addition, the LEVEL5 Knowledgebase augments review articles in extragalactic astrophysics and cosmology with object names and graphical content within the articles linked directly to relevant database queries. These and other NED capabilities have been described in more detail elsewhere (e.g. Helou et al 1990, Mazzarella et al. 2007, 2014). Refer to Paper I (Mazzarella et al. 2017) for more information about the evolution of NED.

3 NED and Big Data

3.1 The four "V"'s

(The following is extracted from Paper I.)

Many challenges confronted by the NED team pertain to the four "V's of Big Data" - Volume, Variety, Velocity, and Veracity.

Volume: Over the next few years, NED will be growing by more than 10-fold to support research with data joined for about 2 billion objects and more than 20 billion attributes. This is being accomplished by upgrading storage and servers; refactoring the database schema to improve scalability, extensibility, and performance; and supporting queries on complex regions via spatial indexing using PostgreSQL extensions (Q3C, PgSphere).

Variety: NED spans the entire spectral domain from gamma rays through radio frequencies. These challenges are being met by continuously updating data capture and fusion techniques, and serving (meta)data using VO standards for interoperability with analysis tools.

Velocity: Data are being published in sky surveys and the literature at an increasing rate. This challenge is being met by accelerating cross-matching and data integration via parallel processing, replacing legacy data processing tools with a new pipeline developed in Python, Perl and C, as well as a small pilot project to apply machine learning techniques to classification of data types in the literature to streamline ingestion.

Veracity: Data in the literature are not refereed to the same degree as scientific results. Thus, there are often issues that impede efficient processing such as incomplete (meta)data, missing uncertainties or observation time-stamps, or ambiguous object identifiers. To help mitigate these issues, the NED team has published *Best Practices for Data Publication to Facilitate Integration into NED* (Schmitz et al. 2014]) [2] as a reference guide for authors and referees. The publishers of MNRAS have included a link to this document in their Instructions to Authors, and the AAS journals (ApJ, AJ) intend to do so soon.

[2]http://ned.ipac.caltech.edu/docs/BPDP/NED_BPDP.pdf

3.2 Software improvements

Since 1990 the NED team have been gaining experience in cross-matching sources in astronomical catalogs in the context of integrating selected data. This expertise has recently been codified and extended to include a probabilistic algorithm in an in-house computer program called MatchEx. The local density of objects is used to estimate the background contamination rate and Poisson statistics are used to balance completeness versus reliability of matches. Scientific vetting is applied to the results of trial runs in order to tune parameters and refine the algorithms. Further information is presented by Ogle et al. (2015). Ongoing and future enhancements will go beyond proximity, adding comparisons of redshifts, object classifications, sizes, and fluxes to the matching decisions.

A recent implementation of MatchEx with parallel processing enables effective and efficient cross-matching of surveys with tens of millions of sources, in addition to maintaining applicability to small data sets extracted from the literature. Most recently, MatchEx was used to process 42 million sources from the Spitzer Enhanced Imaging Products Source List, (SEIP [3]) yielding 37 million new objects and 5 million cross-ids to prior NED objects. Over 360 million photometric measurements in four IRAC bands and the MIPS 24micron band have been integrated into the SEDs for NED objects.

3.3 Database Content additions

In addition to the basic information that NED has traditionally contained (see Section 2), the next few years will see new data and meta-data added in order to enhance data discovery. Galaxy memberships in published pairs, groups, and clusters will be systematically added. Details about the observations (such as the telescope, instrument, and filters used) will be searchable. Survey coverage (footprints) of major surveys will be included.

3.4 Improvements of User Access to the Database

Recent improvements to the user interface include support for asynchronous queries with long run times, connectivity to the IRSA Finder Chart [4] service from NED image query reports, and access to the image archive through the VO Simple Image Access (SIA) protocol. Implementation of a VO Table Access Protocol (TAP) service is in progress, which will enable 'power users' to run ADQL queries against the NED object directory.

Source list uploads and improvements to user-specified tabular output will be implemented. We are improving capabilities of science queries on observations joined across missions, along with derived physical quantities, while also streamlining NED access from popular analytics environments such as Python, R and visualization tools.

4 World-wide Coordination

NED is a resource which relies on coordinated efforts with other resources worldwide.

The authors who provide the published and peer-reviewed data are essential to NED. Additional involvement will include requests that the authors provide NED with data ready for ingestion and to help more in the validation of published data.

[3] http://irsa.ipac.caltech.edu/data/SPITZER/Enhanced/SEIP/docs/seip_explanatory_supplement_v3.pdf
[4] http://irsa.ipac.caltech.edu/applications/FinderChart

NED and astronomy librarians have always had a close working relationship. Plans are underway to incorporate telescope bibliographies [5] and the Unified Astronomical Thesaurus (UAT [6]) which have been coordinated by the librarians.

Journal publishers have been and continue to be helpful in allowing NED access to the original journal articles as well as the data tables and supplementary materials contained within them. Mutual linking between the publishers and NED makes access to each more efficient for the researcher.

Three decades of close coordination of NED with CDS and ADS has proven beneficial to all three projects. NED and CDS share common challenges and have often worked together to determine appropriate solutions. These collaborative exchanges will continue to be helpful in the future.

NED's involvement with the NASA Astronomical Virtual Observatories (NAVO [7]) and the International Virtual Observatory Alliance (IVOA [8]) via contributions and implementations will continue.

5 Summary

NED usage had long been dominated by scientists interactively looking up a few facts about their favorite galaxies one at a time. Although this type of access will continue, current usage is dominated by programmed queries. The NED team is continuously evolving the hardware, software, and data content in response to advances in astronomy and informatics. The system is growing to serve data fused across the spectrum for billions of galaxies, and delivering new capabilities to exploit this unique resource for scientific discovery.

NED is operated by the California Institute of Technology, under contract with NASA. Current NED team members are Kay Baker, Ben Chan, Tracy Chen, Rick Ebert, Cren Frayer, George Helou, Jeff Jacobson, Tak Lo, Barry Madore, Joseph Mazzarella, Olga Pevunova, Ian Steer, Marion Schmitz, and Scott Terek.

References

[1] G. Helou, B.F. Madore, M.D. Bicay et al. , *Windows on Galaxies* edited by G. Fabbiano, J. S. Gallagher, A. Renzini (1990), Vol. 160 of Astrophysics & Space Science Library, p. 109

[2] J.M. Mazzarella, and NED Team, *NED for a New Era* in *Astronomical Data Analysis Software and Systems XVI* edited by R.A. Shaw, F. Hill, D.J. Bell (2007), Vol. 376 of Astronomical Society of the Pacific Conference Series, p. 153–162

[3] J.M. Mazzarella, P.M. Ogle, D. Fadda et al., *xplosive Growth and Advancement of the NASA/IPAC Extragalactic Database (NED)* in *American Astronomical Society Meeting Abstracts #223* (2014), Vol. 223, #302.04

[4] J.M. Mazzarella, and NED Team, *Evolution of the NASA/IPAC Extragalactic Database (NED) into a Data Mining Discovery Engine* in *Astroinformatics* edited by M. Brescia, S.G. Djorgovski, E.D. Feigelson (2017), Vol. 325 of IAU Symposium, p. 379–384 (Paper I)

[5] P.M. Ogle, J. Mazzarella, R. Ebert et al., *Rule-based Cross-matching of Very Large Catalogs* in *Astronomical Data Analysis Software an Systems XXIV (ADASS XXIV)* edited by A.R. Taylor, E. Rosolowsky (2015), Vol. 495 of Astronomical Society of the Pacific Conference Series, p. 25–35

[6] M. Schmitz, J.M.. Mazzarella, B.F. Madore et al., *Best Practices for Data Publication to Facilitate Integration into NED: A Reference Guide for Authors* in *American Astronomical Society Meeting Abstracts #223* (2014), Vol. 223, #302.05

[5]http://aspbooks.org/custom/publications/paper/492-0099.html
[6]http://astrothesaurus.org/
[7]https://heasarc.gsfc.nasa.gov/vo/summary/
[8]http://www.ivoa.net/

Coffee Breaks

Posters session with M. Kurtz

S. Prosser, D. Bouquin and K. Frey

B.M. Meera and F. Genova

(Photos: M. Wenger)

EPJ Web of Conferences **186**, 03002 (2018) https://doi.org/10.1051/epjconf/201818603002
LISA VIII

Research Data Management in India: A Pilot Study

Nishtha Anilkumar[1,*]

[1] *Physical Research Laboratory, Ahmedabad, India*

Abstract. Data is a by-product of the research process where published results are the output. More and more research institutes are getting interested in this by-product. Data could be in the form of statistics, experimental results, observational data, interview recordings, etc. Organizing the varied forms of data is a challenge for any institute. This is particularly so in the present scenario of constantly changing technologies for data storage and retrieval. The funding agencies are making it mandatory to archive these datasets so that these can be preserved for the posterity and / or re-used by others. Libraries, as a part of the research institutes, seem to be well equipped to organize and manage these datasets. The author undertook the present study to find the level of involvement of libraries in 'data management' in India. A survey was done to assess the awareness about data curation, data archival policies, infrastructure required, technologies used, etc. The survey sample consisted of 15 national research / academic institutes in India. The study showed that libraries' role in data management in research / academic institutes was still at a very early stage of development in India.

1 Introduction

Research data management (RDM), also sometimes known as research data curation, essentially involves collecting and organising data which is part of the research outcome in such a manner so as to facilitate easy access and re-use. Research data covers a broad range of types of information like documents, spreadsheets, field notebooks, diaries, audio tapes, video tapes, images, spectra, models, algorithms, scripts, protocols, workflows, software, standard operating procedures, methods, etc. [1]. Research data can be defined as "the recorded factual material commonly accepted in the scientific community as necessary to validate research findings" [2]. It does not include preliminary analysis, drafts of scientific papers, and plans of future research, peer reviews, communication with colleagues, trade secrets, commercial information, personnel and medical information. Research data management also includes record management of items like project files, grant applications, ethics applications, technical reports, research reports, signed consent forms, correspondence regarding these matters.

2 Earlier studies

As a result of data gaining importance, academic and research institutes are developing infrastructures and services to support the management of research data on their campuses [3]. In 2012 data

*e-mail: nishtha@prl.res.in ORCID 0000-0002-6091-2074

curation was identified as one of the top trends in academic libraries [4]. A study carried out by Akers, et al [5] surveyed eight universities in the United States about how they developed programs for supporting RDM. The study concentrated on the role of the library in educating and assisting the researchers in managing the data before, during, and after their research projects are done. It identified similarities and differences in initial motivation to provide RDM, collaborative relationships with other sections/divisions, approaches to assess the users' needs and changes in competencies required in library staff to carry out RDM. Another study carried out by Norman & Stanton [6] illustrates an evolution toward a strategic vision for library leadership in supporting RDM at the University of Sydney by exploring three stories. Each story throws light on key ingredients that characterize RDM support – researcher engagement, partnerships with other divisions and the complementary roles of policy and practice.

3 Need to preserve/manage research data

Research data is an important and expensive output of the scholarly research process across all disciplines. It is an essential part of the evidence necessary to evaluate research results, and to reconstruct the events and processes leading to them. Its value increases as it gets aggregated into collections and it becomes more available for re-use to address new and challenging research questions. Without proper organization, this value is greatly diminished.

With increasing numbers of 'born digital' research publications, there are new possibilities to store and preserve data with the benefits of making research more usable and replicable, saving the resources, and fostering inter-disciplinary research. The future lies in data being easily and effectively stored, preserved, shared, discovered and re-used in support of researchers and for advancement of science.

Most data are produced or gathered as part of publicly funded research, so it needs to be transparent, accountable and available. Currently in many countries, major funding agencies mandate that applicants submit a data management plan (DMP) as part of their research proposal. The data life cycle provides an overview of the stages involved in successful management and preservation of data for use and reuse. The following image of the data cycle clearly shows the importance of data curation for preserving and re-using and taking the research forward.

4 Stakeholders in RDM

RDM consists of a number of different activities and processes associated with the data lifecycle involving the design and creation of data, its storage, security, preservation, retrieval, sharing and reuse. A number of professional services, including academic support, library services and computing services, have an important role to play in RDM at the institutional level. Erway [7] has identified three main stakeholders in carrying out RDM in any institute. These are Researchers, Library & Information Services and IT Services.

4.1 Researchers

Researchers as the producers of the research data are the first stakeholders. The researchers are usually faced with a mix of requirements – managing their data and open access mandates from their funding agency or parent organization. This makes them reluctant to engage in data sharing. Any endeavour to share that data will depend on the trust they have with the RDM unit. Researchers should be explicitly informed of resulting decisions and procedures so that the trust level rises and they become more open to sharing their data for preservation and access.

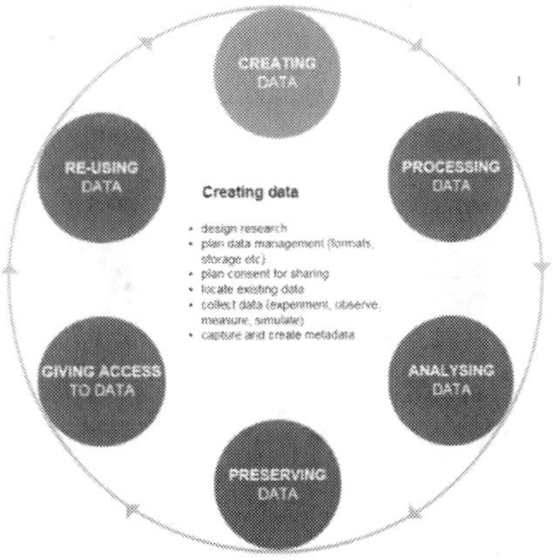

Figure 1. Data life cycle according to UK Data Archives - Posted by: Matthias Töwe, 21/04/2016

4.2 IT Services

In present times most research institutes have robust IT infrastructure which could support advanced data acquisition, storage, management, security, integration, mining and visualization as well as other information processing services. Currently, most researchers keep irreplaceable data on personal storage devices without documentation, version control or back up. Even where data is preserved effectively, it is not being made available to others for re-use. All infrastructure must include systems for documenting, depositing, managing, archiving and preserving data facilitating efficient search and retrieval and providing access.

4.3 Library & Information Services

Libraries are very well positioned to carry out RDM because they understand the need for standards-based information organization and because of the information management skills that librarians have, in particular assigning metadata to the information item for easy retrieval. Besides, libraries have wide experience in collection building, creating and maintaining institutional repositories, preservation and access to library material, including print as well as digital. There is also a potential connection between RDM and the open access agenda that libraries are actively promoting. In addition, libraries have extensive networks with other academic / research divisions within the institute and also with libraries of other institutes. Libraries have earned a trusting relationship with researchers, as their publications are now being preserved in the institutional repository by the library. The general perception of the library is as a safe, sustained and trusted unit for long term document/data preservation. The infrastructure used for an IR may be scaled up to accommodate the data sets. Also, most libraries have experience with copyright issues related to ownership of both source materials and research publications. Libraries are also best suited for the dissemination of information. Libraries can also provide help to researchers with depositing their data to international subject repositories.

An important component of RDM is advice and training of users apart from the technical infrastructure. This helps the librarians to actively re-invent their role and develop while supporting RDM.

Through these activities libraries can play a key role in carrying out RDM effectively. Auckland [8] has identified the following activities which libraries can undertake for data management:

- offering advice / information on funding sources
- conducting literature reviews or CA alerts for research projects
- bibliometric and impact measurement
- reference management software training like Mendeley, Zotero, etc.
- advocacy for open access / institutional repository
- data analysis advice
- advice on copyright issues
- advice on archiving of research records such as correspondence

5 Challenges in RDM

The use of powerful computing technology across disciplines now means that a greater number of researchers generate and use large datasets as part of the research process [9]. Storing this data in a form that can be easily accessed, processed and analysed is a very challenging activity for any research/academic institute. Also, the datasets are potentially fragile, being vulnerable to storage failures and technological obsolescence. It may also be sensitive containing personnel or health information, and so it needs to be managed with proper security measures. The other activities, like reformatting for analysis in various software packages, processing them for potential re-use and carrying out various preservation activities on them, also pose challenges. The responsibility for addressing these challenges then falls on higher education institutes or research institutes. Existing technical infrastructure can be coordinated to support data management. Robust infrastructure needs to be developed to support researchers to manage their data effectively.

One of the major challenges for librarians in taking up RDM for researchers is capacity and workload on the existing, and shrinking, staff. Other key issues to consider are the cost of RDM, including infrastructure, persuading the scientists to deposit their data for preservation and sharing, and convincing the management that their information handling skills are relevant for data curation.

6 Data Policy

Making data available for preservation and re-use adds value to published articles. This preserved data may be used for entirely new purposes in the future, thus increasing the Return on Investment (ROI) of the funding allocated to a particular research project. It could lead to collaborations and new areas of research. Hence, most of the funding agencies have now mandated submission of data management plans along with the research proposal submissions by researchers. Many universities which have no funder mandates for DMPs still carry out data curation and have a data policy in place simply because it is a good practice. Having a data policy helps in clarifying many issues and acts as a guide for the staff carrying out RDM. Before framing a data policy, an institute should constitute a committee comprising of various stakeholders and chaired by a director/dean. Briney, et al [10] have iterated that following points need to be explicitly mentioned while framing a data policy:

1. Why data repository is important?
2. who owns the data?

3. Who is responsible preserving and organising the research data?
4. In case of collaborative research, who will preserve the data?
5. what data should be retained? Who decides which data to keep? No institute can or should retain all research data generated by its researchers. As data curation requires extensive staff time and financial resources, it should be ensured that these are spent on data worth keeping only.
6. How open should the data be? Will it be available immediately or will there be an embargo of 2-3 years and then made open?
7. How long the data is to be preserved? Data may have long-term scientific or institutional value, but all preserved data should be reviewed. When an agreed upon retention period is due to expire, how will it be decided, whether it should be extended?
8. When a dataset is considered no longer to be worth keeping in the repository, who should be notified? Should these weeded out datasets be offered to others or destroyed? What records should be kept to document the weeding out of the dataset?
9. How should digital data be preserved? Each dataset will have unique preservation need.
10. What steps will be taken to preserve the data so that hardware obsolescence does not lead to data loss?
11. Should the DMP be kept with data? What other information should be provided, such as project and personnel records or documentation of instrument calibration?
12. Which kind of metadata is required for different types of data sets?
13. Which software will be used for retrieving the stored data sets?
14. Are the file formats of the data supported by the repository? What descriptors should be applied?
15. What are the implications of cloud storage for data preservation?
16. What are the ethical issues?
    - how will sensitive data be identified and contained?
    - have the consent forms been designed for allowing to share the data after the embargo period say 2-3 years?
17. How is the data accessed?
    - is it necessary to only make the metadata discoverable, with links to the data files?
    - how will the repository monitor access to ensure that restrictions are being enforced?
18. How will the costs be managed? RDM will incur substantial costs.
    - where will the funding come from? if the funding is project based, then how will long term preservation be supported?
19. What happens when the primary researcher leaves the institution?
20. How will the data be acknowledged and cited?

Table 1. Data archiving at various institutes

Institute	Data archiving	Section responsible	Data policy
Bose Institute, Kolkata	Yes	Library	being planned
INFLIBNET, Gandhinagar	Yes	other	being planned
IIAP, Bangalore	Yes	other	no policy
IIM, Ahmedabad	No	Library (being planned)	no policy
IISER, Pune	No	No	no policy
IIT, Gandhinagar	No	No	no policy
IPR, Gandhinagar	No	Library (being planned)	no policy
IUCAA, Pune	No	No	no policy
NIO, Goa	Yes	other	Yes
NRSC, Hyderabad	No	No	no policy
RRI, Bangalore	Yes	Library	no policy
SAC, Ahmedabad	Yes	other	no policy
SINP, Kolkata	No	No	no policy
TIFR, Mumbai	No	No	no policy
Nirma University, Ahmedabad	No	No	no policy

7 Present study

The Physical Research Laboratory, Ahmedabad, to which I am affiliated, wished to start a data curation project for posterity. To assess the level of data curation in India, I decided to do a survey of mainly research institutes and a few institutes of higher education. I sent a questionnaire to 25 research institutes out of the 166 research institutes in India. These 25 institutes were representative members of FORSA, DST, DAE and ISRO Consortium. Out of 25 research institutes contacted 11 have responded. These are Bose Institute, Kolkata, INFLIBNET, Gandhinagar, IIAP (Indian Institute of Astrophysics), IPR (Institute of Plasma Research), IUCAA (Inter-University Centre for Astronomy & Astrophysics), NIO (National Institute of Oceanography), NRSC (National Remote Sensing Centre), RRI (Raman Research Institute), SAC (Space Application Centre), SINP (Saha Institute of Nuclear Physics), TIFR (Tata Institute of Fundamental Research). In addition, I also sent the questionnaire to five academic institutes and out of these four responded - IIMA (Indian Institute of Management, Ahmedabad), IITGN (Indian Institute of Technology, Gandhinagar), IISER, Pune and Nirma University, Ahmedabad. The data compiled from the questionnaire provided information about whether data was being archived at all in the institute. If it was, information was given about which section was responsible for archiving it and whether a data policy was formed to carry out data curation. The following table throws light on these indicators:

Out of the fifteen responses received, nine librarians responded that data is not being archived by the library, but data is maintained by individual researchers. Out of these nine institutes, two libraries plan to start this service for researchers very soon (IIMA & IPR).

In four institutes, data is archived by other divisions like Data & Information Division. Looking at these 4 institutes, Inflibnet, IIAP, NIO and SAC, only NIO has a data policy in place.

In two institutes, data is archived in the library but at a nascent stage. A data policy is being formed at present (Bose Institute & RRI).

The above survey results show that in India RDM carried out by libraries is almost non-existent or at the very least on the idea plane only. As part of the study, I also tried to find standalone data centers funded by the Government of India. I found quite a few data centers in India dedicated to different subject disciplines and catering to researchers, the funding agencies and the general public. These are:

1. ESSO-INCOIS, Hyderabad http://www.incois.gov.in/
ESSO-INCOIS was established as an autonomous body in 1999 under the Ministry of Earth Sciences (MoES) and is a unit of the Earth System Science Organization (ESSO). ESSO- INCOIS is mandated to provide the best possible ocean information and advisory services to society, industry, government agencies and the scientific community through sustained ocean observations and constant improvements through systematic and focused research. ESSO-INCOIS has been designated as the National Oceanographic Data Centre by IOC/IODE of UNESCO and is also identified as the Regional Argo Data Centre for the Indian Ocean.]ESSO-INCOIS was established as an autonomous body in 1999 under the Ministry of Earth Sciences (MoES) and is a unit of the Earth System Science Organization (ESSO). ESSO- INCOIS is mandated to provide the best possible ocean information and advisory services to society, industry, government agencies and the scientific community through sustained ocean observations and constant improvements through systematic and focused research. ESSO-INCOIS has been designated as the National Oceanographic Data Centre by IOC/IODE of UNESCO and is also identified as the Regional Argo Data Centre for the Indian Ocean. Its main activities are:
   - Provides round-the-clock monitoring and warning services for the coastal population on tsunamis, storm surges, high waves, etc. through the in-house Indian Tsunami Early Warning Centre (ITWEC). The Intergovernmental Oceanographic Commission (IOC) of UNESCO designated ITWEC as a Regional Tsunami Service Provider (RTSP) to provide tsunami warnings to countries on the Indian Ocean Rim.
   - Short term (3-7 days) Ocean State Forecasts (waves, currents, sea surface temperature, etc.) are issued daily to fisher folk, the shipping industry, the oil and natural gas industry, the Navy, the Coast Guard, etc. These forecasts inform users about the expected sea conditions during the next few days and help them to plan their activities at sea.
   - Deploys and maintains a suite of Ocean Observing Systems in the Indian Ocean to collect data on various oceanic parameters to understand the processes in the ocean and to predict their changes.
   - Conducts systematic quality checks and archives all observational, satellite and other oceanic data at the ESSO-INCOIS Data Centre and then makes such data available to students, researchers and any other users.

2. ICRISAT Dataverse Network http://dataverse.icrisat.org/
ICRISAT performs crop improvement research, using conventional as well as methods derived from biotechnology, on the following crops: Chickpea, Pigeon pea, Groundnut, Pearl millet, Sorghum and Small millets. ICRISAT's data repository collects, preserves and facilitates access to the datasets produced by ICRISAT researchers to all users who are interested in. Data includes Phenotypic, Genotypic, Social Science, and Spatial data, Soil and Weather.

3. ICSSR Data Service (INFLIBNET) http://www.icssrdataservice.in/
The "ICSSR Data Service" is culmination of signing of Memorandum of Understanding (MoU) between Indian Council of Social Science Research (ICSSR) and Ministry of Statistics and Programme Implementation (MoSPI). The MoU provides for setting-up of "ICSSR Data Service: Social Science Data Repository" and host NSS and ASI datasets generated by MoSPI. The ICSSR Data Service is set-up with an aim to support researchers, teachers and policymakers who heavily rely on high-quality social and economic data for their research. The ICSSR Data

Service extracts and transforms the data from raw datasets before uploading it in the data repository with necessary documentation for the benefit of researcher. Users can explore collection of datasets accompanied with user guides and supporting materials using search interface of the ICSSR Data Service. All datasets documentation and resources are freely available through this platform. The project on setting-up of ICSSR Data Service is being executed by the INFLIBNET Centre with funding from ICSSR.

4. India Meteorological Department http://www.imd.gov.in/
   India Weather Portal has an objective to take meteorological observations and provide current and forecast meteorological information for optimum operation of weather-sensitive activities like agriculture, irrigation, shipping, aviation, offshore oil explorations, etc. Users can check weather of various Indian states, national weather, and weather forecast for agriculture, aviation, and ocean services, etc. It gives warning against severe weather phenomena like tropical cyclones, dust storms, heavy rains and snow, cold and heat waves, etc. which cause destruction of life and property.

5. Indian Oceanographic Data Centre (NIO) http://www.nio.org/iodc/
   The Indian Oceanographic Data Centre (IODC) was established in 1964 in NIO, Goa. The IODC assists national and international users in developing and enlarging their competence in the field of marine science. IODC plays a dual role, one by dissemination of data / information to the user communities and the other is to assist the data personnel in processing, validating, reformatting different types of data generated from the Indian Ocean region. The main responsibilities of IODC are:

   - acquire oceanographic data and information for the Indian Ocean
   - reformat, and perform quality control checks on data
   - develop /update databases
   - develop value -added data / information products
   - disseminate data / information to users' community, and
   - provide training in oceanographic data/ information management

6. Indian Space Science Data Center (ISSDC) https://www.issdc.gov.in/
   Indian Space Science Data Center (ISSDC) is the primary data centre for the payload data archives of Indian Space Science Missions. This data center, located at the IDSN campus in Bangalore, is responsible for the Ingest, Archive, and Dissemination of the payload data and related ancillary data for Space Science missions like Chandrayaan, Astrosat, etc. The Latest news related to the space missions and other activities are also available. The primary users of this facility will be the principal investigators of the science payloads. In addition to them, the data will be made accessible to scientists from other institutions and also to the general public.

7. Kodaikanal Solar Observatory (IIAP) https://www.iiap.res.in/Kodaikanal_Solar_Observatory
   The Kodaikanal Observatory of the Indian Institute of Astrophysics is located in the beautiful Palani range of hills in Southern India. It was established in 1899. Solar observations at this observatory over the last 100+ years provide one of the longest continuous series of solar data. Apart from that, simultaneous observations in different wavelengths make this data a unique one and suitable for multi-wavelength studies. The Kodaikanal observatory has been obtaining solar images since 1904 in broad band white light, narrow band Ca-K 393.37 nm and Ha 656.3nm wavelengths. Many of these observations are still continuing. The historical data which were on photographic plates has been digitized. The first level calibration of the Ca-K, white light

and Ha images have been completed and the data is now available through this portal. Final calibrated images and the data will be posted on this portal in near future. The digitised data are available for use by the scientific community.

8. National Informatics Centre http://www.nic.in/
NIC, under the Department of Information Technology of the Government of India, is a premier Science and Technology Organization, at the forefront of the active promotion and implementation of Information and Communication Technology (ICT) solutions in the government. NIC has spearheaded the e-Governance drive in the country for the last three decades building a strong foundation for better and more transparent governance and assisting the government's endeavour to reach the unreached. With the increased expectations from citizens for online services and the number of e-Governance Projects being launched by the Government, the Data Centre requirements are growing exponentially. There is a need to set up strategic infrastructure that facilitates high availability, quick scalability, efficient management and optimized utilization of resources. To fulfil this requirement, NIC has set up state-of-the-art National Data Centres at NIC Hqs, Delhi, Pune and Hyderabad and 30 small data centres at various state capitals to provide services to the Government at all levels. These Data Centres combine round-the-clock operations and management of systems with onsite skilled personnel.

9. World Data Centre for Geomagnetism, Mumbai http://wdciig.res.in/
This data centre has functioned as a division of the Indian Institute of Geomagnetism, Navi Mumbai since its activities commenced in 1991 in coordination with the International Council of Scientific Unions (ICSU) Panel on World Data Centres. It is responsible for the compilation of final hourly absolute values from nine of the Indian magnetic observatories and deposition of this data to the World Data Centre. In recent years, the centre has prioritized its activities related to digital preservation to ensure digital archiving of magnetic data from the traditional media and also digital conservation of very old hand written/printed data volumes and magnetograms. In view of the scientific importance of data from the Colaba-Alibag Magnetic Observatory, old magnetograms and data volumes are being converted to digital images for long term preservation. In the digital preservation process, the creation of metadata has become an important component in storing information related to old and current scientific records for future use. The centre also hosts a database driven website to make datasets available online to the global scientific community [11].

8 Findings

The survey shows that in India, RDM carried out by libraries is still at the nascent stage of development and will take a few years more to become an integral part of RDM activity in research and academic institutes. The survey results also show that although institutional policy is not in place currently, it is in the process of being formed soon. This indicates that the organizational perspective is changing. However, the prospect of dedicated data centers in different subject fields looks to be quite promising.

9 Conclusion

We as librarians need to convince researchers that data preservation and data access are inter-linked. Preservation helps in access, and active use of data is the best reason for continued preservation. As we all know, researchers are under extreme time pressure. So any new solution would need to be very easy to use and preferably integrated into their existing workflow to have greater chance of being used. By devising an improved information management solution around research activity, some of the problems that researchers themselves regard as most pressing, could be solved. According to Chad & Suzanne [12] creating better records of researchers and research projects would be important building blocks for a successful RDM system. This study helped to crystallize the idea that forming the data policy is the first step to begin data curation for posterity in any research institute. PRL library has presented a proposal for setting up the infrastructure for data preservation. We are in the process of forming a data policy and are engaging with the researchers to discuss the benefits of data archiving for data access. We have a long way to go, but the beginning is made....

References

[1] C.L. Borgman, Journal of the American Society for Information Science and Technology, **63(6)**, 1059–1078, (2012)

[2] OMB Circular 110, Office of Management & Budget, https://www.whitehouse.gov/omb/circulars_a110

[3] C. Tenopir, B. Birch, S. Allard, ACRL White Paper, (2012)

[4] ACRL Research Planning and Review Committee. *College & Research Libraries News, 73*(6), 311–320, (2012)

[5] K.G. Akers, F.C. Sferdean, N.H. Nicholls, J.A. Green, International Journal of Digital Curation, **9(2)**, 171–191, (2014)

[6] B.Norman, K.V. Stanton, International Journal of Digital Curation, **9(1)**, 253–262, (2014)

[7] R. Erway, Research Data Management Policy. Dublin, Ohio: OCLC Research, (2013)

[8] M. Auckland, RLUK Report, (2012), http://www.rluk.ac.uk/wp-content/uploads/2014/02/RLUK-Re-skilling.pdf

[9] S. Pinfield, A.M. Cox, J. Smith, PLoS ONE, **9(12)**, e114734, (2014)

[10] K. Briney, A. Goben, L. Zilinski, Journal of Librarianship and Scholarly Communication, **3(2)**, p.eP1232, (2015)

[11] M. Doiphode, R. Nimje, S. Alex, Data Science Journal, **12**, WDS85--WDS88, (2013)

[12] K. Chad, & S. Enright, Insights: The UKSG Journal, **27(2)**, 147–153, (2014)

EPJ Web of Conferences **186**, 03003 (2018) https://doi.org/10.1051/epjconf/201818603003
LISA VIII

The Astrolabe Project: Identifying and Curating Astronomical 'Dark Data' through Development of Cyberinfrastructure Resources

Gretchen Stahlman[1,*], *P. Bryan* Heidorn[1,**], and *Julie* Steffen[2,***]

[1] *University of Arizona School of Information, 1103 E. 2nd St, Tucson, AZ*

[2] *American Astronomical Society, 667 K Street NW, Suite 800, Washington, DC*

Abstract. As research datasets and analyses grow in complexity, data that could be valuable to other researchers and to support the integrity of published work remain uncurated across disciplines. These data are especially concentrated in the "Long Tail" of funded research, where curation resources and related expertise are often inaccessible. In the domain of astronomy, it is undisputed that uncurated "dark data" exist, but the scope of the problem remains uncertain. The "Astrolabe" Project is a collaboration between University of Arizona researchers, the CyVerse cyberinfrastructure environment, and the American Astronomical Society, with a mission to identify and ingest previously-uncurated astronomical data, and to provide a robust computational environment for analysis and sharing of data, as well as services for authors wishing to deposit data associated with publications. Following expert feedback obtained through two workshops held in 2015 and 2016, Astrolabe is funded in part by National Science Foundation. The system is being actively developed within CyVerse, and Astrolabe collaborators are soliciting heterogeneous datasets and potential users for the prototype system. Astrolabe team members are currently working to characterize the properties of uncurated astronomical data, and to develop automated methods for locating potentially-useful data to be targeted for ingest into Astrolabe, while cultivating a user community for the new data management system.

1 Introduction

As a partnership between the American Astronomical Society (AAS) and the University of Arizona School of Information, Astrolabe was formed in 2015 to address a demonstrated need for additional curation resources in astronomy. Based at the University of Arizona (UA), Astrolabe has thus far been funded by the University's Research, Discovery & Innovation (RDI) division and the National Science Foundation (NSF) Advanced Cyberinfrastructure Directorate to leverage existing cyberinfrastructure to create a new repository and computational environment for astronomical data that are not adequately curated in mission repositories or elsewhere. Additional collaborators at the UA include the Department of Astronomy and Steward Observatory, the University Libraries, and CyVerse (formerly

*e-mail: gstahlman@email.arizona.edu ORCID: 0000-0001-8814-863X
**e-mail: heidorn@email.arizona.edu
***e-mail: julie.steffen@aas.org

known as the NSF-funded cyberinfrastructure project, the iPlant Collaborative). The organizational structure of this collaboration is further strengthened by the University of Arizona's proximity to intellectual and technical resources, and its geographic centrality as a major hub for astronomy and information science research. The core mission of Astrolabe is to collect, preserve, and make accessible astronomical data that are not already successfully managed. The Astrolabe team is also developing tools for analysis and to encourage data sharing, as well as curation resources for authors of data and associated publications. Overall, the Astrolabe Project aims to help researchers expose data at key stages of the research lifecycle, such as the point of submitting a publication, for example, or when collecting new data for an existing project.

2 Background

Popular data lifecycle models highlight the complexity of research across disciplines, particularly considering the deliberate re-use of scientific findings for subsequent research through established channels of review and dissemination as a key aspect of data lifecycles and an important priority for many disciplines and for scholarly publishing [1]-[4]. Additionally, with a tremendous amount of data being created, analyzed, and transformed in astronomy and contemporary science in general, the need for innovative methods of publishing and curating these data is well-documented, including both the technical and social aspects of data management challenges [5]-[10]. Pennock (2007) defines curation as "the active management and appraisal of digital information over its entire life cycle" [1]. While curation tends to bridge disciplinary boundaries, research has shown that curation also requires insightful knowledge of the data themselves - and of the scientific communities, technologies, analyses and activities that produce them - to facilitate best practices and provide researchers with appropriate scholarly and cyber-infrastructure. Furthermore, publication of both research and data is an important component of the data curation lifecycle, and it is critical to develop resources that support the needs of the community and the advancement of the discipline.

2.1 "Dark Data" in the Long Tail of Astronomy

Much of astronomy's data are curated in archives designed by and for large missions. However, much other data are known to be uncurated. Heidorn (2008) hypothesizes that large projects have well-planned data stores, while large amounts of data remain uncurated from the more numerous, small projects: "Like dark matter, this dark data on the basis of volume may be more important than that which can be easily seen" [11]. Heidorn's exploration of research grants funded by NSF across disciplines shows that the largest 20 percent of projects funded in 2007 received more than 50 percent of total funds awarded in that year. Heidorn suggests that this top 20 percent has well-curated data, primarily due to access to curation resources and budgeting for data management, and that it is the uncurated dark data distributed throughout the Long Tail - small projects of this distribution - which represents a major potential for accumulated knowledge if brought to light. Similar distribution of funding and presumably data curation resources are found in other fields such as ecology [12].

A similar analysis of active awards in the Astronomy and Astrophysics Division of NSF in 2016 illustrates that while the characteristics of astronomy's Long Tail are unique reflecting the extremely high funding given to a small number of large missions, a diversity of projects and levels of funding nonetheless exist that may require data curation services, to aggregate knowledge currently hidden from other researchers, and to support published research for the benefit of the astronomical community. Essentially, 20 percent of NSF-funded research projects in astronomy receive 90 percent of funding.

Total Grants	855	
Over $500	$1,893,911,112	
	20% by number of grants	80% by number of grants
Number of grants	171	684
Total Dollars	$1,703,997,476	$189,373,212
Range	$3,500,000-$540,424	$539,648-$5,000
	20% by total value = $378,782,222	80% by total value = $1,515,128,890
Number of grants	1	854
Range	$350,000,000	$305,900,000-$5,000

Table 1. Grant Size Distribution of NSF Astronomy and Astrophysics Active Awards, running as of July 2016

Astrolabe is intended to provide a repository and computational environment for smaller datasets and projects with less funding, and to make easy-to-use interfaces for uploading and interacting with data, and eventually integrating into the publication process.

Curating data in the Long Tail of any discipline is a challenge due to heterogeneity of data and a tendency towards lack of standard metadata. Astronomical data types and formats correspond to particular research projects and observing techniques that vary widely across sub-disciplines and facilities. It is also often necessary to collect many observations over time, and astronomy faces growing challenges associated with generating, publishing, and utilizing very large and numerous datasets. Reprocessed or derived datasets and data collected by smaller instruments frequently remain uncurated. While the majority of data in astronomy conform to the Flexible Image Transport System (FITS) file format, with headers including coordinate information and details about instrumentation, many possible descriptive headers can exist, reintroducing heterogeneity. However, due to relatively successful adoption of FITS and other standards (including standards for interoperability developed by the International Virtual Observatory Alliance - IVOA - across the discipline and its global communities) astronomy is known for best practices in data management, and for widely supporting open access to scientific research.

2.2 Astrolabe System Design

Astrolabe aims to support analysis and re-use of data in addition to collection and preservation services. Astrolabe is being constructed in partnership with CyVerse cyberinfrastructure, a platform for data sharing and analysis that was originally designed to support research in life sciences, but has expanded to other disciplines. CyVerse provides longevity to support data management planning, easy access to supercomputing resources and community-developed research tools, along with data federation through iRODS technology for portability and discoverability, and minting of permanent identifiers for static published data [13]. While anyone can obtain a CyVerse account and upload data, Astrolabe functions as a liaison between CyVerse and the astronomy community, facilitating development of functionality to make the system useful for astronomy data, and building on a foundation of tools and storage that include:

- The "Discovery Environment"for convenient data and application sharing;

- "Atmosphere" to launch virtual instances for analysis using High Performance Computing (HPC) resources; and
- A "Data Store" with ample free data storage (100GB per user).

A substantial user community has formed within CyVerse, with more than 31,000 active users and nearly 3,000 Terabytes of user data, and projects representing a variety of disciplines across Natural and Life Science domains, all now based in CyVerse technology [14].

Stakeholder-collaborators in the Astrolabe project are united by mutual need and interests and provide distinct assets to the organizational design of the project. This strategic collaboration includes: a School of Information (University of Arizona iSchool), representatives of a scientific discipline (astronomy), an academic library system (University of Arizona Libraries), a scholarly society (American Astronomical Society), and a cyberinfrastructure platform for cloud-based data storage and computation (CyVerse). As the publisher of four major journals in astronomy and astrophysics, with a global author community and with thousands of members from the United States astronomical community, the AAS stewards and supports other key elements of a broader vision to bring literature and data together and to progressively facilitate research and astronomy outreach, representing a key strength of the project. An advisory board has been established and convenes periodically to provide external guidance and to assist with establishing relationships with potential Astrolabe users and additional stakeholders.

3 Preliminary Research: Astrolabe Workshops

To begin planning and development for Astrolabe, two workshops were held in 2015 and 2016 to obtain expert feedback and to develop functional requirements for the system. Participants in the "Astrolabe" astronomy cyberinfrastructure workshops were primarily from the University of Arizona, National Optical Astronomy Observatory (NOAO), the American Astronomical Society (AAS) and CyVerse staff. Data collected from both workshops include: field notes taken by workshop organizers using a collaborative note-taking platform; preliminary surveys administered online using Qualtrics survey software; audio recordings of presentations, group discussions and breakout sessions; and transcription of key dialogue. This material was analyzed qualitatively by coding for key themes and sub-themes.

3.1 2015 Workshop Description and Outcomes

This workshop was held in Tucson in July 2015 with 32 participants and organizers, including astronomy and data science experts. Final recommendations of participants included:

- Identify a mission statement and clear science use cases;
- Take advantage of CyVerse cyberinfrastructure and the longevity of University of Arizona as a home for this new repository;
- Obtain community buy-in and manage expectations for the system;
- Focus on "low-hanging fruit" such as data not curated elsewhere and data behind figures in journals; and
- Develop a follow-on workshop for additional feedback.

3.2 2016 Workshop Description and Outcomes

The 2016 "Astrolabe"" workshop included 22 local and nonlocal participants. The group was comprised of research astronomers, University of Arizona and CyVerse representatives, and other expertise related to astronomy data and publishing, along with five workshop organizers. The following four topic areas were recommended for the immediate attention of Astrolabe:

- Physical format of dark data (i.e. historical data stored on tapes);
- Author websites archiving data (not typically long-lived);
- LSST time domain and serendipitous data cases (such as follow-up to LSST observations and discovery through historical data); and
- Searching the literature for references to dark data (for indicative text, broken links, etcetera).

4 Current Status and Ongoing Development of Astrolabe

Current activities of the Astrolabe Project include:

- Soliciting pilot datasets to develop workflows, and cultivating a user community;
- Searching for uncurated or "at-risk" data by mining the literature, and by contacting authors individually based on the Astrolabe team's review of particular types of publications;
- Working towards the objectives specified in a NSF grant (award #1642446) to develop WorldWide Telescope (WWT) software as a repository front-end and visualization tool;
- Submitting additional funding proposals for system development, including to broadly create protocols for migrating data from obsolete media into Astrolabe;
- Collaborating with CyVerse to develop and optimize interfaces, applications, and indexing;
- Developing a metadata pipeline, including an application for automatic extraction of some metadata from FITS headers, and creating a standard metadata template following the IVOA ObsCore model [15];
- Incorporating the Unified Astronomy Thesaurus (UAT) as a controlled vocabulary [16]; and
- Designing a web portal as an interface to the CyVerse data store, to conveniently facilitate data deposition and reuse.

4.1 WorldWide Telescope Interface

Originally developed by Microsoft Research but currently managed by the AAS, WorldWide Telescope (WWT) is an open-source tool for visualization of astronomical data. With WWT, Astrolabe will provide users with a convenient option for automatically visualizing compatible image data within the repository environment and making Astrolabe datasets visible and accessible to WWT users. This new visualization functionality and cloud-based processing of images and all-sky and catalog data for display in WWT clients will be implemented over the next three years.

4.2 Searching for "Dark Data" in the Literature

Attempts to locate publications of interest for Astrolabe follow-up have resulted in combinations of search terms in the Astrophysics Data System (ADS) that narrow down the literature for manageable human review and selective follow-up with authors who may possess uncurated data. A list of papers without existing data links in the ADS index has been generated using the following search parameters:

- 24 articles contain the terms "new survey" or "new catalog" with a URL
- 82 articles contain the term "SDSS-III"
- 188 articles contain the term "online data"
- 1399 articles published between 2006 and 2016 contain a URL

This spreadsheet provides additional information about the identified articles, including ADS bibcode identifiers, associated grant information, and number of times read online. As Astrolabe initially aimed to seek data from highly-cited Arizona authors, a second spreadsheet of articles lists 1142 articles associated with Arizona who are first-authors and published between 2005 and 2015. A third study has identified broken links in journals published by AAS (a likely indicator of hidden data associated with a publication). Efforts continue to locate relevant papers within and beyond the four astronomy journals published by AAS. Machine learning and/or statistical modeling to determine significant variables associated with inaccessible data could assist the Astrolabe team with honing searching strategies to more accurately locate papers of interest, simultaneously providing insight into data sharing behavior and trends over time.

5 Conclusion

The Astrolabe project builds on existing technology as much as is practical to provide preservation and access to otherwise uncurated data. The Astrolabe team is working with the scientific community to identify both the most critical datasets as well as the functionality that is needed for a new easy to use data and computing environment for the astronomical community.

References

[1] M. Pennock, Library and Archives **January** (2007)

[2] J.C. Wallis, A. Pepe, M.S. Mayernik, C.L. Borgman, *An Exploration of the Life Cycle of eScience Collaboratory Data*, iSchools Conference (2008) http://hdl.handle.net/2142/15122

[3] J. Greenberg, Cataloging and Classification Quarterly **47(3-4)**, (2009)

[4] P. Darch, A. Sands, textitBeyond Big or Little Science: Understanding Data Lifecycles in Astronomy and the Deep Subseafloor Biosphere, iConference (2015) http://hdl.handle.net/2142/73655

[5] T. Hey, S. Tansley, K. Tolle (Eds.), *The Fourth Paradigm: Data-Intensive Scientific Discovery* (Microsoft Research, 2009)

[6] A. Accomazzi, R. Dave, *Semantic Interlinking of Resources in the Virtual Observatory Era* in *Astronomical Data Analysis Software and Systems XX*, edited by I.N. Evans, A. Accomazzi, D.J. Mink, A.H. Rots (2011), Vol. 442 of Astronomical Society of the Pacific Conference Series, p. 415-424

[7] C.L. Borgman, Journal of the American Society for Information Science and Technology **63**, 1059–1078 (2012)

[8] C.L. Borgman, *Big Data, Little Data, No Data: Scholarship in the Networked World.* (MIT Press, Cambridge, MA, 2015), ISBN 9780262028561

[9] E. Henneken, Bulletin of the Association for Information Science and Technology **41**, 40-43 (2015)

[10] A. Goodman, J. Fay, A. Muench, A. Pepe, P. Udomprasert, C. Wong, *WorldWide Telescope in Research and Education* in *Astronomical Data Analysis Software and Systems XXI*, edited by P. Ballester, D. Egret, N.P.F. Lorente (2012), Vol. 461 of Astronomical Society of the Pacific Conference Series, p. 267-270 https://arxiv.org/abs/1201.1285 (2012)

[11] P.B. Heidorn, Library Trends **57**, 280-299 (2008)

[12] W.C.Lenhardt, M. Conway, E. Scott, B.O. Blanton, A. Krishnamurthy, M. Hadzikadic, M. Vouk, A. Wilson, *Cross-institutional research cyberinfrastructure for data intensive science* in J2016 IEEE High Performance Extreme Computing Conference (HPEC) (2016)

[13] S.E. Hampton, C.A. Strasser, J.J. Tewksbury, W.K. Gram, A.E. Budden, A.L. Batcheller, C. Duke, J.H. Porter, Frontiers in Ecology and the Environment **11**, 156-162 (2013)

[14] CyVerse, http://www.cyverse.org/learning-center (2017)

[15] M. Louys, D. Tody, P. Dowler, D. Durand, L. Michel, F. Bonnarel, A. Micol, http://ivoa.net/documents/ObsCore/index.html (2017)

[16] A. Accomazzi, N. Gray, C. Erdmann, C. Biemesderfer, K. Frey, J. Soles, *The Unified Astronomy Thesaurus.* in textitAstronomical Data Analysis Software and Systems XXIII, edited by N. Manset, P.Forshay (2014) Vol. 485, p. 461-464 https://arxiv.org/abs/1403.6656

The conference's room (Photo: M. Wenger)

EPJ Web of Conferences **186**, 04001 (2018) https://doi.org/10.1051/epjconf/201818604001
LISA VIII

Growing a Bibliography

Sherry Winkelman[1,*], *Arnold* Rots[1], *Raffaele* D'Abrusco[1], *Glenn* Becker[1], *Sinh* Thong[1], and *Michael* McCollough[1]
[1] *Harvard-Smithsonian Center for Astrophysics*

Abstract. The Chandra Data Archive (CDA) has been tracking publications based on Chandra observations in journals and on-line conference proceedings since early in the mission. Our goals are two-fold: 1) provide a means for Chandra users to search literature on Chandra-related papers to further their scientific research; and 2) provide a means for measuring the science produced from Chandra data. Over the years the database and its associated tools have expanded dramatically. In this paper I will give a history of the development of the bibliography with a focus on the human capital involved, along with the skill sets and management structures developed which allow us to maintain a very rich and extensive bibliography with a limited number of full time employees (FTEs). I will also cover how the diverse metadata collected has made the Chandra bibliography an essential resource in managing the Chandra X-ray Center.
This work has been supported by NASA under contract NAS 8-03060 to the Smithsonian Astrophysical Observatory for operation of the Chandra X-ray Center. It depends critically on the services provided by the ADS.

1 Introduction

The Chandra X-ray Observatory (CXO) was launched on July 23, 1999 and is the third of NASA's Great Observatories. It has two instruments which can be combined with two gratings to create 6 instrument modes. Chandra's data archive is a repository for Chandra data as well as a source of information for all Chandra-related activities such as: proposal submission and management; mission planning; data processing, including the Chandra Source Catalog; and data retrieval. The Chandra archive operations group is responsible for ensuring the integrity of data in the CDA; maintaining the Chandra Observation Catalog; curating the Chandra Bibliography; and tracking downloads of Chandra data. The focus of this paper is the growth, development, and use of the Chandra bibliography: Section 2 describes the development of the Chandra bibliography; Section 3 reviews the management tools and skills sets needed to develop and curate the bibliography; Section 4 explores some of the ways the Chandra bibliography has been used; and Section 5 gives some suggestions of what will be next for the Chandra bibliography.

2 Chandra Bibliography

The purpose of the Chandra Bibliography has always been two-fold: 1) provide a means for astronomers to search literature on Chandra-related papers to further their scientific research; and 2)

*e-mail: swinkelman@cfa.harvard.edu ORCID: 0000-0001-7354-6221

provide tools for measuring the science produced from Chandra data in the context of the astronomical literature as a whole. Since its inception, the Chandra Bibliography has linked Chandra data to the papers which analyze the data and shared the links with ADS. Throughout the mission, we have been curating the bibliography and providing database and application development for the bibliography with <= 1 FTE. A plot of FTEs required to develop and curate the Chandra Bibliography over the course of the mission is shown in Figure 1.

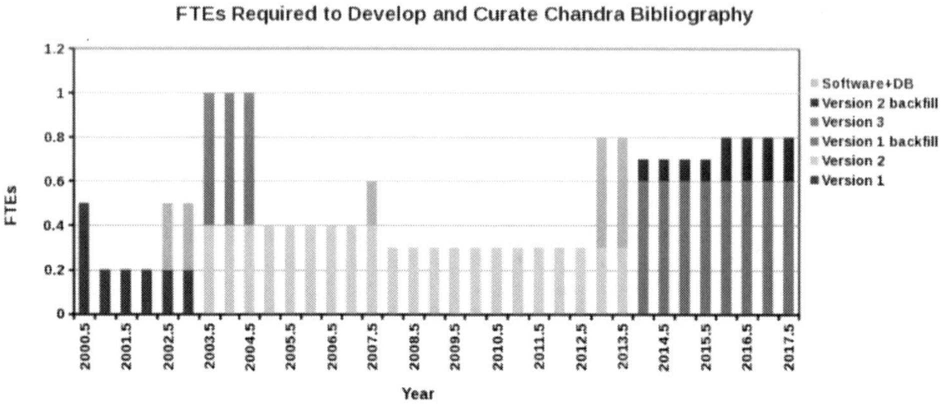

Figure 1. FTEs required to develop and curate the Chandra bibliography.

The bibliography has four basic categories of Chandra related papers. Each category has metadata which describes the Chandra relation within the paper in more detail. The general categories are:

- **Chandra Science Paper (CSP):** Chandra data contributes significantly to the science in the paper
- **Chandra Observatory Paper (CXO):** Chandra instruments, software, or operations
- **Miscellaneous Chandra Papers (Misc CXC):** Papers which refer to the Chandra X-ray Observatory in some way but do not rise to the level of CSP or CXO
- **Not Chandra Related:** Chandra/CXO is in the text of the paper, but the paper does not fit into any other category

Over the years the Chandra bibliography has undergone two major expansions. Each expansion in metadata was accompanied by new classification tools; added complexity to the bibliography database; and a backfill effort to bring the entire bibliography up to the new version. As can be seen in Figure 1, backfilling for Version 2 took 1.5 years operating at the 1 FTE level. Due to the size of the bibliography and the complexity of Version 3, the backfill efforts for this new version will be much greater. To date, we have spent an average of 0.17 FTEs per year in the backfill effort for four years and have completed the metadata for $\sim$ 10% of the backfill. We will need to enlist the help of more FTE's to complete this backfill effort.

The Chandra bibliography has a very rich set of metadata which are described in detail in Table 1 for each version of the bibliography. A GUI was developed to aid classifiers in compiling the metadata attached to papers. Figure 2 provides screenshots of several aspects of the current bibliography classification interface. The left panel shows the publication and status information window which also contains buttons for accessing the rest of the metadata options for a paper. The right panel shows the summary of all flags selected for the paper.

Table 1. Description of the three versions of the Chandra bibliography

Version 1: September 2000 - June 2004
Paper Related Flags:
pub_form: article
Chandra Related Flags:
CSP with direct analysis of data
Data links for all papers

Version 2: April 2003 - present
Paper Related Flags:
pub_form: title, abstract, article
pub_type: journal, proceeding, review, thesis, circular
refereed: Y or N
publication date as noted by ADS
citation count
date found on ADS
Chandra Related Flags:
CSP with direct analysis of data including data links and links to defined datasets
CSP with indirect analysis of data, but no data links
Misc CXC and CXO papers included
Instrument flags: ACIS, HRC, HETG, LETG, HRMA, PCAD, EPHIN
Software and operations flags
Chandra Source Catalog added in 2009
Chandra science context: theory, follow-up analysis, multi-observatory
Status Related Flags:
Person responsible for classifying
Status: in progress, published, request to author
Comment field

Version 3: October 2013 - present
Paper Related Flags: same as V2
Chandra Related Flags: Same as V2 plus
7-9 defined justification flags for each category plus ability to add in a custom justification
Chandra surveys used in paper
Status Related Flags: Same as V2 plus
a detailed history section to annotate the classification process
Data Related Flags: Attached to each ObsId
How data was located: ObsId given, ChaSer, contacted author, reference in paper, etc.
Where in paper lead was located: main body of text, table, figure caption, etc.
Type of analysis: direct, indirect: multiwave, indirect: follow-up, indirect: theory
Reference bibcode if ObsId location came from another paper
Acknowledgment Related Flags:
Grant source: CXC with (in)valid grant, NASA with (in)valid grant, SAO with (in)valid grant, etc.
Data: Chandra data, archive or source catalog
Team: Chandra Director's Office, Mission Planning, science team, etc.
Software: CIAO, ChIPs, Sherpa, Chart
Miscellaneous: support from CXC, acknowledgment with no Chandra connection, no acknowledgment section
Grant Related Flags: Attached to each grant
Grant number as stated in text
Official grant number
Related to proposal: direct, substantial, tangential, questionable, none
General Content Related Flags:
Content flags: data analysis, theory and/or computation, new survey or catalog, etc
Observatories by waveband
Publication Date Flags:
Publisher dates: received, revised, accepted, published, conference date
ArXiv dates: up four versions

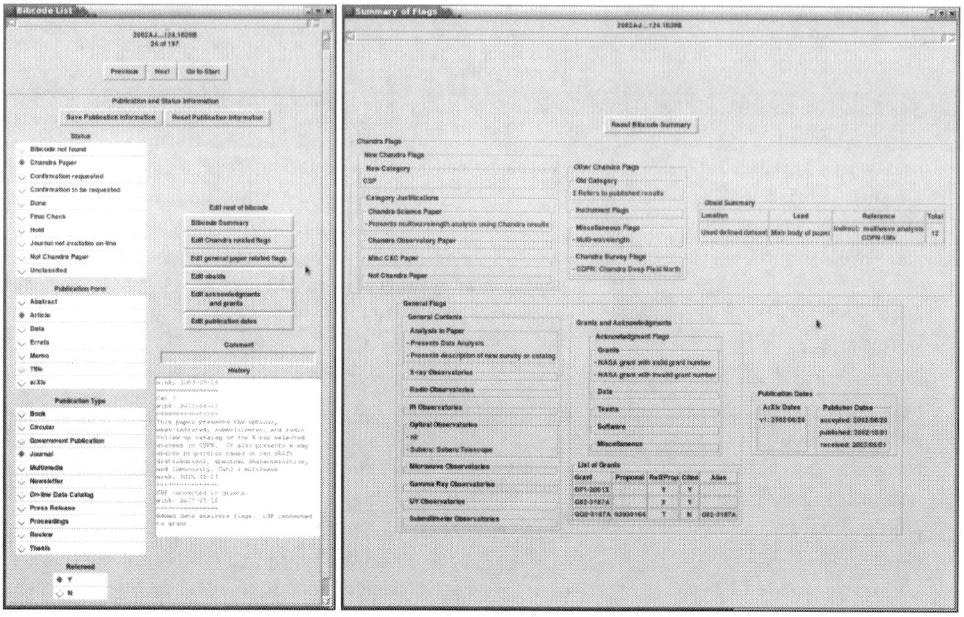

Figure 2. Screenshots of the bibliography editing GUI. The left panel is the publication and status information window. The right panel shows the summary window.

3 Management Tools and Skill Sets

Accuracy and consistency throughout the bibliography is critical. An essential component for maintaining that consistency as classifiers change is providing well documented metadata descriptions and classification procedures. Curators also need good tools to support their efforts and these tools may need to be provided by the curators themselves. We have developed applications which cover: preprocessing of bibcodes so we do not look at papers which are not relevant to us; classifying bibcodes so we can visualize the metadata we are attaching to papers; and post-processing of the bibliography to ensure data integrity and to produce metrics and other products useful to our users.

Data linking also requires good tools to search for data as they are described in the literature to match with the data as they are presented by the data repository. These tools are often available from the repository, but curators need to understand how to use those search tools effectively to match data descriptions in papers with data descriptions in the repository.

The most important skills for classifying and publishing papers are having: a good understanding of the tools available for curating papers; an acute attention to detail; and an understanding of the scientific process as it is described in astronomical papers. Applications development requires knowledge of: a programming language such as Perl, Python, or Java; database design and query structures; and web-application development tools. It is important that management understand and buy into the fact that both elements need to be supported in order for a bibliography to be maintained and effective.

In the 15+ years that the Chandra bibliography has been around, we have had twelve different classifiers/developers with the current Chandra bibliography team consisting of three classifiers and a manager/publisher/applications developer. Two of the classifiers can also provide application development support, while the manager/publisher also provides classification support when needed.

As our bibliography has grown and become more complex, we have found the following management structures to be of great use for curating the Chandra bibliography:

- defining separate roles for publishing and classifying papers
- automating actions with scripts whenever possible and as soon as possible
- providing feedback loops for maintaining and improving documentation and classification procedures
- setting reasonable schedules and priorities for backfill when adding new metadata fields
- consulting with a person with an astronomy research background to clarify how to classify papers which do not fit easily into your classification scheme
- advocating the inclusion of the bibliography in observatory tools and services
- anticipating future needs of users of your bibliography, particularly management, so you can prepare to handle more complex and detailed queries to your bibliography (they will come).

4 Uses of the Chandra Bibliography

As stated earlier, the Chandra Bibliography has two objectives: assist astronomers in their research and provide measures on the science output of the observatory. Implementation of these goals can be separated into three categories: interfaces; metrics; and assessing observatory program allocations.

4.1 Interfaces with the Chandra Bibliography

In addition to providing data links to ADS, the CDA offers a number of services to the astronomy community which tie into the Chandra bibliography: 1) the Bibliography Search Pages (http://cxc.harvard.edu/cgi-gen/cda/bibliography) for querying the bibliography (Figure 3); 2) a CXO Paper Repository (http://cxc.harvard.edu/cda/cxo_papers/cxo_papers.html) providing access to papers on the instruments, software and operations of the observatory (Figure 3); 3) the Chandra Proposal Search pages (http://cxc.harvard.edu/soft/propsearch/prop_search.html) to aid in proposal preparation (Figure 4); and 4) ChaSer (http://cda.harvard.edu/chaser/) for searching the CDA (Figure 4).

4.2 Metrics

We use the Chandra Bibliography to provide a number of metrics to various levels of management to give a sense of the science productivity of the observatory based on the number of Chandra Science Papers published, number of citations to those papers, etc. We have also developed metrics based on the data links in our bibliography. These metrics give us insight into the archival use of Chandra data and are less sensitive to observatory-specific characteristics than are the traditional metrics. Reference [1] gives a complete description of the data-centric metrics we have been using to measure speed of publication, fraction of observing time published, and archival usage. Recently we have been working on identifying high impact astronomy papers which are based on Chandra data and have a paper in this volume discussing our findings [3].

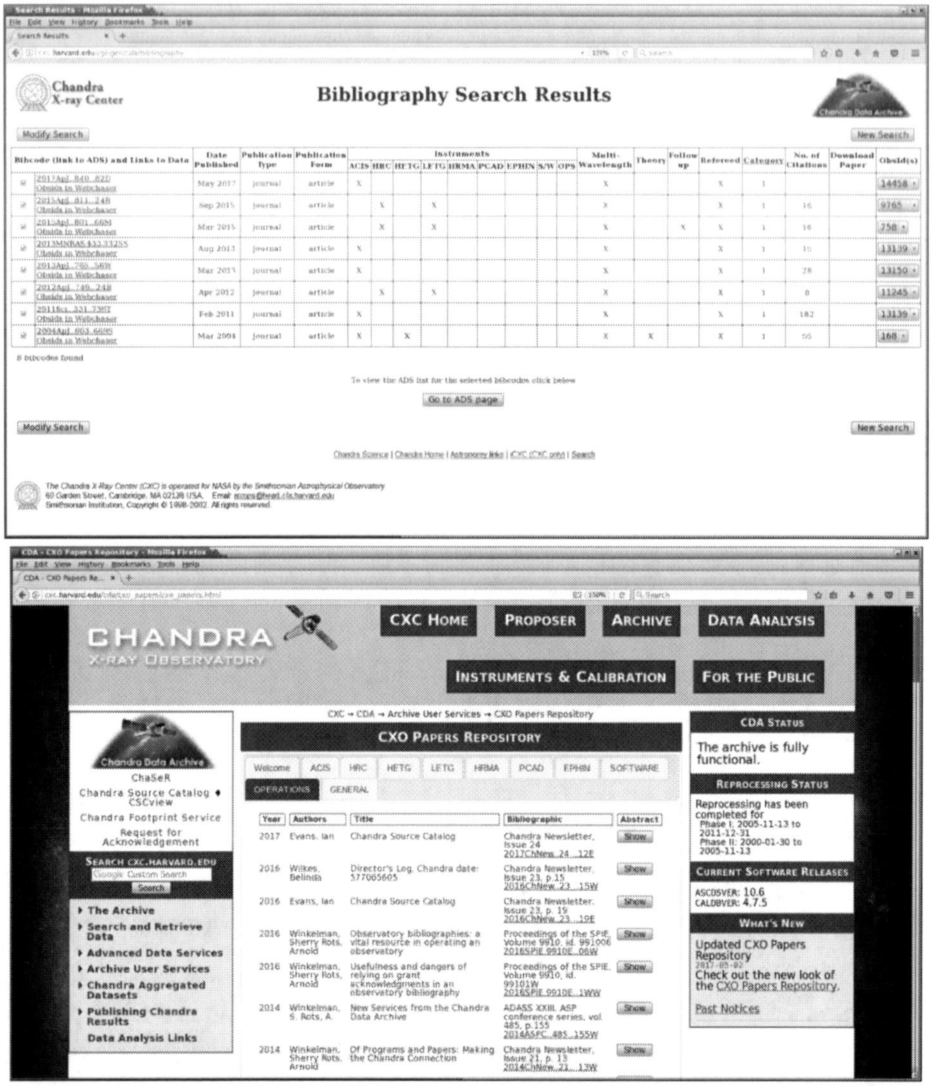

Figure 3. Screenshots of interfaces which access the Chandra Bibliography. The top panel shows search results from the Bibliography Search Pages for refereed journal articles which are CSPs with multi-observatory analysis of the Crab Nebula. The bottom panel displays the CXO papers focused on observatory operations.

4.3 Assessing Observatory Program Allocations

The CDA is routinely asked to provide input to the Chandra Users Committee (CUC), an advisory committee to the Chandra Director's Office representing the principal groups concerned with Chandra development and the general Chandra user community, regarding the productivity of various types of observing and funding programs offered by the CXC. The Chandra bibliography is heavily relied

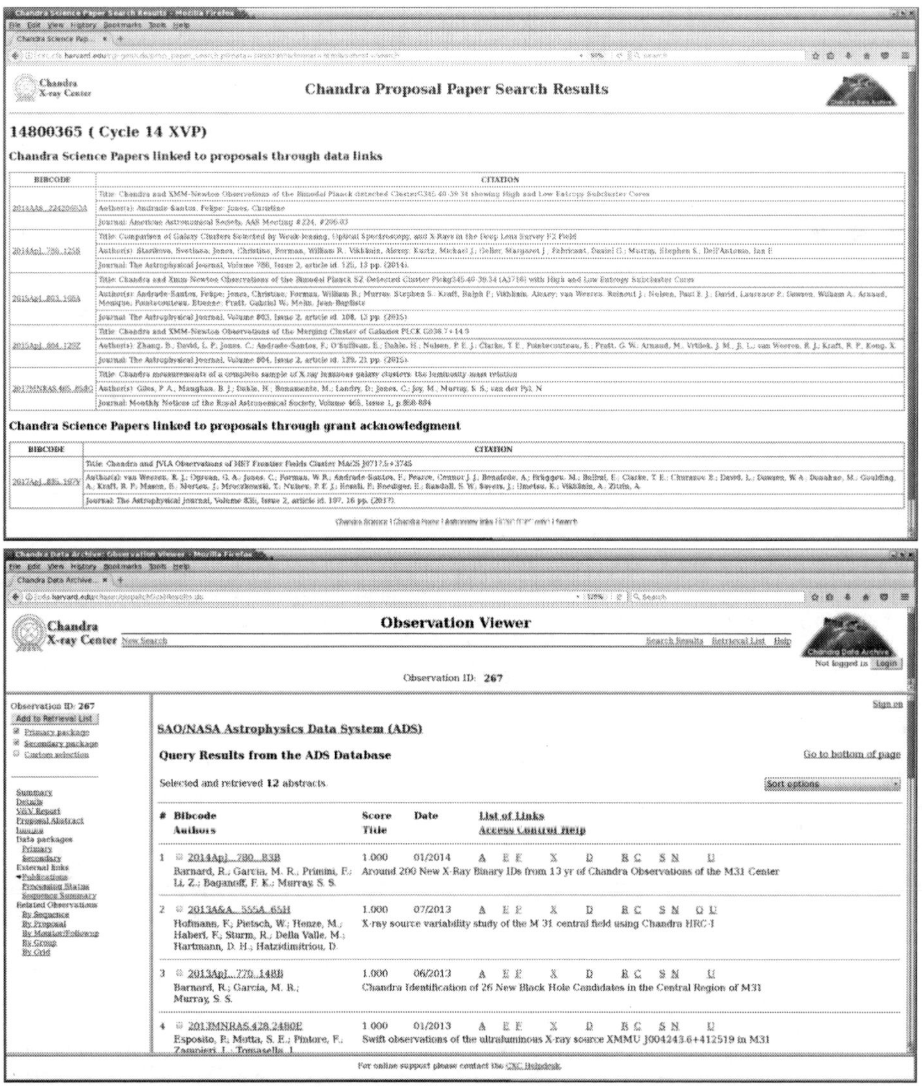

Figure 4. Screenshots of interfaces which access the Chandra Bibliography. The top panel shows search results from the Chandra Proposal Search pages for the XVP proposal for the Chandra COSMOS Legacy Program. The bottom panel displays the publications results within ChaSer for a search on M31 and then selecting ObsId 267.

upon for these assessments and in at least one instance the results lead to new metadata being added to the bibliography.

The first major assessment was with regards to the productivity of archive and theory awards granted by the CXC. These programs are money awards only, so data links were of limited use here. Our initial research focused on locating papers by searching ADS for Chandra archive and theory grants. However, we were eventually given access to all grant numbers associated with all Chandra proposals. As a result, grant links are now a part of the bibliography and we can use this information for determining whether a paper is associated with an observing proposal. We have also found severe limitations in using grant information which are detailed in reference [4].

In observing cycles 13-16, the Chandra Call for Proposals solicited proposals for X-ray Visionary Projects (XVP) which described major coherent science programs to address key, high-impact scientific questions in current astrophysics. The time allocated to XVPs largely came from an increase in the total time available to the observatory due to changes in Chandra's orbit, so other time allocations were not affected. Future calls for XVPs will require a change in allocation to other observing categories. Because the XVPs are too young to have a reliable publication history, the CDA was asked to assess the science impact of XVP-proxies to provide insight into offering another XVP round. Our approach to the problem was to define aggregated observing programs from the archive to create pseudo-XVPs and using the Chandra bibliography to apply publication statistics to the aggregated sets. Details of the study are in reference [5].

5 What Next?

Expectations are that the Chandra Observatory will remain in operation for the next 10-15 years. As the archive grows, the needs of the astronomy community for accessing archival Chandra data will likely expand. As the observatory ages, program changes will likely be implemented. The Chandra bibliography will play a role in both of those developments. In anticipation of future needs, the CDA has a few projects planned to lead us into the next decade of Chandra science:

- Continue backfill efforts to migrate to Version 3 to take full advantage of the new metadata
- Add SIMBAD descriptions to the Observation Catalog to provide another set of flags for searching for Chandra data and Chandra papers
- Transition from Dataset Identifiers to DOIs [2] to aid in citing Chandra data
- Update Chandra Bibliography Search application to include expanded flags
- Add new flags for High Impact Papers [3]
- Explore crowd sourcing as a means of engaging citizen scientists to add flags to papers
- Migrate database infrastructure and applications to the Archive Development Team

6 Conclusion

In summary, it is possible to create and maintain a very rich bibliography for an observatory with <=1 FTE, but a diverse set of skills and tools are required. Processes should be automated where ever possible and as early as possible; clear documentation of metadata and classification procedures is essential; and feedback loops are necessary to fit the needs of classifiers.

As a bibliography grows in size and complexity, the uses of the bibliography will also expand. Interfaces for accessing the archive should link into the bibliography, making the bibliography an integral part of the observatory. As the observatory matures, the bibliography will be seen a resource for managing the observatory, so plan to handle more complex questions from upper management.

Once a bibliography is considered an integral part of an observatory, new projects will present themselves, sometimes at a far too rapid clip but that is part of the fun.

This research was supported by NASA contract NAS8-03060 (CXC) and relies heavily on the services provided by the Astrophysics Data System (ADS). We gratefully acknowledge the work by all past and current members

 https://doi.org/10.1051/epjconf/201818604001

of the Chandra Data Archive Operations Team, who have helped build and maintain our bibliographic database. Without their efforts there would be no bibliography.

References

[1] A. Rots, S. Winkelman, G. Becker, Publications of the Astronomical Society of Pacific **124**, 391–399 (2012).

[2] A. Rots, R. D'Abrusco, S. Winkelman, *A Model for Using DataCite DOIs in Observatory Bibliographies*, These Proceedings (2018)

[3] S. Winkelman, A. Rots, R. D'Abrusco, *Evaluating High Impact Papers: Are We Missing Something?*, These Proceedings (2018)

[4] S. Winkelman, A. Rots, *Usefulness and dangers of relying on grant acknowledgments in an observatory bibliography* in *Observatory Operations: Strategies, Processes, and Systems VI* (2016) Vol. 9910, p. 99101W

[5] S. Winkelman, A. Rots, *Observatory bibliographies: a vital resource in operating an observatory* in *Observatory Operations: Strategies, Processes, and Systems VI* (2016) Vol. 9910, p.991006

Andràs Holl (Photo: M. Wenger)

EPJ Web of Conferences **186**, 04002 (2018) https://doi.org/10.1051/epjconf/201818604002
LISA VIII

Shared nomenclature and identifiers for telescopes and instruments

Emmanuelle Perret[1,*], *Mireille* Louys[1,2], *Mihaela* Buga[1], and *Soizick* Lesteven[1]

[1] *Université de Strasbourg, CNRS, Observatoire astronomique de Strasbourg, UMR 7550, F-67000 Strasbourg, France*
[2] *Université de Strasbourg, CNRS, ICube Laboratory, UMR 7357, F-67412 Illkirch, France*

Abstract. In the context of sharing public data, science results are expected to be reproducible and therefore we need full traceability of the origin of the data. On the documentalist side, there is a need to relate instrumental origins to the published data.
We propose to define a shared nomenclature to index each publication with unique designations for facilities, telescopes and instruments which could benefit from the Virtual Observatory work on semantics.
This would help the documentalists to check the consistency of the instrument description in publications or make it more explicit. Observation period, data quality and spectral coverage, for instance, may be checked by referencing a global instrumentation service which gathers the nominal observation parameters for the telescope/facility/instrument involved.
Based on this indexation mechanism, then the bibliographic metrics for telescope/instrument usage would be easy to compute, and tracking services like the ESO telescope bibliography database (TelBib) or others would be easier to feed. This paper traces the existing initiatives and gives the example of a facility description framework reusing Virtual Observatory metadata which could be fed by the community.

1 Introduction

Sharing public data is now well adopted in the sciences, especially in astronomy, and supported by public policies – e.g. the G8 Science Ministers Statement in 2013 [1]. Reusability and reproducibility of the data that researchers have to share are key features.
Particularly in astronomy, where science questions are driven by the need for multi-wavelength, multi-messenger observations, it is crucial to attach metadata to observations in order to describe facilities, instruments, wavelengths and other particular information related to the data. By doing so, those observations become discoverable and understandable and so can be reused by the astronomer community.
Another interesting point related to data sharing is the interoperability provided by metadata standards which allow querying multiple archives at the same time – e.g. as provided in the Virtual Observatory framework.
To this day, there are individual usages of the descriptions of facilities but no common effort from the

*e-mail: emmanuelle.perret@astro.unistra.fr ORCID: 0000-0002-4068-8175 community to coordinate them. The aim of this paper is to review the existing initiatives in terms of facility designations and to initiate a discussion on a possible common strategy.

1.1 Definition points

The terms facility, observatory or even instrument can have different meanings according to the usages – e.g. Voyager 2 or HST are both observatories when querying the SPASE registry [2] whereas HST is a telescope in the WISeREP database [3]. In this paper, we propose to define the following terms:
- Facility: a global term to designate any installation containing a telescope, a satellite or an observatory.
- Observatory: the ground-based buildings equipped with a telescope.
- Telescope: any platform ground-based or spacecraft.
- Instrument: a physical device that collects data.

2 Existing initiatives

We conducted a survey of various repositories describing facilities on the Web. All these sites deal with astronomical data and belong to the following categories:

- Agencies webpages which describe telescopes, space missions and instruments (how data are produced)
- Publishing services which distribute scientific papers (how data is analyzed)
- Astronomical services providing data (how data is distributed: by domain of interests or data collections)

2.1 Agencies

Agencies describe their missions and facilities online with more or less details. Although most agencies provide the names and descriptions of their facilities, they do so in an uncoordinated manner.
Besides, agencies have more specific services which use facility designations. For instance, the NASA Space Science Data Coordinated Archive (NSSDCA) provides a master catalog to search for spacecraft. It also provides a search through the SPASE Registry [2] for NASA observatories and instruments.
Furthermore, the Jet Propulsion Laboratory (JPL) provides the Navigation and Ancillary Information Facility (NAIF) integer ID codes [4] which are part of an observation geometry information system to help scientists to interpret scientific observations from space-based instruments aboard robotic spacecraft.
In both cases, alternative names of instruments are managed by a unique identifier designating all the synonyms. But, if you search Voyager 2 on SPASE, its code is "1977-076A" whereas the NAIF identifier is "-32". So no uniform index currently exists.

2.2 Publishing services

As a matter of fact, the American Astronomical Society (AAS) which publishes three of the main journals in astronomy, initiated the use of a list of controlled facility keywords 12 years ago [5] in order to encourage authors to tag their publications with these facility tag labels. However, authors adoption did not spread enough and facility names are still searched in free text.
In order to index articles with names of the facilities which have provided observational data, a common reference list among journals and the astronomical community at large would be very helpful.

2.3 Astronomical services

An interesting example of an online repository fed by the supernovae community worldwide is the Weizmann Interactive Supernova data REPository (WISeREP) [6] which gathers observational data on supernovae. This data repository provides a list of telescopes, instruments and gives the wavelength domains. Its purpose is to select appropriate datasets. It also provides filter names and operational status information, for example updates and decommissioning dates.
Besides, comparing the spectral coverage of various filters helps to tag associated data correctly, select datasets and combine measurements together. For instance, this is done when a spectral energy distribution profile is built from catalog data.
The SVO filter profile service [7] is a good example of an initiative to gather all the information concerning filters coming from diverse facilities. A constrained but rich list of instruments and facilities is proposed for the search of the filters. Furthermore, the service provides the transmission plots and reference parameters for the filters like the effective wavelengths and widths.

In order to synthesize the observed behaviors of the various initiatives in our survey, we suggest registering the following features as columns in Table 1:

- "Goal" describes the main objective of the site
- "Facility type" indicates the subclass of facility as: instruments 'Inst", telescopes (ground or space) as "Tel" or observatories "Obs", if provided
- "Facility description" indicates the presence of a free text description
- "Instrument type" indicates "F" for full, if a precise list of instrument is given, or "P" for partial if an indication is given (e.g. spectroscopic or photometric in WISeREP [3])
- "Time range" indicates if the period when the facility is operational is given - usually it is not given but some sites indirectly provides some time information (TelBib [8] via program ID and VESPA [9] via time intervals).
- "Spectral coverage" shows "F" as full, "P" as Partial or "N" as no, depending on whether the site gives the range of wavelength/frequency/energy covered by the instrument
- "Alternative names" indicates "Y" (yes) if alternative names can be resolved on the web site API
- "Reference features" mentions if additional parameters of the instrument are given like the spectral or spatial resolution for instance
- "Spatial location" can be "Y" for yes if the coordinates location is given or "P" for partial if the observatory site or the country only is mentioned
- "Number of entries" gives the number of facility rows for each site

Table 1. Examples of websites which propose diverse identifiers for facilities (F=Fully described, P=Partially described, Y=Yes, N=No)

Website	Goal	Facility type	Facility Description	Instrument type	Time range	Spectral coverage	Other names	Reference features	Spatial location	Number of entries	Comment
NSSDC SPASE Query (NASA) [2]	All NASA facilities	Inst - Obs	F	F	N	F	Y	Y	N	?	Registry query with description sheets of each instruments
WISeREP [3]	SNe data and instrumental profiles	Tel - Inst	N	P	N	P	N	N	P	109 - 182	Gives also the URLs of instruments and diameters
NAIF (NASA) [4]	Planetary science facility index	Tel	N	N	N	N	Y	N	N	151	Integer IDs used for an observation geometry information system named "SPICE"
Facility Keywords (AAS) [5]	Indexing publications	All	N	N	N	F	N	N	P	500	The list of facilities contains general type of facility if solar.
Filter Profile Service (SVO) [7]	Filters data	Tel - Inst	N	N	N	F	N	Y	N	54 - 95	Gives specific details of all filters of an instrument like the effective widths
TelBib (ESO) [8]	Indexing ESO publications	Tel - Inst	N	N	P	P	N	Y	N	9 — 54	Provides the links to datasets which give the properties of observations
VESPA (Paris observatory) [9]	Planetary sciences	Inst	N	N	P	F	N	N	N	?	A lot more data like targets, exposure time, phases...
ADS facility identifiers (SAO/NASA) [10]	Naming convention for datasets	Obs	N	N	N	N	N	N	N	258	A list of codes for naming facilities in the dataset identifiers
Minor Planet Center (IAU) [11]	Planetary sciences	Obs	N	N	N	N	N	N	Y	2038	Locations of observatories along with an integer code; also searchable on GAVO [12]
NA1-Matrix: Europlanet [13]	Planetary sciences	Obs - Tel	N	F	N	F	N	N	P	305	Gives also locations (with a map), area of interest, diameters
JKT's list (personal Keele Univ. page of John Taylor) [14]	Telescope availability for UK astronomers	Tel - Inst	F	P	N	F	N	Y	N	33 - 113	Gives URLs of instruments, comments, application deadlines. All instruments available via OPTICON [15]
AstroWeb (CDS) [16]	Interesting URLs in astronomy	Obs - Tel	P	N	N	N	N	N	N	375	Not maintained since 2010
Planetary Data System (JPL - NASA) [17]	Planetary Sciences	Tel	F	F	F	F	N	Y	N	389?	Archive all planetary data obtained with NASA facilities and ESA also
Space Assigned Numbers Authority (CCSDS) [18]	Spacecraft registry	Tel	N	N	N	Y	N	N	N	1172	Protocol registry created under the Consultative Committee for Space Data Systems (CCSDS) for spacecraft identifiers
OSCAR tool (WMO) [19]	Satellites registry	Tel	F	F	F	N	N	N	F	683	Observing Systems Capability Analysis and Review Tool (OSCAR) from the World Meteorological Organization

3 Classifying the requirements

As we can see in Table 1, various level of details are provided for facility descriptions in order to serve diverse needs.

3.1 Simple list

For some usages, a list of all the unique, common and long-lasting identifiers of existing facilities would be enough. For example, in the case of the naming convention proposed by the SAO/NASA Astrophysics Data System (ADS) [10] to build dataset identifiers from facility names, the users only have the list of the unique identifiers – based on human-readable acronyms – of 258 facilities.
However, even for a simple list, the problem of alternative names needs to be solved. Among the examples given in Table 1, we can cite the NAIF codes [4] which give for one integer code, diverse synonyms of the same spacecraft. In the case of SPASE [2], you also have unique identifiers as we have already seen in section 2.1. but here, the alternative names are given in a description sheet and they can be queried as well as their identifiers.
Finally, in the case of Telbib [8] or the SVO filter profile service [7], the lists of queriable instruments and facilities are constrained, so alternative names are not searchable. But, in those cases, the alternative names have to be managed internally, e.g. the librarian at ESO keeps an updated list of the ESO facilities with alternate designations.

3.2 Detailed facility profile

Going a little further, according to the usages, diverse degrees of details related to the facility are required. In Table 1, indeed, all examples other than ADS give more information than a simple list of identifiers.

For some efforts, information about the instrument, where it is installed (telescope, observatory name and location) or to which space mission it belongs, is provided (SPASE [2]). For others, the dependency is not tracked, or it is described in a non standard way (AAS [5]).
The instrument type is also an important information which is provided at diverse levels. From the simple distinction between photometric or spectroscopic (e.g. WISeREP [3]) to a detailed list of instrument types with the ability to search among them (e.g. Europlanet [13], SPASE).
Another aspect, which is particularly important in planetary science, concerns the observational periods of the instruments. For this parameter, we can have the global commissioning dates, the global observational dates corresponding to a specific version of the instrument or we can have the specific dates corresponding to a given observation (e.g. indicated in the program identifier of an observation). For instance, we already note in section 2.3 that WISeREP provides comments on instrument upgrades or decommissioning dates. This leads to the versioning notion, which is indeed something to keep in mind when we want to describe observational data.
Finally, in the case of planetary science, the location of observatories is really important. For example the Minor Planet Center (MPC [11]) provides unique codes along with the precise coordinates for more than 2,000 observatories. This kind of information is also given in the case of VESPA [9]. To a less extent, the country of observatories hosting telescopes is also often provided (e.g. AAS, WISeREP, Europlanet).
From those specific usages of facility designations, we can define three levels of facility profiles:

- precise : these are the exact parameters for a specific observation

- intermediate: the usual values – e.g. spectral coverage, recorded for a data collection
- global : generic values for the facility – e.g. when no other information is provided for the spectral range of a data set, the minimum and maximum spectral range of the instrument can be used if there is a link toward the agency's instrument web-page description.

4 Discussions and Feedbacks

4.1 Specific usages

As we have seen in section 2.2, a list of unique identifiers would allow tagging of the scientific publications in a homogeneous way.
Considering the new technological tools, tagging afterwards by automated extraction of facility names thanks to text mining and machine-learning tools, could help both ways by:

- Adding synonymous identifiers to a built list of standardized facility identifiers
- Assisting authors to rightfully tag the publication with a standardized facility identifier

It is worth noting that those kind of tags would be useful for librarians and would allow the tracking of papers more easily according to the facilities used to obtain the scientific results. For instance, a consistent tag in publications could be used by ADS to allow a search among papers using specific facilities.
For the VizieR associated data service [20] which archives images and spectra associated to publications, a simple list of facilities would be a great help because this list, once integrated in the indexation tool, could help the documentalists and the authors to attach the correct telescope/instrument to the corresponding observations.
Another level of usage is reached with the ESO Telbib service [8] which allows the retrieval of all the scientific papers in which the ESO facilities have been used for observations.
This service is really beneficial for the agency, because all the metrics of this archive can help the decision makers. For instance, when an instrument is less cited in papers, the agency can think of improvements to make to this instrument [21].
However, to provide such a service, the librarians need to parse the publications in order to find the citations of the ESO facilities and program identifiers, and this is a real time-consuming effort. Even if curation will always be necessary, it would be easier to find the relevant papers if the publications have homogeneous facility tags.
Another use-case example can be the VizieR Photometry Viewer tool [22] which offers to plot the fluxes for one source from the photometry gathered among all the VizieR catalogs (currently 16,200). In order to do so, precise data on filters must be collected and attached to the VizieR datasets. Those metadata are carefully enriched by the whole team of astronomers, engineers and documentalists. Having unique identifiers for all facilities along with the links to instruments and filters descriptions provided by the agencies could be a good basis to enrich the metadata.

Furthermore, by tagging publications, other usages could be implemented. For instance, in SIMBAD, each data measurement is linked to its bibcode which displays the title of the corresponding scientific paper. And so, with facility-tagged publications, each parameter could eventually show the facilities used to make the observations for that paper.

4.2 Feedback

From all the existing initiatives and use cases, we can conclude that there is a need for a common nomenclature, with unique and long-lasting identifiers for facilities. Should this telescope/instruments

list be as complete as possible, this would emphasize interoperability and would be very beneficial for the community. This conclusion has been enforced by the LISA VIII meeting at the end of which a working group started to discuss all these aspects, towards a definition of a nomenclature. This interest working group currently includes more than twenty people from more than ten institutes (including NED, ADS, ESO, Keck, NOAO, CfA, etc.)
The interest comes from librarians and data center service developers who need to tag the facilities in their descriptions or publications. Curation of facilities is especially hard because, on one hand all the alternative names, and on a second hand all the facilities which share the same names, but are different. This is true for the organizations, i.e. SAO (Cambridge) versus SAO (Moscow) but also for the instruments, i.e. MegaCam can be the instrument of diverse telescopes (CFHT or MMT and Magellan).
Also the difficulties in defining all the terms we are talking about (see section 1.1) and the level of facility profiles (see section 3.2) are some illustrations of the complexity of sharing a common description. All the diversity of usages suggest that we need a common base on which more layers can be linked from diverse origins.

4.3 On the way to having a general facility representation

In the framework of the IVOA Semantics group, we initiated an effort to homogenize the notation and description of facilities properties and link it to existing web pages maintained by agencies in charge of building them. This representation is spread among 6 basic concepts: Facility, Telescope, Observatory, Space Mission, Spacecraft, and Instrument.

The Facility Class is derived from a VOResource [23] metadata structure which gathers basic curation parameters. It handles a list of alternate names for the Facility, and a start time and stop time interval related to the operational period of the Facility.
The other classes are derived from the Facility class. Two subclasses are defined from the Observatory Class: a 'ground-based' Observatory has location coordinates (longitude, latitude, altitude), a SpaceCraft is an Observatory subclass related to a Space Mission instance. All facilities usually have an Agency name attribute and are described by a webpage or document by each agency. A more detailed description is available in [24].
As we examined other projects, namely in Space Science, the PDS4 [25] data model representation offers a richer description of these items, with an efficient binding of the various instrumental classes together with the data collections. This perspective is pursued currently in our group to produce a clear relational diagram of the various concepts and derive an Instrument Class with precise instrument types. This is an ongoing effort.

5 Conclusion

We have presented a survey of various repositories describing facilities available on the Web. We have categorized those examples by the origin of data (how data are produced, analyzed and distributed) and by the type of information requested according to diverse needs. From the reviewed simple list of facility identifiers to detailed facility profiles, we can conclude that there is a need for a common nomenclature, with unique and standardized identifiers for facilities. A common and basic description would help to maintain and build upon diverse astronomical services. More detailed facility descriptions as proposed in the PDS4 data model suggest a way to go for developing a standardized facility description repository with shared contributions from the whole astronomical community.

Acknowledgements

We thank U. Grothkopf and G. Schwarz for fruitful discussions before launching the survey and thank all LISA participants who helped to expose their current practice for tagging facility names in datasets, articles, and other metadata.

References

[1] *G8 science ministers statement* (2013), http://www.gov.uk/government/news/g8-science-ministers-statement
[2] *NASA NSSDCA - SPASE registry*, http://nssdc.gsfc.nasa.gov/spase/
[3] *Weizmann interactive supernova data repository*, http://wiserep.weizmann.ac.il/
[4] *NAIF integer id codes*, http://naif.jpl.nasa.gov/pub/naif/toolkit_docs/FORTRAN/req/naif_ids.html
[5] *American astronomical society facility keywords*, http://journals.aas.org/authors/aastex/facility.html
[6] O. Yaron, A. Gal-Yam, *Wiserep—an interactive supernova data repository* (2012)
[7] *SVO Filter Profile Service*, http://svo2.cab.inta-csic.es/theory/fps3/
[8] *ESO Telescope Bibliography*, http://telbib.eso.org/
[9] *Virtual European Solar and Planetary Access (VESPA) Query Interface*, http://vespa.obspm.fr/planetary/data/epn/query/all/
[10] *ADS Web services - list of facilities for which datasets are available*, http://vo.ads.harvard.edu/dv/facilities.txt
[11] *IAU Minor Planet Center, List of Observatory Codes* , http://www.minorplanetcenter.org/iau/lists/ObsCodesF.html
[12] *MPC observatory codes on German Astrophysical Virtual Observatory (GAVO)*, http://dc.zah.uni-heidelberg.de/obscode/q/query/form
[13] *Europlanet NA1-Matrix of Ground-based facilities and space missions*, http://europlanet-na1.oeaw.ac.at/matrix/index.php
[14] J. Southworth, *JKT's list of available telescopes*, http://www.astro.keele.ac.uk/jkt/telescopes.html
[15] *Optical Infrared Coordination Network for Astronomy (OPTICON)*, http://www.astro-opticon.org/
[16] *Astroweb: space telescopes and observatories*, web page, http://cdsweb.u-strasbg.fr/astroWeb/astroweb/telspace.html
[17] *The Planetary Data System (PDS): data search*, http://pds.jpl.nasa.gov/tools/data-search/
[18] *Space Assigned Numbers Authority -SANA- Spacecraft Identifier*, http://sanaregistry.org/r/spacecraftid/spacecraftid.html
[19] *World Meteorological Organization (WMO) Observing Systems Capability Analysis and Review Tool (OSCAR)*, http://www.wmo-sat.info/oscar/instruments
[20] *VizieR Associated Data*, http://cdsarc.u-strasbg.fr/assocdata/
[21] U. Grothkopf, S. Meakins1, D. Bordelon, *Use Cases of the ESO telescope bibliography*, in *These Proceedings* (2018)
[22] *VizieR Photometry viewer*, http://vizier.u-strasbg.fr/vizier/sed/

[23] R. Plante, M. Demleitner, the IVOA Resource Registry Working Group, *Voresource: an xml encoding schema for resource metadata, version 1.1*, IVOA Recommendation (2017), http://www.ivoa.net/documents/VOResource/20170425/index.html
[24] M. Louys, B. Ceconni, D. S., P. Le Sidaner, *Unique identifiers for facilities and instruments in astronomy and planetary science*, in *Astronomical Data Analysis Software and Systems XXV, 2015* (in press), Astronomical Society of the Pacific Conference Series
[25] PDS4 Information Model Specification Team, *PDS4 Information Model Specification* (2017), https://pds.jpl.nasa.gov/pds4/doc/im/current/index_1800.html

Strasbourg Observatory (Photo: Pierre-Alain Duc)

EPJ Web of Conferences **186**, 04003 (2018) https://doi.org/10.1051/epjconf/201818604003
LISA VIII

First Light for DOIs at ESO

Dominic Bordelon[1,*], *Uta* Grothkopf[1], and *Silvia* Meakins[1]

[1] *European Southern Observatory, Karl-Schwarzschild-Str. 2, 85748 Garching near Munich, Germany, library@eso.org*

Abstract. Digital Object Identifiers (DOIs) are a helpful tool for academic research, and their use continues to grow. DOIs assist in citing and identifying research data, as well as in combating link rot. As a producer of large amounts of astronomical data, ESO has become interested in creating DOIs for its datasets. However, there are significant considerations to be undertaken in reaching this goal, and some local infrastructure is required. The ESO Library has planned, designed, and implemented a local solution for creating DOIs, consisting of an application, the ESO DOI Service, which other departments can use. This paper discusses the requirements, development process, and features of the ESO DOI Service.

1 Introduction

Digital Object Identifiers (DOIs) are an increasingly useful tool in research and scholarly communications. As a producer and provider of large amounts of astronomical data, ESO has become interested in DOIs for enhancement of the usability and visibility of its data. Within ESO, beginning in mid-2015, the Library has led the effort in planning, designing, and implementing a local solution for creating DOIs. In early 2017, the first ESO DOIs were created, and work continues towards the creation of more DOIs and better software features.

This paper describes how DOIs have come to be used at ESO, the software employed to do so, and future plans for DOIs at ESO.

2 Digital Object Identifiers

Digital Object Identifiers (DOIs) are persistent and globally unique identifiers designed to facilitate long-term access to resources. Although DOIs have many potential applications, they have been used particularly in a scholarly context since their inception in 2000. As of March 2017, roughly 143 million DOI names have been assigned.[1] A DOI is created (or "minted") by a client via a Registration Agency (RA), which in turn is a member of the International DOI Foundation (IDF). The IDF is a nonprofit organization which acts as the central authority and maintenance agency for the DOI system. DOIs and the DOI system are described by ISO Standard 26324:2012[1], and the DOI Handbook published by the IDF is the authoritative source for information about DOIs.[2]

* e-mail: dominic.bordelon@eso.org ORCID: 0000-0002-8382-1300

[1] http://www.doi.org/factsheets/DOIKeyFacts.html (accessed July 2017)

A DOI is conceptualized as a *digital identifier of an object* rather than an *identifier of a digital object*. It consists of the DOI name—the identifier string itself—as well as accompanying metadata which describes the resource. DOI names have a syntax (see Figure 1) which follows the pattern `10.1234/abc-678/90.1`, where: `10.` is common to all DOIs; `1234` represents the registrant of the DOI in question; / (forward slash) is a separator between the prefix and suffix; and `abc-678/90.1`, the suffix, is an arbitrary (but unique) string determined by the registrant. The accompanying metadata helps users—human or machine—make use of the described resource, and is implemented through a minimal metadata schema, the DOI Kernel, which Registration Agencies are free to supplement according to specific needs or practices.

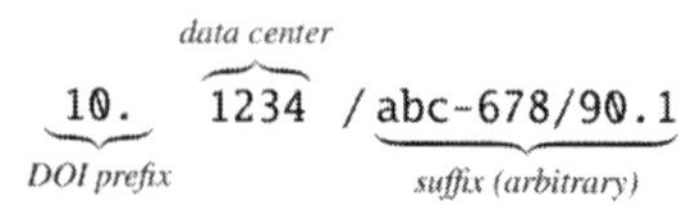

Figure 1. DOI name structure

Besides serving as a unique, unambiguous identifier for a resource, DOIs are also "actionable" or "resolvable" in the sense that a resource can be retrieved via the internet by prepending `https://doi.org/` to its DOI name. When the user requests this URL, the doi.org proxy server responds with the resource's actual URL, and the user is seamlessly redirected. This resolution therefore promises a persistent link to the resource, which is ensured through the system's technical infrastructure as well as the obligations undertaken by registrants. In practice, the resolution via doi.org does not return the resource (i.e., data file) to the user, but instead a "landing page." This landing page displays useful metadata, along with a link to the resource itself (e.g., a PDF file for a journal article) if available, perhaps in multiple file formats. The address of the landing page and the resource file(s) themselves may change, but the registrant is obligated to maintain up-to-date information with the DOI system so that resolution continues to function properly. Persistence is a key goal not only of the DOI name, but also of the resolution service, in an effort to combat the problem of link rot (see e.g., Klein et al. 2014[3]).

The metadata component of the DOI system enables interoperability and the creation of additional services. For example, CrossRef, a well established RA for scholarly publishers, offers services such as reference linking, "Cited-by" (tracking resources that have cited the resource in question), and a registry of research funders. DataCite, an RA specialized for research data, offers similar services, and collaborates with CrossRef to link metadata between their respective repositories.

Organizations wishing to mint their own DOIs must fulfill the following requirements: become members/clients of an RA (which is a member of the IDF); upload formatted metadata for each resource to the RA; serve a landing page for every resource; and maintain both the resource's record in the DOI system and the target landing page URL, to which the resolver redirects. It is important to note that, by minting a DOI, the organization is making a promise of persistence of the resource, its associated metadata, and the URL association. Organizations should also expect pay for the RA's service, e.g., a per-DOI fee.

3 ESO's interest in DOIs

In 2015, the ESO Library began to take interest in adopting DOI creation at ESO. Such adoption would facilitate data usage and sharing, assist the organization in tracking usage of its data, and further advance a commitment to long-term data availability. Two areas that would seem to benefit were astronomical data and publications.

As a highly productive research data center, ESO produces and shares a large amount of astronomical data, amounting to a current total of over 1.5 million images and spectra, occupying approximately 65 terabytes of disk space.[2] The ESO Science Archive Facility offers both raw data and "data products," packages of pre-processed data, to the scientific community.

ESO's Education and Outreach Department (ePOD) produces a quarterly publication, *The Messenger*, which presents ESO's activities to the public.[3] Each issue contains roughly a dozen articles, covering the topics of instrumentation, science, and organizational news.

The Library initiated conversations with these departments, and they agreed that minting DOIs would be beneficial. The next step was to make an arrangement with an RA. Since ESO anticipated using DOIs primarily for datasets (as opposed to publications), DataCite was chosen due to its prominence in this area. This choice was further facilitated by the arrangement between the Technische Informationsbibliothek (TIB; a German national library for sciences, engineering, and math based in Hannover) and DataCite: TIB allows research centers in Germany proxy membership to DataCite. The Library was eager for ESO's signatory to be a member of the upper administration, to represent an institution-wide commitment to the responsibilities engendered by DOI minting. In March 2016, a contract was signed between ESO's administration and TIB for the creation of DOIs.

4 Planning and points of consideration

In planning to mint DOIs, several aspects had to be considered. In addition to typical concerns of project management (e.g., time allocation, schedule, and quality assurance) and software development (e.g., choices of programming language and data storage), the local situation at ESO also raised further questions. In particular, more than one department was interested in minting DOIs, and others could arise in the future. Therefore the first question was who would have ultimate responsibility for DOIs organization-wide; since the Library initiated use of DOIs, it adopted this role. But other questions that had to be answered were:

- Who is responsible for DOIs in each "client" department?
- How would the Library deal with diverse technologies—e.g., different databases—and needs—e.g., different landing page appearance—among departments?
- How would the Library translate each department's metadata, describing different kinds of resources, into a common format?
- How could the Library make a system that future, unknown clients would be able to use?

Each of these questions was of key importance in creating a solution.

5 Implementation: architecture and process

In August 2016, the Library began development of a software solution for minting DOIs for ESO. In February 2017, a minimum viable product was achieved, and the following month the service launched with the first batch of DOIs (articles for vol. 167 of *The Messenger*).

The Library created an application, called the ESO DOI Service, which is responsible for (1) communicating with DataCite for create–update–delete operations, and (2) serving landing pages.

[2] http://www.eso.org/public/about-eso/esoglance/ (accessed July 2017)
[3] https://www.eso.org/sci/publications/messenger/

The service's clients are various departments which want to mint DOIs, such as the Data Archive or ePOD. A client sends a structured request for a new DOI, containing the desired DOI name and resource metadata, to the service via an HTTP API; the service creates its own record and attempts to register the DOI with DataCite; and finally the service gives the client a response indicating whether creation was successful or not. Editing and deaccessioning DOIs each follow a similar pattern. As soon as the DOI is registered with DataCite, the ESO DOI Service serves a landing page at `http://doi.eso.org/<DOI>` (where <DOI> is replaced by the DOI name), displaying the metadata that the service has for that resource. This architecture is depicted in Figure 2.

The ESO DOI Service was developed using Test-Driven Development (TDD) and using common application design patterns such as Model–View–Controller and Data Mapper.[4] The decision was also made early to decouple the application from those of other departments, and to communicate with clients only via the HTTP interface, to avoid any unnecessary dependency. As a result, the ESO DOI Service and a client application can be "black boxes" to one another: internal details can safely change as long as the agreed-upon HTTP interface is maintained. The ESO DOI Service does not need to "know" which database system (for example) a client department uses and vice versa.

The central data model of the ESO DOI Service is the DataCite Metadata Schema (DMS).[5] This decision was made to answer the question of how to deal with multiple types of resources. It also serves the goal of producing the XML required by DataCite for DOI minting, which is the key output of the system. Semantic mapping takes place outside of the scope of the application; each client determines, in consultation with the Library, how its data needs to be translated to fit the metadata scheme. A simple example is that *Messenger* authors and principal investigators of Archive datasets both become DMS `creators`. Generally these mappings are not difficult, but they are needed to mint DOIs with DataCite.

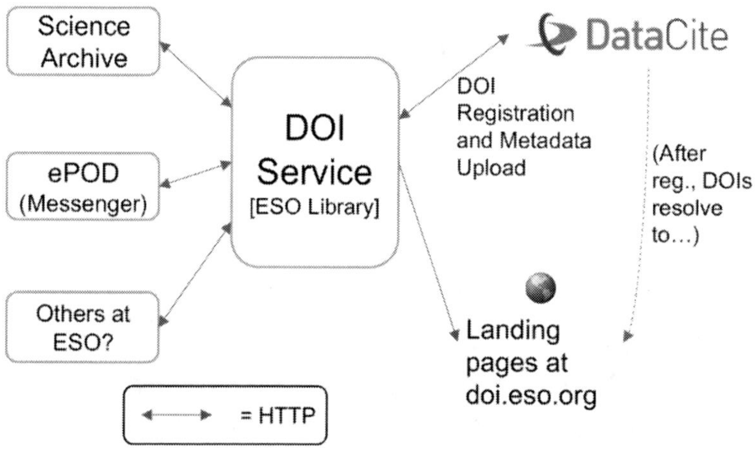

Figure 2. DOI Service architecture within ESO.

6 Features

In addition to the ESO DOI Service's core functionality described above, a few more features have already been implemented.

6.1 Namespacing & extensibility

ESO DOIs begin their suffix with a namespace, such as `/0722-6691/` in the DOI `10.18727/0722-6691/5010`. Although this namespace is, by design, meaningless to the global DOI system, the ESO DOI Service uses it for next-consecutive-integer DOI naming. In the example above, the next DOI minted in the same namespace would end with `5011`. This allows a client to request a DOI without having to specify the full name, instead specifying only the namespace.

The ESO DOI Service also uses the namespace to "know" more about what kind of resource is being described. The basic data model can be extended for the sake of richer landing page content, and the visual design of the landing page itself can change based on the namespace. This has already been implemented for *Messenger* articles, whose landing pages display volume/section/page information (which is outside of the core data model), and which display labels like "author" instead of "creator." Such behavior is possible for as many namespaces as desired.

6.2 Machine-readable embedded metadata

FORCE11, a multidisciplinary organization centered upon advances in scholarly communication, released in 2014 a "Joint Declaration of Data Citation Principles," whose purpose is to help provide "a foundation of robust, accessible data" for "sound, reproducible scholarship" through the expression of eight principles.[6] Although a full discussion of the Declaraction is outside the scope of this article, the principle of "Access" through metadata is particularly relevant to the ESO DOI Service. In turn, Fenner et al. have produced an extremely helpful "Road Map" to fulfilling these principles.[7]

One recommendation by Fenner et al. is the use of `<meta>` tags in landing pages, with Dublin Core attributes. The purpose of these tags is for landing pages to interoperate with reference manager applications, since this is how they extract metadata. In a similar vein, ESO DOI Service landing pages also contain JSON-LD using Schema.org, which improves discoverability (for example by search engines).

6.3 "Cite this article"

ESO DOI Service landing pages also have a simple "Cite this article" (or "Cite this dataset") feature. A user can copy-paste the pregenerated reference, so citing the resource is quite easy—and guaranteed to include the DOI, if the user does it this way.

7 Outlook and conclusion

Returning to the questions posed in Section 4, the chosen approach seems to have addressed these challenges. The Library deals with diverse technologies by offering its own service, accessible via a common, language-agnostic interface (HTTP), but whose internals are isolated from the applications of other departments. The adoption of the DataCite Metadata Schema (DMS) as the core data model provides a common target for the metadata of various departments and resource types, and also allows for easy translation to the XML required for registering the DOI with DataCite. Finally, the use of namespaces and a data model extensible beyond the DMS enables the application to serve diverse wishes.

However, this local solution—like the DOI system as a whole—is not merely technological, but must also rely upon critical social components in order to succeed. The Library must continue to communicate with client departments to begin minting DOIs for Archive datasets, and the ESO DOI

Service's metadata schema and API must be well conveyed for clients to use them effectively. Any technical problems that arise will also require interdepartmental coordination.

Although the ESO DOI Service is an operational application, there are several "nice-to-have" features which have yet to be implemented. These include: search of ESO DOIs; automated handling of inverse relations between resources; downloadable bibliographic formats for landing-page users; regular checks for resource persistence; and perhaps others. The Library hopes to accomplish these in the short-to-medium term.

DOIs are an important component in contemporary research and scholarly communications, but they require a global infrastructure, and each participating organization must somehow integrate its data, using its own infrastructure. For ESO, the ESO DOI Service provides that integration. In crafting such a solution, the Library has found another way to continue its mission of supporting researchers and data providers for the benefit of the astronomy community as a whole.

References

[1] International Standards Organization, *ISO 26324:2012 Information and documentation — Digital object identifier system* (ISO, Geneva, 2012)

[2] International DOI Foundation, *DOI® Handbook*, https://doi.org/10.1000/182 (accessed July 2017)

[3] M. Klein, H. Van de Sompel, R. Sanderson et al., PLoS ONE **9(12)**, e115253 (2014), https://doi.org/10.1371/journal.pone.0115253

[4] M. Fowler, D. Rice, M. Foemmel et al., *Patterns of Enterprise Application Architecture*, Addison-Wesley, Boston, (2002), ISBN 0-321-12742-0

[5] DataCite Metadata Working Group, *DataCite Metadata Schema Documentation for the Publication and Citation of Research Data* Version 4.0 (DataCite e.V., 2016) https://doi.org/10.5438/0012

[6] Data Citation Synthesis Group, "Joint Declaration of Data Citation Principles" (San Diego, FORCE11, 2014) https://www.force11.org/group/joint-declaration-data-citation-principles-final (accessed July 2017)

[7] M. Fenner, M. Crosa, J. Grethe et al., *A Data Citation Roadmap for Scholarly Data Repositories*, bioRxiv **97196**, (2016) https://doi.org/10.1101/097196

[8] N. Paskin, "Digital Object Identifier (DOI®) System", *Encyclopedia of Library and Information Sciences, Third Edition* (CRC Press, 2009) 1586-1592

EPJ Web of Conferences **186**, 05001 (2018) https://doi.org/10.1051/epjconf/201818605001
LISA VIII

A Bibliometric Analysis of Observatory Publications 2011-2015

Dennis Crabtree[1,*]

[1]*National Research Council Canada, Dominion Astrophysical Observatory, 5071 West Saanich Road, Victoria, BC V9E 2E7 Canada*

Abstract. Bibliometrics are increasingly used as metrics to measure the performance of individuals, institutions and countries. Refereed publications are the primary output of modern observatories. In this paper, I use bibliometric techniques to examine the performance of astronomical observatories.

1 Introduction

Bibliometrics is increasingly used as an evaluation tool in science. Papers in refereed journals represent the key product of modern astrophysical observatories. These publications represent the observatories' contribution to knowledge and represent the *return on investment* for the partners. Observatories carefully track the papers based on data from their telescopes as this is of great interest to the various governance bodies and other stake holders.

The number of refereed papers each year is a measure of an observatory's annual productivity. A telescope's productivity usually ramps up over a period of 5-10 years following the start of operations. While a telescope's productivity is one metric of a telescope's performance, the relevance, or impact, of the publications is another important metric.

Citation counts are a standard measure of the impact of a paper, i.e., the more citations a paper receives, the more relevant the work is to the broad astronomical community.

In this paper, I examine the productivity and impact of most optical telescopes with an aperture large than 3.5-meters. A small number of other telescopes, such as the Hubble Space Telescope (HST) and JCMT are also included in this study.

2 Observatory Publications

Observatories maintain bibliographies of refereed papers that are based on analysis of data from their observatory. To be considered an observatory publication a paper must base some or all of its conclusions on data from that observatory. IAU Commission 5 has developed a document [1] describing the best practices for creating such a bibliography. At least one observatory now includes papers based on analyses of the catalogs of objects identified from large scale surveys undertaken by their telescope. Papers based on catalogs already form a large part of the SDDS bibliography and will also be a significant fraction of LSST's bibliography.

*e-mail: Dennis.Crabtree@nrc-cnrc.gc.ca ORCID: 0000-0002-0162-6086

I have retrieved the bibliography for each telescope in this study from individual observatory websites, directly from the librarians or other individuals in a few cases or via other sources such as personal ADS libraries. This data is usually in the form of a list of bibcodes. In all cases, I rely on each of the observatories to determine whether a paper meets the definition of an observatory paper. I do minimal filtering, removing papers in Annual Review of Astronomy and Astrophysics that are included by some observatories.

The basic publication information as well as the bibcode for each paper is stored in a relational database (Microsoft Access). The bibcode for each paper is validated via ADS and corrections made to the data in the case of an invalid bibcode. Once a paper's bibcode is verified, further information including the number of citations to the paper is retrieved from ADS and added to the database.

3 Impact Metric

The cumulative citation count for a paper increases as the paper ages. It takes at least a year before the citation count is meaningful, and this is the reason this study includes papers only up to 2015. Generally, a paper's citation rate (number of citations per year) peaks 2 years after publication although each paper's citation history can exhibit quite different behavior.

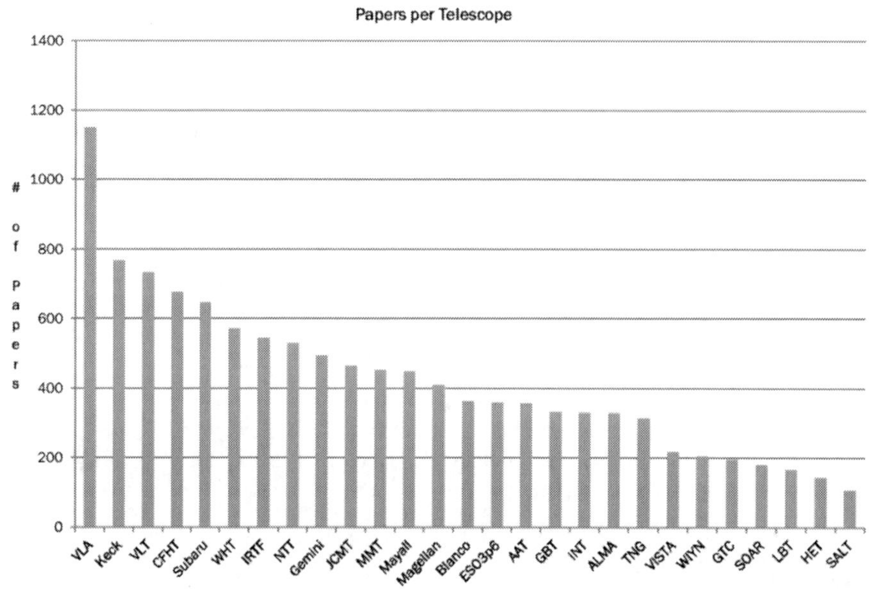

Figure 1. Productivity: Papers per Telescope: 2011 - 2015

Since citation counts grow with time and I will be comparing papers published over a 5-year period, it is necessary to correct the citation counts for this aging effect. My approach is to normalize each paper's citation count by the citation count for the median AJ paper (based on number of citations) of the same year. I refer to this normalized citation count as the *impact* of a paper.

4 Productivity

The productivity of each telescope in this study for the period 2011-2015 is shown in Figure 1. In the case of a telescope actually representing multiple physical telescopes (e.g., Keck has two, VLT has four) I have normalized the paper count to produce a productivity per telescope.

Note that HST, which produced 4081 papers in the period, is not included in this figure as it would distort the scale of the plot. There is a large range of productivity between the most productive (VLA, Keck) and the least productive (LBT, SALT) telescopes. The low productivity of the LBT and SALT is due to technical issues that have kept the telescopes off-line for periods of time. A variety of factors affect a telescope's productivity. This includes the quality of the site, quality, and aperture, of the telescope, instrumental capabilities and the user community.

The data used in this study allows an examination of other aspects of refereed publishing in the field of astronomy. The growth in the size of research teams, and the corresponding increase in the number of authors on papers has often been noted. I examined the long-term trend in the number of authors per paper using the data in my observatory publication database which extends back to the early 1990s.

Figure 2 shows the average number of authors on each paper for the period 1991-2015. The smooth line is a fit to the data with an extrapolation to the year 2025. The error bars represent the standard error of the mean. The data prior to 2000 is based on a smaller number of telescopes and the errors are larger. The data indicates that the increase in the number of authors per paper was relatively flat during most of the 1990's and then began to increase rapidly after 1997. An extrapolation of the data indicates there will be an average of 18 authors on each paper in 2025!

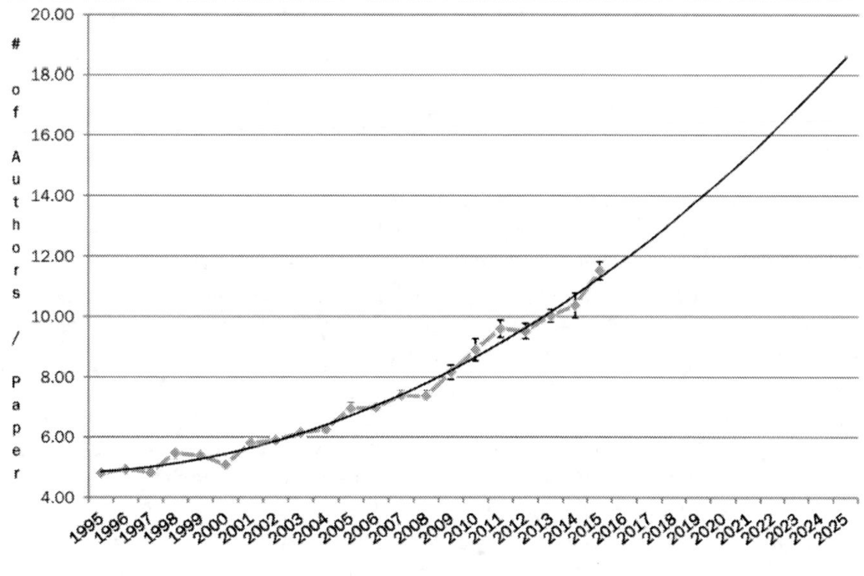

Figure 2. Average # of Authors per paper by # of authors

5 Impact

While the productivity of an observatory is an important metric of the performance of an observatory, the impact of the published research is also an important metric. The number of citations to a paper is considered a good quantitative measure of a paper's impact. Impact is not to be confused with the quality of the research. Rather, impact is a measure of the relevance of the paper to other research and researchers in the field. Of course the number of citations is influenced by other factors such as area of research and the citation culture of each sub-field. However, since we are studying large aggregates of papers and not comparing individual authors, the effect of these factors should be minimal.

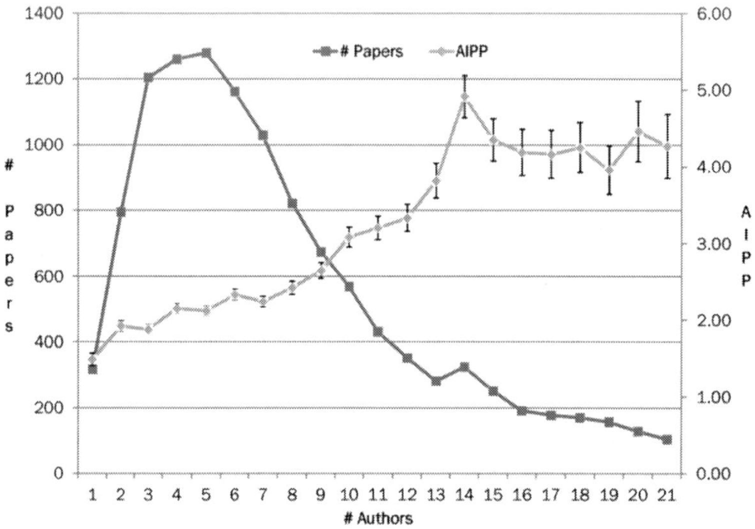

Figure 3. Average impact per paper and the number of papers by year

Before investigating the impact of the observatory papers in this study, I will look at how the impact of a paper correlates with the number of authors. As discussed earlier research teams are increasing in size (indicated by the increasing number of authors per paper).

Figure 3 shows the average impact per paper (AIPP), as well as the number of papers, as a function of the number of authors for all papers in this study. The AIPP curve has error bars that represent the standard error of the mean. Note that there are more papers with eleven authors than there are single author papers.

Note the dramatic increase in impact with the number of authors. The AIPP increases until there are 14 authors per paper and then more or less levels out. This clearly indicates that on average, papers produced by larger teams have higher impact, i.e., higher number of citations.

I also investigated the average impact per paper (AIPP) by observatory. Figure 4 shows the AIPP as well as the productivity for each observatory by year. HST is not included in the sample but produced 4081 papers during this period with an AIPP of 3.22. There are considerably smaller differences in impact between observatories than there was in productivity. While Keck is still near the top (#2), ALMA now holds the honor of having the highest average impact per paper for this period. Also

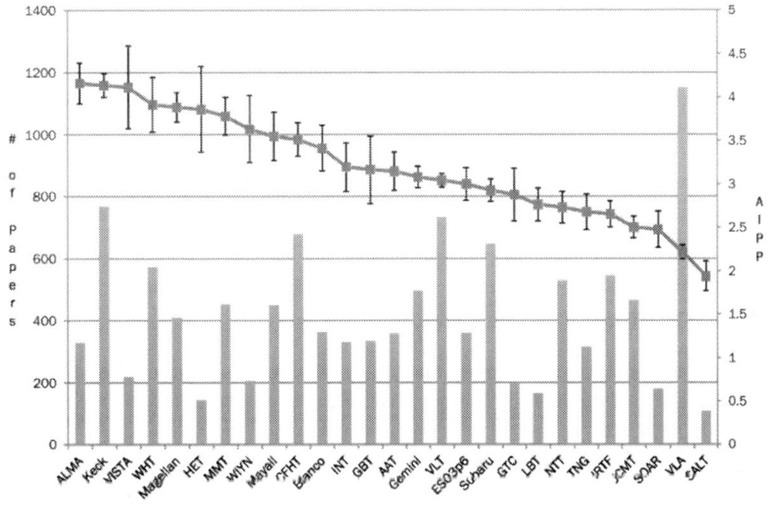

Figure 4. Average impact per paper and the number of papers by observatory

worth noting is that HST's AIPP falls in the middle of the pack. So while HST produces significantly more publications than ground-based telescopes, the impact of most of its papers are quite average.

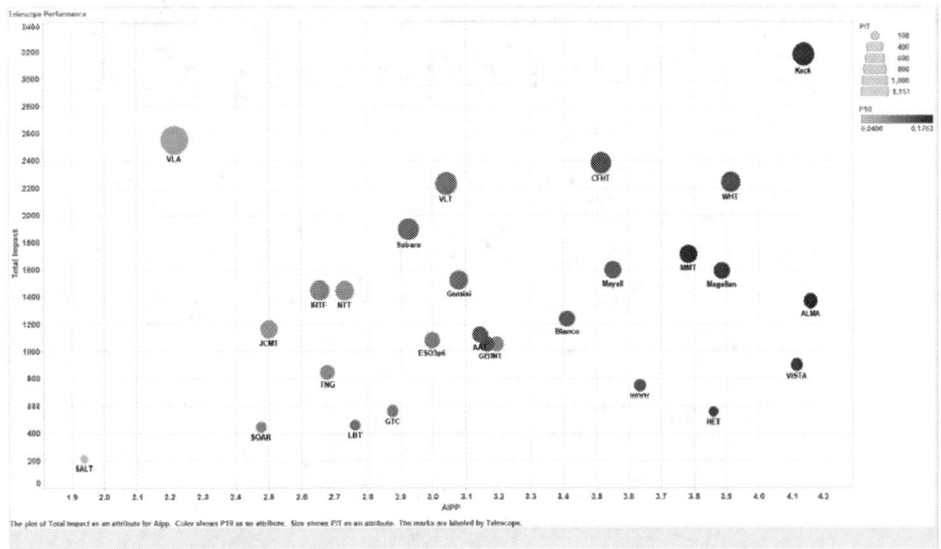

Figure 5. Performance plot of all observatories.

Another bibliometric that is often used is the number, or fraction, of papers that are top-cited 10% of all papers(P10). Figure 5 is an overall performance plot that combines the productivity, AIPP, total impact, and fraction of the papers in the top-cited 10%. The size of the circle is proportional to the observatory's productivity, while the shade of the circle indicates P10 with a darker shade, indicating a higher P10 andthe percentage of the observatory's papers that are in the top-cited 10%. The total impact is simply the sum of the impacts of an observatory's papers, or the AIPP times the number of papers.

Increased performance indicators (higher productivity, higher AIPP, and higher P10) tend to move together for an observatory. That is, if a telescopes is high in one of these indicators the others will also tend to be higher. However, this is not ironclad. For example, the MMT has a very high P10 for its location in the plot.

Acknowledgments

Thank you to observatory librarians and other staff who maintain their observatory bibliographies. This work would not have been possible without their dedication. This work has made use of NASA's Astrophysics Data System.

References

[1] Lagerstrom, J., *Best Practices for Creating an Observatory or Telescope Bibliography from the IAU Commission 5 Working Group on Libraries* in *Open Science at the Frontiers of Librarianship*, edited by András Holl, Soizick Lesteven, Dianne Dietrich, and Antonella Gasperini (2015), Vol. 492 of Astronomical Society of the Pacific Conference Series, p. 99–103

Open session : Mark Allen (Photo: M. Wenger)

EPJ Web of Conferences **186**, 05002 (2018)
LISA VIII
https://doi.org/10.1051/epjconf/201818605002

Astronomy in Chile: Assessment of scientific productivity through a bibliometric analysis

Rodrigo Cortes*, *Denise* Depoortere**, and *Lucina* Malaver***

Bibliotecas UC, Pontificia Universidad Catolica de Chile

Abstract. The skies of Northern Chile are considered among the best in the world for astronomy due to their geographical, climatic and atmospheric conditions. In fact, during the last several decades, a great number of astronomical observatories have been built by space research institutions devoted to space research, turning Chile into one of the countries with the greatest astronomical observation capacity in the world. Consequently, it is relevant to explore and assess the development of astronomy in Chile during the last ten years (2005-2015), carrying out a bibliometric analysis to extract traditional metrics, complemented with alternative metrics, to identify the Chilean production and scientific contribution of research in this field of knowledge. The results from traditional metrics, based on the information obtained in the Web of Science (WoS) database analyzed through InCites for the Astronomy & Astrophysics category indicate a sustained increase of the scientific production for the discipline during the last ten years. In particular, the Normalized Citation Impact of organizations and institutions conducting research in Chile is above the worldwide average. On the other hand, the alternative metrics, including for instance, the altmetrics types presented by the SCOPUS database, reflect that the Chilean research impact is much lower in the context of social networks.

1 Introduction

Chile is quite active in astronomical research given its special sky conditions. The skies of Northern Chile are regarded among the best in the world for astronomy due to their geographical, climatic and atmospheric conditions like the absence of clouds most of the year with 340 clear sky days[1], dry soils, and low light intensity or pollution. During the last decades, a great number of astronomical observatories have been built by institutions devoted to space research, turning Chile into one of the countries with the greatest astronomical observation capacity in the world. In fact, it is estimated that by 2020 70% of all observational activity of the world will be carried out in Chile. Currently, the number is 40%[2], allowing a remarkable advance in scientific production in the area of Astronomy[3].

*e-mail: rcortee@uc.cl
**e-mail: ddepoort@uc.cl
***e-mail: lmalaver@uc.cl

[1]Educarchile. 2006, http://www.educarchile.cl/ech/pro/app/detalle?id=75547

[2]Treister, E. 2016, http://www.emol.com/noticias/Tecnologia/2016/05/04/797699/Chile-tendra-todos-los-telescopios-mas-grandes-del-mundo.html

[3]Espinoza, C. 2016, http://www.latercera.com/noticia/chile-segundo-en-latinoamerica-en-ranking-de-publicaciones-en-revistas-de-excelencia/

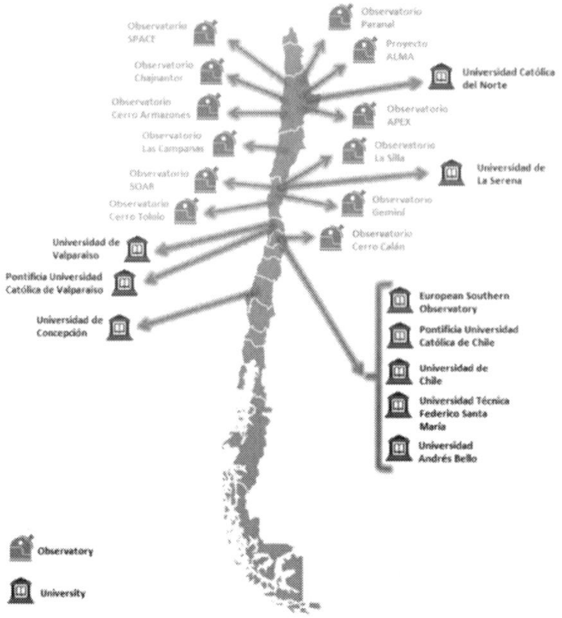

Figure 1. Observatories and Institutions in Chile

Figure 1 displays a map indicating the geographical distribution of observatories and universities related to astronomy in Chile. It can be observed that most of the observatories are located in the North of Chile while the research centers are mostly located in the Central Region.

2 Objectives

The main objective of this article consists of showing the important increase of the astronomical scientific productivity developed in Chile during the last decade (2005-2015), analyzed by means of traditional and alternative metrics. Moreover, the specific goals of this work are: (1) To assess the scientific impact of astronomy in Chile; (2) Show the countries and institutions that carry out most of the scientific astronomical collaboration with Chile; and, (3) Investigate the evolution of altmetrics indexes in the Chilean scientific production in Astronomy.

3 Methodology

The methodology adopted in this analysis for exploring and assessing the development of astronomy in Chile is based on the study of bibliometric indicators provided by InCites[4] and complemented with alternative metrics provided by SCOPUS. The analyzed data correspond to scientific publications of Chile in Astronomy & Astrophysics over a 10 years span. The first focus of this bibliometric analysis is on the Chilean scientific productivity, identified through the analysis of the authors' institutional

[4]InCites is a research analytics tool that allows to analyze institutional productivity and benchmark output against peers worldwide (http://researchanalytics.thomsonreuters.com/m/pdfs/indicators-handbook.pdf)

affiliation. For visualizing research trends and growth of knowledge in this scientific field it is important to identify these scientific publications. The second focus of this bibliometric study is the impact of the published scientific papers, identifying the impact of Chilean scientific production analyzing the times a paper has been cited. The citation of these scientific documents was analyzed based on Normalized Citation Impact (NCI), this measure is used to assess the overall impact of the scientific publications in the category. This type of analysis allows for exploring the scientific context of the field and helps to select meaningful metrics for developing bibliometric methods in local applications.

4 Results and analysis

4.1 Traditional metrics

This bibliometric analysis focuses on productivity, impact and international collaboration of Chile in Astronomy & Astrophysics, analyzing a global perspective and their influence throughout the period under study. Quantity and performance indicators were used, such as: number of publications, number of citations and field normalized citation score (Rehn, Gornitzki, Larsson, & Wadskog 2014).

Year	Web of Science Documents	Productivity rate	Times Cited	% Docs Cited	Normalized Citation Impact
2005	440	-	14.632	93,18	1,24
2006	547	↑ 1,24	20.550	91,77	1,31
2007	522	↓ 0,95	23.006	91,95	1,42
2008	592	↑ 1,13	16.434	84,97	1,30
2009	605	↑ 1,02	21.347	85,95	1,48
2010	699	↑ 1,16	18.033	84,84	1,26
2011	784	↑ 1,12	22.795	91,96	1,58
2012	926	↑ 1,18	24.510	89,31	1,71
2013	880	↓ 0,95	17.666	95,23	1,55
2014	1.152	↑ 1,31	21.698	90,19	2,20
2015	1.172	↑ 1,02	8.663	83,79	1,40

Figure 2. Annual Evolution of Chilean Scientific Production and Impact Indicators

Figure 2 exhibits the data of the Astronomy & Astrophysics annual evolution of Chilean scientific production and impact indicators. In the period under study, the Chilean papers annual productivity has tripled. It stands out the year 2014 where it is exceeded in 30% the productivity with respect to the previous years. In addition, it has a strong impact indicator (NCI) that is above the world average. Therefore, Chile has a high performance within the range of impact of the world scientific community (Greco, Bornmann and Marx 2012).

Figure 3 displays the evolution of papers published annually in this field and the most productive scientific institutions during the last ten years. The European Southern Observatory is the reference for productivity in this area[5]. Pontificia Universidad Catolica de Chile, University of Chile and University of Concepcion are the most productive institutions in this area. This is consistent with the

[5]European Southern Observatory. 2017, http://www.eso.org/public/chile/announcements/ann16019/

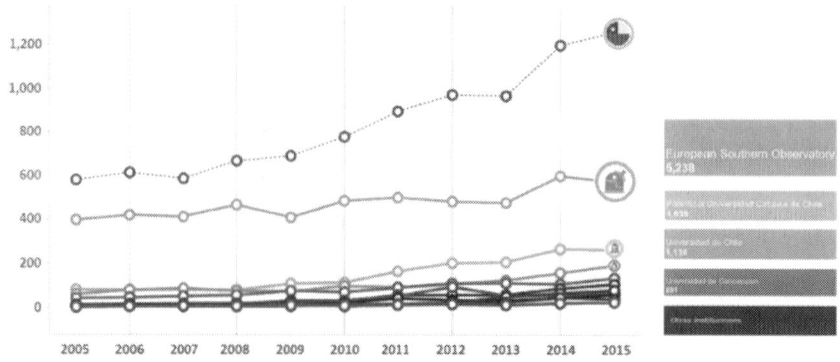

Figure 3. Evolution of Papers Published Annually in Astronomy & Astrophysics

overall results of the national scientific production (Navarrete 2015). The bibliometric analysis shows that Chilean scientific publications exhibits a positive growth that demonstrates the development of academic research in this area.

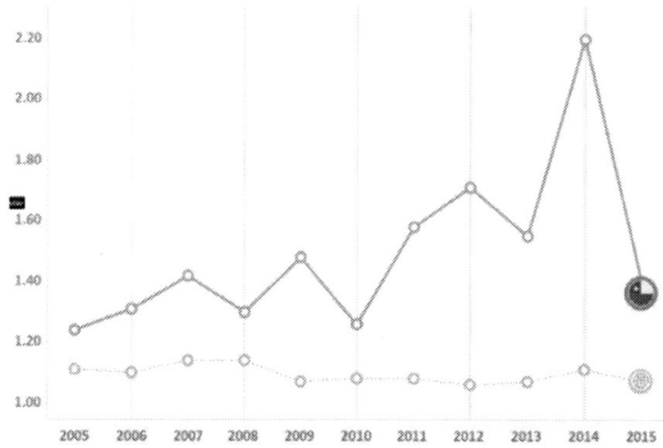

Figure 4. Normalized Citation Impact (NCI)

Figure 4 shows the evolution of Chilean scientific documents citation, analyzed using the Normalized Citation Impact (NCI) in astronomy and astrophysics. Based on these data, the plot indicates that over the last ten years, the Normalized Citation Impact of Chilean scientific documents is higher than the worldwide impact in this field. The figure shows that Chilean papers citation impact in astronomy and astrophysics has been above 20% of the worldwide impact average, during the period studied. The peak is produced by articles with an unusual high impact for that year.

Figure 5 displays the impact of Chilean top institutions publishing scientific documents in astronomy and astrophysics based on the NCI. Most of the institutions conducting research in Chile in this field are above the worldwide average. On the other hand, for exploring Chilean international collaboration by country, this bibliometric analysis focuses on document counting.

https://doi.org/10.1051/epjconf/201818605002

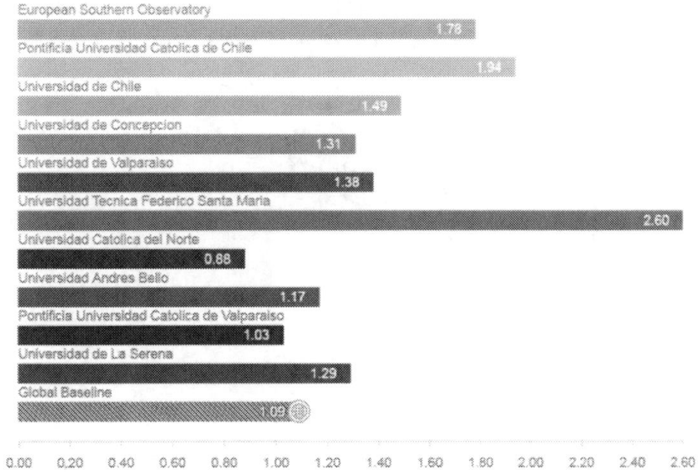

Figure 5. Normalized Citation Impact of Chilean Top Institutions

Ranking	Country	Web of Science Documents	% Collaboration WoS Documents with Chile	Times Cited	% Documents Cited	Normalized Citation Impact
1	USA	4.193	50,40	151.155	94,40	2,15
2	GERMANY	2.877	34,58	103.139	94,23	2,26
3	ENGLAND	2.249	27,03	89.699	96,40	2,60
4	FRANCE	2.134	25,65	79.654	94,05	2,48
5	ITALY	1.786	21,47	67.543	95,52	2,57
6	SPAIN	1.657	19,92	64.739	95,17	2,68
7	CANADA	1.187	14,27	61.560	94,78	3,32
8	NETHERLANDS	1.075	12,92	52.519	97,49	3,34
9	AUSTRALIA	1.057	12,71	45.277	97,45	2,70
10	JAPAN	867	10,42	35.882	93,54	2,76

Figure 6. International Collaboration of Chilean Institutions

Figure 6 presents the data of the astronomy and astrophysics international collaboration network of the scientific production between authors of Chilean institutions and other countries (2005-2015). In this case, we show the top 10 countries that have contributed significantly to develop production and impact in the area.

Figure 7 shows the international collaboration network of the scientific production among authors of Chilean institutions and other countries. The US has collaborated with Chile with more than 4,000 documents in the period studied. For another perspective Germany, England and France have more 2,000 documents each in collaboration with Chile in the same period.

Figure 7. International Collaboration Network

4.2 Alternative metrics

The methodological approach of this study is complementing the traditional metrics with the alternative ones. In this case, the Altmetric tool was used to provide information on the dissemination and discussion of research results, as well as a view of the references in non-conventional sources that complement traditional quotations (Melero 2015).

			Scholarly Activity		Scholar Commentary			Social Activity				Mass Media
Top articles	Publication Date	Times Cited	Mendeley	CiteULike	Blogs	Wikipedia	Q & A sites	Twitter	Facebook	Google+	Reddit	Noticias
1	2007	4201	225	11	1	7	0	2	1	0	0	0
2	2012	3824	52	5	38	5	3	1413	29	13	3	13
3	2014	3690	500	1	21	8	1	41	4	4	0	6
4	2009	2470	0	0	1	1	0	3	0	0	0	0
5	2006	1654	60	4	0	1	0	1	0	0	0	0
6	2014	1126	166	0	14	1	2	12	1	2	0	3
7	2008	990	28	0	0	1	0	0	0	0	0	0
8	2008	848	120	7	1	2	2	0	0	0	0	2
9	2014	704	0	0	10	8	1	35	10	3	0	0
10	2012	628	105	0	0	0	0	1	0	0	0	0

Figure 8. Top Ten Cited Papers in WoS versus Alternative Metrics through SCOPUS

Figure 8 displays the data of the top ten cited papers in Web of Science that were analyzed in SCOPUS, focusing on alternative metrics in these four aspects: Scholarly Activity, Scholarly Commentary, Mass Media and Social Activity. It can be observed that the use of Mendeley, Blogs and Twitter are important tools to enhance scientific publications. When comparing traditional metrics with alternative metrics, as shown in Figure 9, the Altmetric indicators presented by SCOPUS reflect that the impact of Chilean research in Astronomy & Astrophysics is low in the context of social networks in the studied period.

5 Conclusions

The analysis of traditional metrics from the information obtained in Web of Science analyzed through InCites indicates a steady increase of the scientific production in Astronomy & Astrophysics during

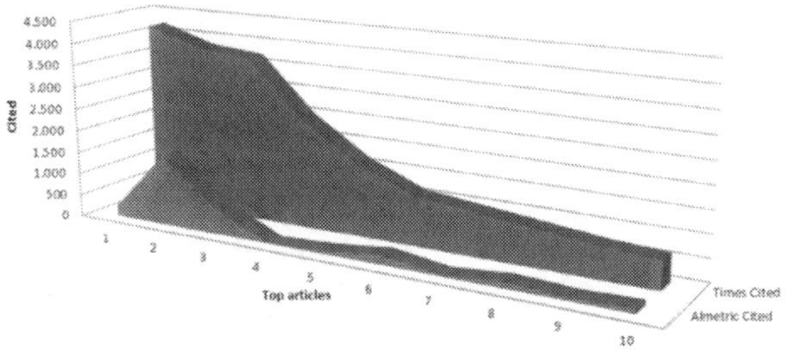

	1	2	3	4	5	6	7	8	9	10
Almetric Cited	247	1.574	586	5	66	201	29	134	67	106
Times Cited	4.201	3.824	3.690	2.470	1.654	1.126	990	848	704	628

Figure 9. Traditional Metrics versus Alternative Metrics

the last 10 years. In addition, the Normalized Citation Impact of the organizations and institutions that carry out research in Chile is above the average at a worldwide level. Furthermore, alternative metrics presented by SCOPUS show that the impact of the research is low in the context of social networks.

Acknowledgements

The authors would like to thank the Pontificia Universidad Catolica de Chile for supporting the development of this work and the attendance at the LISA conference. We are also very grateful to the Friends of LISA for travel and registration funding. In addition, we would like to thank the Librarian Patricio Cortes for his fundamental collaboration in this study and to Prof. Wilfredo Palma for his help with the manuscript edition.

References

[1] A. Greco, L. Bornmann, W. Marx, El profesional de la informacion **21**, 607-612 (2012)

[2] R. Melero, Biochemia Medica **25**, 152-160 (2015), http://dx.doi.org/10.11613/BM.2015.016

[3] J. Navarrete, Plataforma Cientifica, (2015), http://www.plataformacientifica.cl/que-dicen-los-indicadores-sobre-la-produccion-cientifica-en-chile/

[4] C. Rehn, C. Gornitzki, A. Larsson, D. Wadskog, *Bibliometric handbook for Karolinska Institutet* (Thomson Reuters, USA, 2014), https://kib.ki.se/sites/default/files/bibliometric_handbook_2014.pdf

Progress of Astronomy in India: A Scientometric Study

base on paper published during 1991 – 1995 and 2011 – 2015

Vijay Kumar Rai[1,⋆], *K. P. S.* Senger[2], and *Rajesh K* Lohiya[3]

[1] *IUCAA, Ganeshkhind Pune, India*
[2] *IMTECH, Chandigarh, India*
[3] *NEERI, Nagpur, India*

Abstract. Astronomy is the oldest of the natural sciences. It is well known that the period covering about seven centuries, from Aryabhata to Bhaskaracharya II (c. 476-1150), was the golden age of Indian astronomy. In the present study, we identify research trends and the growth of knowledge in the field of astronomy science research. The study is an essential tool to measure the scientific publications in the field. This study analyses the publications of astronomy research in India published during 1991-1995 and 2011-2015. The study assesses how astronomy progresses in India and its impact is reflected in the science citation index over the period 1991-1995 and 2011-2015. The publication output has been analyzed using quantitative and qualitative indicators, such as progress in research article publishing, change in authorship patterns, citation patterns during 1991-1995 and 2011-2015, articles that received the most citations, and international collaboration with other countries. Further, the study investigated highly prolific authors and highly preferred journals. The author is from one of the FORSA institutes, therefore the study includes a separate ranking for FORSA institutes in term of output as well citations.

1 Introduction

The science of astronomy has fascinated humans from the beginning of time. In the early stages of the evolution of ideas, we found conceptions of the Mother Earth being entertained by Father Sky to nourish her and create life. The journey from there to new satellites, such as ASTROSAT, has been fascinating.. In India, astronomy has a long and has a glorious past. There is documented evidence that astronomy was studied in India from the Vedic times dating back to 1500 BC, and then nurtured in the fifth and sixth century AD by the greats like Aryabhatta, Varahamihira, and Brahmagupta.. The advent of telescopes shifted the leadership in astronomy studies from Asia and the Middle East to Europe.

There are numerous institutes in India which are doing frontline research in astronomy and astrophysics, and there are even several telescope facilities now available. They include solar telescopes, radio telescopes, optical telescopes and others. India is marching ahead in astronomy research aided by ground-based, balloon-borne and satellite facilities. The future of astronomy in India is bright and the next decade promises that there will be substantial contributions coming from India.

⋆e-mail: vkrai@iucaa.in ORCID: 0000-0003-4548-403X

In the early 1980s, due to the proliferation of information, library professionals working in the Institutes where astronomy is one of the main areas of research, felt the need to come together and form a forum, which can act as a springboard for sharing and exchanging information. The sheer necessity brought all likeminded astronomy librarians in the country together. The Forum for Resource Sharing in Astronomy and Astrophyics (FORSA) was launched on July 29, 1981, with a vision and mission to share resources held in each library. At present, there are eleven Institute members: Aryabhatta Research Institute of Observational Sciences (ARIES), SN Bose National Centre for Basic Science, CASA-OU, Harish-Chandra Research Institute (HRI), Indian Institute of Astrophysics (IIA), Inter-University Centre for Astronomy and Astrophysics (IUCAA), National Centre for Radio Astronomy (NCRA), Physical Research Laboratory (PRL), Raman Research Institute (RRI), Saha Institute of Nuclear Physics (SNIP), and Tata Institute of Fundamental Research (TIFR) [1].

Research trends and growth of knowledge in a scientific field can be identified by using essential tools to measure the scientific publications in the field. There are various quantitative methods and techniques to use for measuring the impact of scientific work. It has become essential to adopt standard operating procedures and to measure the research work using various methods from historic bibliometric to modern scientometric. There are tools to measure the science and technology, most notably, for published papers in referred journals the scientometric technique is applied. Scientific research productivity usually culminates in scientific writing, which leads to research publications. Scientific literature is considered to be a reliable medium for promoting national and professional interest in one's field of specialization.

2 Past Studies

Astronomy and astrophysics have been subject to a number of scientometric studies in the US and Europe, but in India only a few studies have been conducted. These include a scientometric study of the doctoral theses of aphysical research laboratory done by Nistha Anilkumar [2]. Barve and Gopal-Krishna(2002)[3] found that there is a steady increase in the number of observational papers published by Indian radio astronomers. They found a steep rise in the number appearing as "rapid publication."

In the productivity of Indian Telescopes study by Meera and Manjunath, [4] they examined the productivity of modern Indian telescopes that were installed after India attained independence from colonial rule. The study found that 1,643 papers were published using Indian telescopes. The highest number of citations is associated with the GMRT (Radio Telescope) papers. The first international publication from India was in the "Monthly Notices of Royal Astronomical Society" in 1906. India ranked fifteenth in the Global Aurora research with 182 publications and 1559 papers from astronomy and astrophysics (Dwivedi, 2016)[5]

A scientometric study of astronomy research in IUCAA done by Vijay Kumar Rai and K P S Senger found that with a collaborative approach in research, several papers received an extraordinary number of citations and many papers were published in high impact journals[6]. A similar study of visiting associates of IUCAA found that there is a collaborative approach in research among them, and the degree of collaboration is quite high [7]. Dutta and Rath (2013) studied cosmology research in India and found that it is a mainly collaborative effort with the USA and other European countries. The study found that there are some exorbitantly large teams with more than one hundred members, and sometimes as many as 715. Almost all core journals are published by the USA and European countries and possess relatively high impact factor. The study found that individually IUCAA, IITs and Jadavpur University are the top three affiliating institutions [8].

3 Objectives of Study

The primary aims of the present study are to analyze the broad characteristic features of the publication output of Indian Astronomy during 1991-1995 and 2011-2015, using quantitative and qualitative indicators. In particular, the study focuses on the growth of research output, citation impact, international collaboration, communication channels preferred, and contribution author and highly prolific author.

4 Source Database and Methodology

The study is based on the publication data derived from Web of Science Core Collection. The study is limited to the period of 1991–1995 and 2011–2015.

To retrieve data an advance strategy was adopted. First I applied the term India for Author Address and limited the search results to the period 1991–95 and 2011–2015 respectively. The results which I got, I refine with the Subject Category "Astronomy and Astrophysics."

The study periods, 1991-1995 and 2011-2015 have been chosen purposefully to understand the shift in the research output of Indian astronomy. There has been a substantial expansion in the information technology (IT) infrastructure, budget and researchers in later years. The data was analyzed as per the objectives of the study after the validation.

5 Data Analysis and Interpretation

To identify research trends and the growth of knowledge in astronomy and astrophysics, it is essential to measure the scientific publications in the field. This study analyses the publications of astronomy research in India published during 1991-1995 and 2011-2015. Study analyses the production activity, distribution of journals, citation analysis, and international collaborations.

5.1 Publication Activity

The Indian astronomy and astrophysics research output has increased three folds during 2011-2015 in comparison to 1991–1995. During 1991-95, a total of 1987 articles were published and in 2011-15 this jumped to 5186, which is 2.61 times higher. There is no definite pattern of literature growth ascertained from the data, at least for the years of 1991-95. The study finds that the research output for the period of 2011-2015 is an unexpected decline, but there was a growth of publication observed for the year of 2016.

Indian researchers published a total of 1987 papers, with an average number of publications per year. This number was 397 during 1991-1995. During 2011-15, the maximum number of articles published was in 2011 (1175) and the minimum was in 2015 (893), with an average of 1037 as shown in Table 1. The reason for the decline in the number of articles published is yet to be ascertained. Observation indicates that astronomy articles publication does not have a particular pattern.

5.2 Distribution of journals

During 1991-95, 1640 articles were published in peer-reviewed journals and the rest published in a note, review, individual proceedings, editorial, etc. Out of 1640, 38 articles were published either in proceedings series or book series. Most of the articles were published in international journals. These

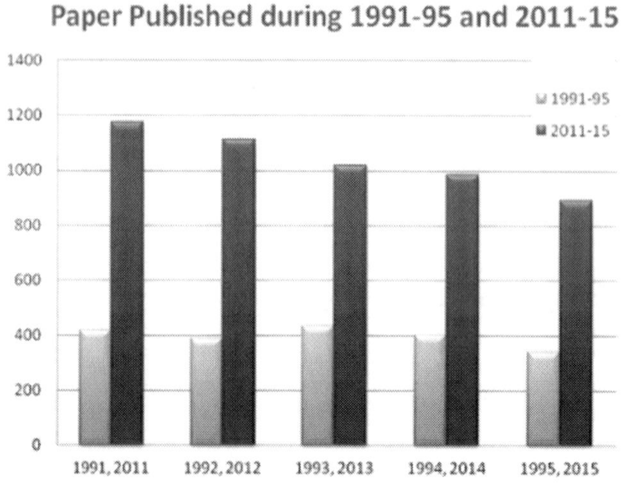

Figure 1. Paper Published during 1991-95 and 2011-15

Table 1. Papers Published and Growth

Year	No. of Papers	Year	No. of Papers	Growth
1991	420	2011	1175	280%
1992	389	2012	1113	286%
1993	434	2013	1019	235%
1994	403	2014	986	245%
1995	341	2015	893	262%
Total	1987		5186	261%

articles were published in 51 journals. Out of these 51 journals, there are only 3 Indian journals, but the highest numbers of articles (252) were published in the Indian Journal of Radio Space Physics. The journals Astrophysics and Space Science (219), Physical review D (208), Physics Letters B (206), Astrophysical Journal (136) published more than 100 articles each. Four journals published between 50-100 articles, 11 journals published between 10-15 articles and 31 journals published ten or fewer articles. While the highest number of papers were published in the Indian Journal of Radio Space Physics, most of the papers were published in well-known international journals.

A total of 4581 articles were published in peer-reviewed journals during 2011-15, including 42 articles published in proceedings and the rest published in correction, editorial material, news item, book chapter, biographical item, etc. The highest numbers of articles were published in Physical Review D (1067). MNRAS (515), Astrophysics and Space Science (490), Physics Letters B (419), and Astrophysical Journal (385) were among the top five journals chosen by researchers to publish articles. These top 5 journals published 57% of the total articles published by Indian researchers. Four journals published between 100-250 articles. Nine journals published between 50-100 articles. 13 journals published between 10-49 articles. 21 journals published less than 10 articles.

It shows that Indian researchers prefer to publish their papers in International journals with high impact factors.

5.3 Citations Analysis

It is clear that Indian astronomers published their papers in internationally known and high impact factor journals. Therefore, Indian astronomy research articles are well cited. As shown in Table 3, during 1991-95, the highest number of papers (859) were cited between 1-10 times. However, 341 papers were not cited even once. One article is cited 1043 times. During 2011–15, 2779 papers are cited between 1–10 times and 1266 papers are cited between 11–50 times.

Table 2. Citation Analysis

Citation received	Paper Published 1991 - 1995	Paper Published 2011 - 2015
0	314	453
1-10	859	2779
11-50	500	1266
51-100	83	122
101-200	25	29
201-300	4	7
301-400	1	3
401-800	1	3
801+	1	4

There are two papers which are cited 4072 times, and one article is cited 3857 times. Citations have consistently increased every year no matter the number of articles published that particular year. There were 1987 articles published during 1991–95. These 1987 papers received 4052 citations in during this period, but overall these papers received 24326 citations until 2017. The average citation per paper was 14.10 and the h-index 68. It was found that the number of publications decline but the number of citations increased for the articles published during 2011–15. Even though a minimum number of articles were published in 2015, these papers got maximum citations. The average citations per paper is 14.56 and the h-index is 83. The cumulative citations graph shows that citations for these papers increased every year and still counting.

5.4 International Collaboration

Indian astronomy research progresses with international collaboration. During 1991-1995, Indian researchers collaborated with researchers from about 50 countries; however, in 2011-2015 collaboration reached 89 countries. There were no major group or project collaborations in 1991-1995. Indian astronomy researchers showed significant progress in collaborative research at an international level during 2011-2015. Even Indian research is part of many international projects like CMS Collaboration (141), D0 Collaboration (83), BABAR Collaboration (78), BELLE Collaboration (73), ATLAS Collaboration (47), PLANCK Collaboration (41), ALICE Collaboration (39), LIGO SCI Collaboration (22), and VIRGO Collaboration (19).

During 1991-1995 USA (261), Germany (116), Italy (110), France (84), and Switzerland(74) were the among the top five countries which are preferred by researchers for collaboration. During 2011-2015, England and Russia were among the top five countries. Researchers from USA (1590), Germany (1120), France (867), England (851) and Russia (808) collaborated the most with Indian researchers.

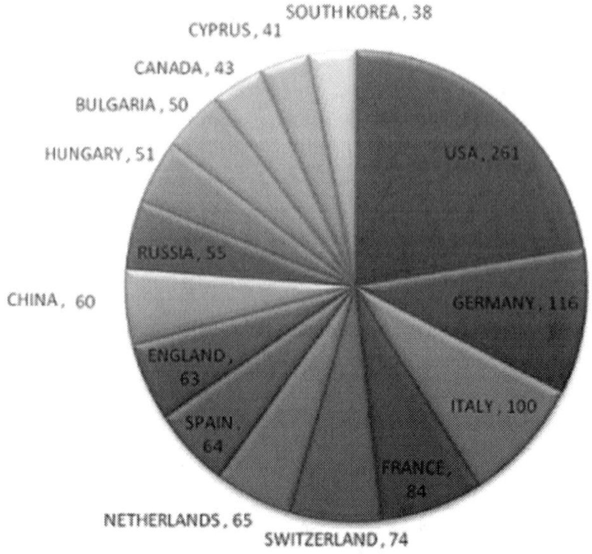

Figure 2. International Collaboration with Top 15 countries

5.5 Highly Productive Author

The study has identified 41 Indian authors who have produced the most papers during 1991–1995. Out of these 41 authors, 8 authors produced between 46–53 papers and these papers were published by a small group of authors. All these 8 authors are affiliated with TIFR, Mumbai. Among them, S C Tonwar (53) and V K Gupta (51) produced the most papers. However, K P Singh (22) from the TIFR is the most prolific author, followed by B M Reddy, U Sarkar and R P Singh who published 21 papers each.

The study finds that during 2011-2015 collaborative research was a trend, and the result is scores of collaborations. In total, 66 collaboration groups were found. The top five group authors produced 599 papers.

Chakrabory, S(61) from IUCAA, Chakrabarti, SK(59) from Indian Ctr Space Phys, Subrahmanyan, R(57) from RRI, Kumar S(55) from BITS Pilani, Banerjee, S(52) from IIT Jodhpur were the top five authors.

6 FORSA Institutes Output Analysis

FORSA institutes have astronomy research as one of its core research areas. The study analyzed output of FORSA institutes separately. Output of TIFR includes NCRA output as well. Two institutes, UPSO and MRI (1991–1995) were renamed Aryabhatta Research Institute of observational sciencES (ARIES) and Harish Chandra Research Institute during 2011-2015 respectively.

6.1 Publication Activity

Among the FORSA institutes SN Bose NCBS and HRI have shown maximum growth but published minimum papers. All FORSA institutes published more than a hundred papers during 2011-15 compared to three institutes during 1991-95. Cumulative output of FORSA institute is 329.44 % higher during 2011-15.

Table 3. Paper Published by FORSA Institute

FORSA Institutes	Paper Published 1991 - 1995	Paper Published 2011 - 2015	Growth of Publications
TIFR	368	929	252.45%
IIA	213	409	192.02%
PRL	155	309	199.35%
IUCAA	90	514	571.11%
RRI	50	225	450.00%
UPSO now ARIES	46	207	450.00%
SINP	30	347	1156.67%
MRI now HRI	9	129	1433.33%
SN Bose NCBS	7	120	1714.29%

FORSA institutes produced around 48% of total papers published by Indian astronomers during 1991-95 and it reached 61% during 2011-15. It shows that FORSA institutes are more focused on astronomy research than other institutes.

6.2 Citation analysis

TIFR published the maximum numbers of papers. Citations received and average citations per year by TIFR papers are also the maximum during 1991-95. Papers published by FORSA institutes during 1991-95 have received 18300 citations up to June 2017 with the average citations per item 20.09. The highest citations totaling 1605 were received in 1996 for these papers, and then the number of citations declines gradually. One paper published in 1994 got 1044 citation, and 20 articles published during this period got 100 or more citations. Cumulative citations increased until 1996 for the papers published during 1991-95.

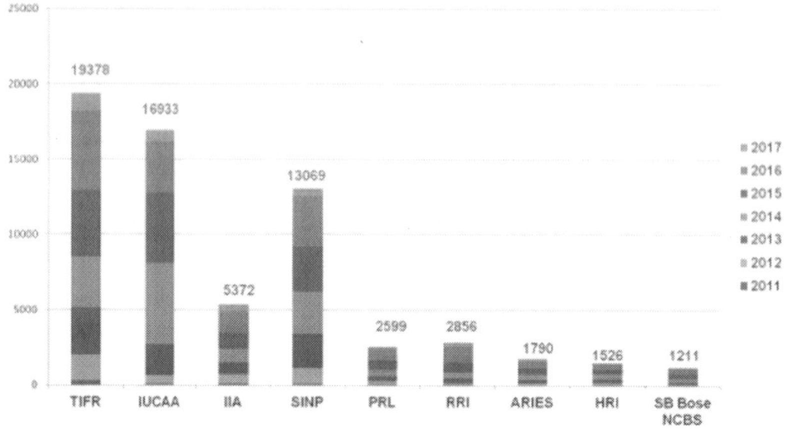

Figure 3. FORSA Institute's Citations

In 2011-15 TIFR received the maximum citations as well. The average citations for TIFR is 2768.28 per year. IUCAA jumped to second in terms of paper publications and its papers received 2418.86 average citations per year.

7 Conclusion

It is found that astronomy research papers publication has never had a precise pattern. In comparison to 1991-95 studies, output has been increased three fold in 2011-15. Although this study finds a decline in the number of research output during 2011-2015 for Indian Astronomers, citations have consistently increased every year. Tata Institute of Fundamental Research, Mumbai is the leading institute and IUCAA is an emerging Institute in terms of number of articles published and citations in the field of astronomy and astrophysics research in Indian. It was found that there is a decline in research output during 2011-2015 for Indian astronomers but FORSA institutes have shown progress. Indian studies in the area of astronomy are more collaborative than individual studies. Indian researchers are making contributions in some exorbitantly large international collaborative research projects. Indian researchers prefer to publish their research papers in international journals with the highest impact factors. To conclude, it is clear that Indian astronomy research has a bright future.

References

[1] *Forum for Resource Sharing in Astronomy and Astrophysis* http://www.rri.res.in/htmls/library/forsa.html

[2] N. Anilkumar, *Scientometric Study of Doctoral Theses of the Physical Research Laboratory*, in *21st Century Astronomy Librarianship, From New Ideas to Action*, edited by Eva Isaksson, Jill Lagerstrom, András Holl, and Nirupama Bawdekar (2010), Vol. 433 of Astronomical Society of the Pacific Conference Series, p. 102–108

[3] S. Barve, Gopal-Krishna, *nalysis of the Publication Pattern of Radio Astronomers from India during 1990–2001* in *Library an Information Services in Astronomy IV (LISA IV - Prague)* edited by B.G. Corbin, E.P. Bryson, M. Wolf, (2003), p. 298–301

[4] B.M. Meera, M. Manjunath, *Productivity of Indian Telescopes: Impact Analysis through Scientometric Methods* in *Open Science at the Frontiers of Librarianship*, edited by András Holl, Soizick Lesteven, Dianne Dietrich, and Antonella Gasperini (2015), Vol. 492 of Astronomical Society of the Pacific Conference Series, p. 118–122

[5] Dwivedi, International Journal of Research in Library Science, **2(3)**, 155(2016)

[6] V.K. Rai, K.P.S. Senger, *Astronomy Research in IUCAA: A Scientometric Study of the Papers Published by IUCAA Academics during 1993-2013* in *Open Science at the Frontiers of Librarianship*, edited by András Holl, Soizick Lesteven, Dianne Dietrich, and Antonella Gasperini (2015), Vol. 492 of Astronomical Society of the Pacific Conference Series, p. 128–132

[7] V.K. Rai, K.P.S. Senger, S.K. Pathak, *A Scientometric Study of Research Papers Published by Visiting Associates of IUCAA, Pune, India during 2003-2013* in *Open Science at the Frontiers of Librarianship*, edited by András Holl, Soizick Lesteven, Dianne Dietrich, and Antonella Gasperini (2015), Vol. 492 of Astronomical Society of the Pacific Conference Series, p. 123–127

[8] B. Dutta, D.S. Rath, Library Philosophy and Practice (e-journal), (2013) https://digitalcommons.unl.edu/libphilprac/996/

G. Schwarz, G. Landais and A. Muench (photo: M. Wenger)

EPJ Web of Conferences **186**, 06001 (2018) https://doi.org/10.1051/epjconf/201818606001
LISA VIII

Use Cases of the ESO Telescope Bibliography

Uta Grothkopf[1,*], *Silvia* Meakins[1] and *Dominic* Bordelon[1]

[1] *European Southern Observatory, Karl-Schwarzschild-Str. 2, Garching near Munich, Germany, library@eso.org*

Abstract. ESO's mission is to enable front-line research in astrophysics by operating and maintaining a wide range of world-class telescopes and state-of-the-art instruments. Various key performance indicators are used to understand whether ESO is achieving these ambitious goals. In order to assist the ESO Management, the Library builds and maintains the Telescope Bibliography (telbib). telbib provides insights into the performance of ESO's facilities and helps to understand publishing trends among the user community. This paper will highlight some use cases and recently added enhancements.

1 Introduction

In this paper, we will show various use cases of the ESO telescope bibliography (telbib)[1] and will highlight some recent additions. The ESO librarians spend a lot of time, energy, and ideas on the development of telbib, our database of refereed papers that use ESO data. By now, telbib has developed into a sophisticated tool that does not only list all the papers, but links them to the observations stored in the ESO Science Archive and provides the opportunity to add additional tags and identifiers. A multitude of reports and statistics can be derived from telbib, allowing stakeholders to answer a large variety of questions about how observational data are used in papers. The ESO Library continuously enhances telbib in order to enable users to obtain further metrics.

The overview of this paper is as follows: In section 2 we will explain why it is important that telescope bibliographies are curated. In section 3 selected use cases will be presented. We conclude with a description of some questions that cannot be answered at this time.

2 Curated bibliographies

Why do we need curated telescope bibliographies? In order to understand the importance, we take a look at the ESO telbib curation process. Our workflow starts with running a text mining software which finds scientific articles that contain at least one ESO-related keyword. Very importantly, each and every paper is visually inspected by the librarians. This inspection is the first evidence of human curation. It is essential in order to make sure that all telbib papers fulfill our main selection criterion,

*e-mail: uta.grothkopf@eso.org ORCID: 0000-0001-6830-0702
[1] http://telbib.eso.org

i.e., they must use (study, analyze) ESO data. If papers only quote results from the literature, or describe future observations, or even if they discuss instrumentation, they are not part of telbib. Further information about telbib can be found on the web[2] and in various publications.[3]

Once the decision is taken that a paper should be included, a record is created in telbib. Bibliographic details obtained from the SAO/NASA Astrophysics Data System (ADS) are the basis of all records. In a second step, the Librarians assign a multitude of tags, labels, and annotations to characterize each entry, for instance regarding author affiliations, observational surveys from which data originated, or ESO press releases that featured specific papers. However, in particular, we add metadata regarding the observing programs, including the program ID(s), observing mode and type, and the instrument(s) used for the observations. Very importantly, we also track whether the data were originally observed by the authors or whether they were obtained from the ESO Science Archive.

As a result, reports and statistics can be obtained for this multitude of parameters. telbib is the only place where scientific papers and ESO's observational data are linked.

Observational details in telbib are identified through the program IDs. However, only approximately 70% of the authors mention these identifiers in their papers. For the remaining 30%, the IDs need to be retrieved through other ways, for instance by querying the Archive for observed objects, instruments, and observing dates, or through communication with the authors. Without this human curation, the ESO data used in 3 out of 10 papers could not be identified, and the papers would not be linked to the observations.

3 telbib use cases

Which questions can we answer using telbib? What are the main use cases? The statistics we provide can be divided in four categories: (1) Productivity and impact, (2) Author demographics, (3) Publishing trends, and (4) Instrumentation management. In the following sections, we will show examples for each category.

3.1 Productivity and impact

Productivity and impact are the most common performance measures that are derived from telescope bibliographies, and very likely everybody who maintains bibliographies provides such statistics.

Productivity is often measured by looking at the number of papers that were published, and impact by the number of citations. This can be the total number of papers or citations, or it can be a group, for instance the fraction per instrument, telescope, observing program, observing type, etc. Since productivity and impact are the most obvious and frequently applied metrics, we will only mention them briefly here.

One example is to study the number of papers per journal (Fig. 1). In the case of the ESO telescope bibliography, more than 60% of the papers were published in the journal *A&A* in the early years of VLT (Very Large Telescope) data papers. Over time, this fraction has gone down to (still remarkable) 40%, while more papers were published in particular in *MNRAS*. One explanation for this shift might be that the UK joined ESO in 2002, and therefore publications in a UK-based journal have increased. It is always important to put information retrieved from bibliographies in context with what is happening in the organization at large.

In a similar way, citations can be investigated for specific entities, for instance for ESO program types. ESO observations are assigned to various program types, including Large Programs (LP),

[2] http://www.eso.org/sci/libraries/telbib_info.html

[3] http://www.eso.org/sci/libraries/useful_links/publications.html

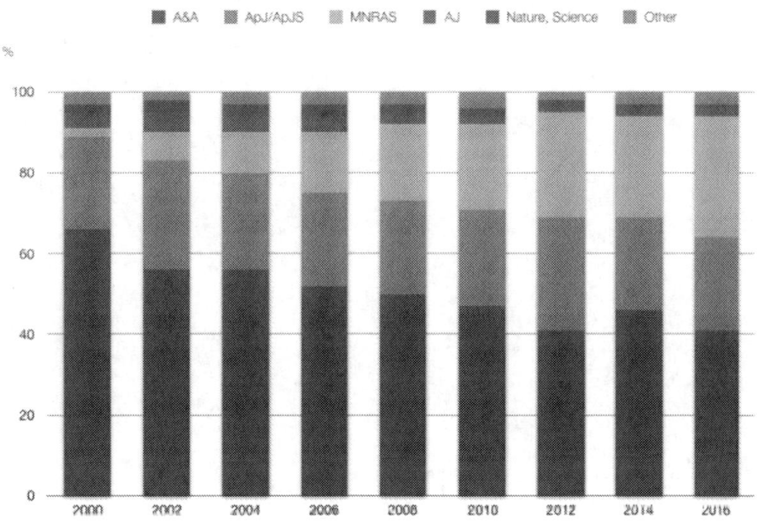

Figure 1. telbib papers by journal 2000-2016

Guaranteed Time Observations (GTO), Target of Opportunity observations (ToO), and others. Sterzik et al. (2015) [1] investigated the role of different observing types and modes in detail, using telbib data for an in-depth analysis of productivity and impact of VLT observations.

3.2 Author demographics

The second area for which statistics are compiled from telbib focuses on author demographics. Central questions are: How many distinct authors publish with ESO data, and what are their nationalities? In this context, we define nationalities as the country of the author's affiliation.

We studied the number of distinct authors of telbib papers since 1999. A steady increase can be seen from approximately 1,300 authors in 1999 to almost 17,000 in 2016, so within approximately 15 years the number has increased by a factor of 13. These numbers are cumulative.

What are the nationalities of these approximately 17,000 authors? We have created a world map of 1st authors of telbib papers (Fig. 2) and found that they have affiliations from 64 countries (marked in dark blue). This constitutes approx. a third of all countries on this planet, so we can say that the ESO user community is truly global.

3.3 Publishing trends

The ESO Management is often very interested in learning about publishing trends. Lately, the Library investigated three topics: (1) Can we see any trends in the team size of authors, and has this changed over the years? (2) What is the fraction of papers that use archival data? and (3) Are there any synergies of ESO observations with data from other observatories?

Looking at the number of authors that publish with ESO data, we found that the first paper with over 100 authors was added to telbib in 2005. In 2011, where a slightly steeper increase can be found

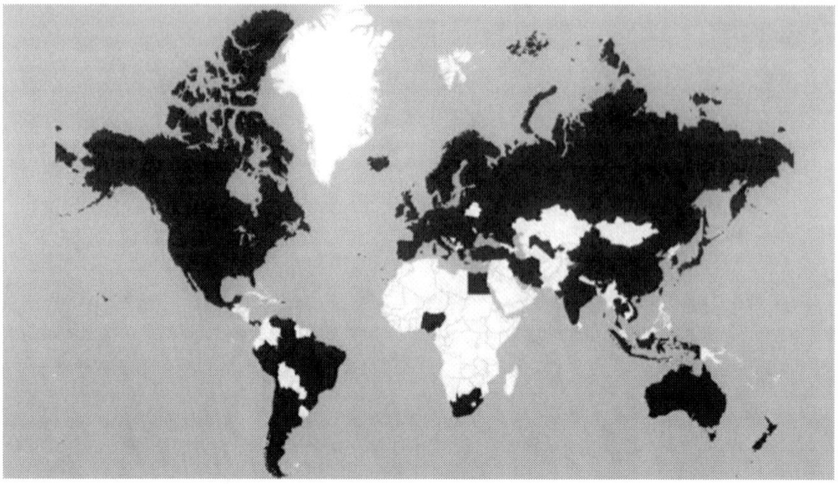

Figure 2. World map of 1st author nationalities of VLT data papers

in comparison with previous years, four papers with more than 100 authors were added to telbib. Two years later, in 2016, three papers had even more than 600 authors, explaining the noticeable upward swing from 2015 to 2016. For author team sizes, measured by the average number of authors per telbib paper from 2000 until 2016, a duplication from 6.5 to almost 14 co-authors can be seen. The number for publishing year 2016 might be slightly inflated due to the three papers with more than 600 authors mentioned above. However, the general trend towards multi-author papers remains. In that context, collaborative writing along with supporting platforms like Authorea, Overleaf, and Writing Studio is gaining importance.

If the average number of authors increases, it is interesting to see whether the number of ESO programs used per paper has also changed. Indeed, for telbib papers, we found an increase by a factor of 2: in the early years of this millennium, authors used on average 1.5 programs in their papers. By 2016, this number has increased to over three programs that provide data for a given paper. A general trend can be seen towards papers with more authors that deploy observations from more programs.

An important aspect for ESO is how archival data are used. Observational data obtained with ESO's facilities are stored in the Science Archive from where data are made available to astronomers worldwide on request. Therefore, the Archive is often seen as an observing facility of its own. It is essential for ESO to understand how many papers are published that use archival data. Statistics derived from telbib show that in particular during the past five years, the fraction of papers that use archival VLT data (exclusively or along with ESO data obtained by the authors) has been on average 23%, or almost a quarter of all VLT papers. Archival papers are divided almost equally between those that use exclusively archival data (12%) and those that deploy archival data in addition to proprietary data (11%).

Are ESO data mostly used alone in papers or in conjunction with data from other observatories? We investigated VLT data papers published between 2012 and 2016. Using information provided by other observatories and available through the ADS (see e.g., Accomazzi et al., in press [2],) we found that 24% of the VLT papers are also included in the HST bibliography. Also Gemini (8%), Keck (8%), or Subaru (3%) data are used in papers along with VLT data. Interestingly, on average 1% of the papers use VLT together with ALMA data. This trend seems to be growing as in 2016, the overlap increased to 4%. In total, 36% of the VLT papers published in the past five years also deployed data from one or several of the other five observatories, while for 64% of the VLT data papers no overlap was found. These papers still may have used data from observatories other than HST, Keck, Gemini, Subaru, and ALMA (Fig. 3).

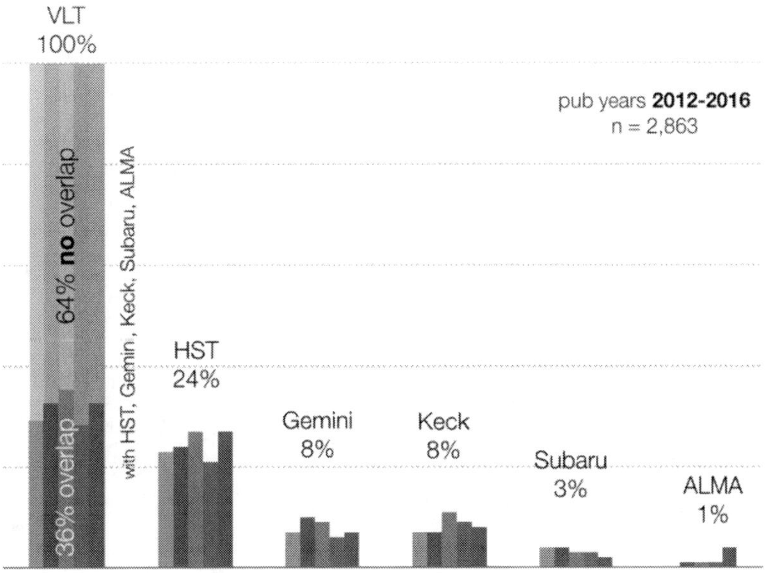

Figure 3. VLT data papers using also data from HST, Keck, Gemini, Subaru and/or ALMA

3.4 Instrumentation management

Instrumentation management is of key importance to ESO. The organization uses a variety of tools and metrics to make decisions about whether instruments need to be upgraded, refurbished, or even retired and replaced. Such tools include for instance the general knowledge of the community's needs or the oversubscription rate, i.e., the demand for observing time. This information is coming from other sources, not telbib.

telbib information in this context focuses mostly on the (total or average) number of papers and citations. In addition, lately the Library adopted a new metric originally presented by S. Kulkarni [3]. Using this methodology, citations to data papers are grouped by the year of the citing papers. The resulting graphs are called citation flux curves. Kulkarni concludes that instruments with decreasing overall citation counts (citation flux curves) should be upgraded or replaced.

To assist with instrumentation management, the Library calculated citation flux curves for all VLT instruments. Here we present two examples (Fig. 4).

Example 1 studies the X-SHOOTER instrument, a fairly recent instrument that has seen a steep rise of papers as well as citations from the start. However, the paper count now seems to have "plateaued" at approx. 70 papers per year on average. As a consequence, some upgrades are now foreseen to prevent the instrument from aging.

The 2nd example is ISAAC, one of the oldest VLT instruments, which has reached a publication plateau several years ago. Even the citation flux curve has been flat during at least the past five years. Indeed, ESO has decommissioned ISAAC in 2013.

In this way, information from telbib can be used to confirm or even initiate instrumentation management decisions.

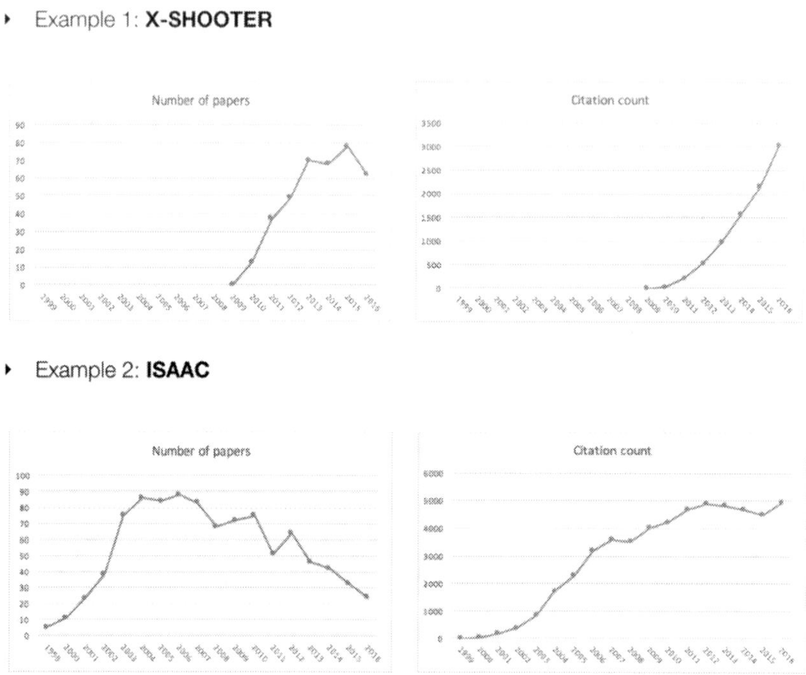

Figure 4. Paper counts and citation flux curves of two VLT instruments

4 Questions we cannot answer (yet)

While there are various strong use cases for the ESO telescope bibliography, there are also some questions that cannot currently be answered. For instance, telbib is not well suited to analyze the subject areas dealt with in papers, or general trends in topics. During previous telbib studies, we experimented with obtaining insights from the keywords authors assigned to their papers, combined with title and/or abstract words, and from the ESO observational categories[4] to which the programs were assigned. However, the results were less meaningful than we had hoped for.

Gender studies are receiving a lot of attention these days, but since we do not record (or even know) the gender of telbib authors, we cannot run any statistics pertaining to this topic.

When papers are ingested into telbib, no "weights" are assigned to reflect the percentage of ESO data in comparison with data from other observatories in the same paper. Regardless of whether only

[4]http://www.eso.org/sci/observing/phase1/p100/opc-categories.html

one spectrum was used or a large data set, all papers are treated (and weighted) equally as far as statistics are concerned.

One considerable complication is that by now telbib contains so much and so varied information that results can become somewhat indistinct or "blurry" at times. This is true in particular for papers that deploy data from many programs with a large variety of different characteristics. At present, it is not possible to disentangle the various parameters.

5 Conclusion

The ESO Telescope Bibliography (telbib) is curated by the ESO Librarians. We interpret the word curation as *to take care of*. Our workflow ensures quality content, including, but not limited to, linking observations to papers, and provides the largest possible warranty for complete and correct data.

Thanks to the detailed annotations and tagging of telbib records, numerous statistics, reports, and visualizations on a large number of parameters can be provided. telbib started in 1996 so that data are available for more than 20 years. The database has therefore become an essential tool to understand publishing trends among the ESO user community.

Despite the many facets of telbib, stakeholders occasionally request statistics that at present cannot be obtained. In order to provide further insights, the ESO Librarians continuously enhance the system with new features.

References

[1] M. Sterzik, C. Dumas, U. Grothkopf et al., The ESO Messenger **no. 162**, 2–8 (2015)

[2] A. Accomazzi, M.J. Kurtz, E.A. Henneken et al., *Aggregation and linking of observational metadata in the ADS*, in Proceedings of the ADASS XXV conference, arXiv:1601.07858, in press

[3] S. Kulkarni, *Instruments on large optical telescopes – a case study*, arXiv:1606.06674 v.2, subm.

EPJ Web of Conferences **186**, 06002 (2018) https://doi.org/10.1051/epjconf/201818606002
LISA VIII

Evaluation of research publications and publication channels in astronomy and astrophysics

Eva Isaksson[1,*] and *Henrik* Vesterinen[2]

[1] *Helsinki University Library, University of Helsinki, Finland*
[2] *University of Tampere, Finland*

Abstract. The astronomy community usually turns to the Astrophysics Data System for bibliometrics. When the context is cross-disciplinary, commercial products like Web of Science and Scopus are used along with related analytics tools instead. The results are often tainted by inherent problems in the chosen classification system. A review of the most common challenges and pitfalls is given. Commercial altmetrics products could be added to the evaluation toolbox in the near future despite the fact that they are best suited for promotion instead of evaluation. Norway, Denmark, and Finland have created journal and publisher ranking systems that are used in national funding models. Differences in how astronomy journals are weighed in these systems night be related to the volume of papers published on a national level.

1 Introduction

The growth of evaluation culture in the 2000s has produced a number of tools and classifications to automate the assessment of university research outputs. While the Astrophysics Data System remains the preferred bibliometric tool for the astronomy community, it is not applicable in evaluations that span a wide range of research areas. University rankings and other evaluation processes use bibliometric data from either Web of Science or Scopus, while universities increasingly choose to purchase related do-it-yourself analytics tools.

In this paper the focus is on mainstream bibliometrics, using commercial bibliometrics products. We have checked how well they can handle evaluation of astronomy and astrophysics publications.

We will cover Web of Science, Scopus, and the analytics tools they use (InCites and SciVal). Our examples are mostly from Nordic countries. We also checked altmetrics, and the so called Nordic model of ranking journals. The time frame used is 2000-2015.

A key question when we turn to other databases and tools is: How do we define astronomy, if we want the results to be useful and realistic. The astronomy community has a pretty good idea as to which journals cover astronomy. That is not so clear when using commercial databases.

The usual way is to pick a research category defined by either Web of Science or Scopus. Both of these have their merits and problems.

*e-mail: Eva.Isaksson@helsinki.fi ORCID: 0000-0002-1488-5757

2 Web of Science and InCites: The HEP dilemma

Web of Science by Clarivate Analytics (formerly Thomson & Reuters Research Intelligence) is the preferred database used in Nordic comparisons of institutes [1]. It uses about 250 research categories. Each journal is listed in one or more categories. There were 63 journals in the astronomy & astrophysics category for 2016. If a journal is listed both in physics, particles and fields (PPF) and in astronomy & astrophysics (AA) categories, all of its articles are included in both, no matter what each article actually contains. Space and planetary physics journals tend to get an additional geosciences (GEO) related category.

To make the process more straightforward, we only looked at 21 top journals, where the Nordic countries (Sweden, Denmark, Finland, Norway) publish 95% of their astronomy papers. Next, we divided the journals into three groups according to which categories they belong:

- **AA category only:** *Astronomy &Astrophysics, Astronomical Journal, Astronomische Nachrichten, The Astrophysical Journal, The Astrophysical Journal Letters, The Astrophysical Journal Supplement, Astrophysics and Space Science, Icarus, Monthly Notices of the RAS, Planetary and Space Science, Solar Physics, Space Science Review* (12 journals)
- **AA & PPF categories:** *Physical Review D, Physics Letters B, Journal of Cosmology and Astroparticle Physics, Classical & Quantum Gravity, Astroparticle Physics* (5 journals)
- **AA & GEO categories:** *Journal of Geophysical Research: Space Physics, Annales Geophysica, Advances in Space Research, Radio Science* (4 journals)

Looking at the number publications, it turns out that there are some differences between Nordic countries regarding these groups. The AA group was strongest (58% of the total Nordic output), except for Norway, where the PPF was on top. After *Physical Review D* was added to the category in 2005, the share of the PPF group has become significant (28% of the output). The GEO group was the smallest everywhere (14%), and it was particularly small in Denmark. What are the effects on such differences?

To see the impact of a set of research publications, one usually normalizes the citations so to reflect the typical number of citations for a particular research category in a chosen time frame. The world average for each category is set to one (1). Impacts higher than 1 are said to be above world average for a particular category.

Figure 1 shows the category normalized results for Sweden, Denmark, Finland and Norway. The starting point is year 2000 and the publications are smoothed into four year time frames.

The results for the AA and physics group of journals differ considerably from the geosciences group. Astronomy and physics journal groups are trending upward, well above the world average. This is probably caused by participation in the very big, high impact collaborations like Planck. In comparison, the trend for the geosciences journal group does not look too good. That does not necessarily mean that the astronomy published in these journals has very little impact. Instead, they are being compared to journals with a different citation rate.

One should keep in mind that we only see some general trends, as InCites does not fractionalize the numbers to reflect the varying number of countries, institutions, groups or authors.

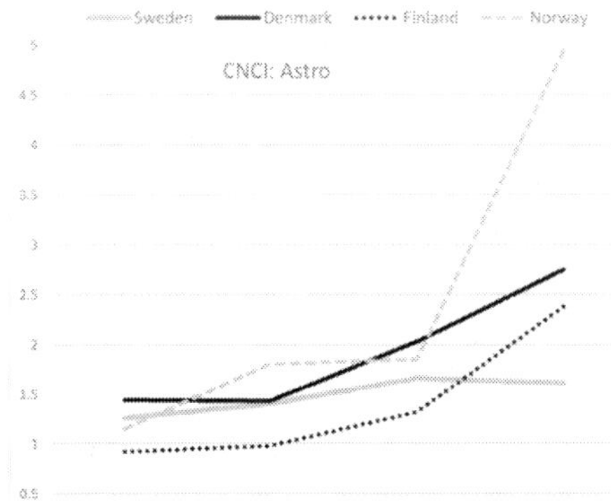

Figure 1. Category Normalized Citation Impact (CNCI) for Nordic Countries for the astronomy, physics and geosciences journal subgroups of the Web of Science Astronomy & Astrophysics research category, 2000–2015. World average is set to 1.
 InCites dataset updated 2017-05-13. Includes Web of Science content indexed through 2017-03-31.

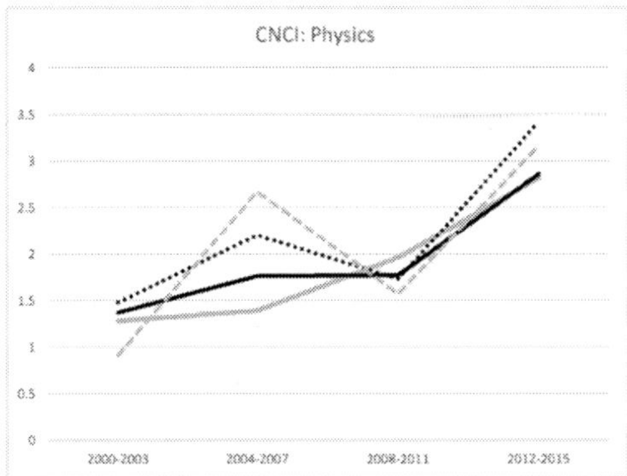

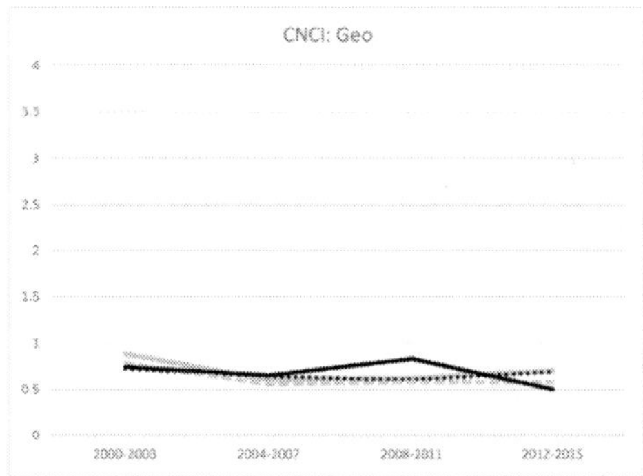

3 Trend Visualizations with VOSviewer Maps

Is there a way to check what these journals really contain? Are the journal groups really different from each other? One way to look inside is to use term maps. With VOSviewer,[1] you can map terms from titles and abstracts of publications.

Figures 2 and 3 contain term maps (clusters) for the selected journals categorized as astronomy by Web of Science. The underlying data consist of the titles and abstracts of all the papers published in the journals in Finland in the time period 2000-2015.

The term maps in Figure 2 are from the start of the time period we are looking at. We can see distinct clusters for the astronomy, PPF and geosciences journal terms, but they are not that much apart. Figure 3 shows the same group of journals but many years later. The astronomy and geophysics groups are now closer together. The PPF content has drifted into a direction of its own.

[1]http://www.vosviewer.com

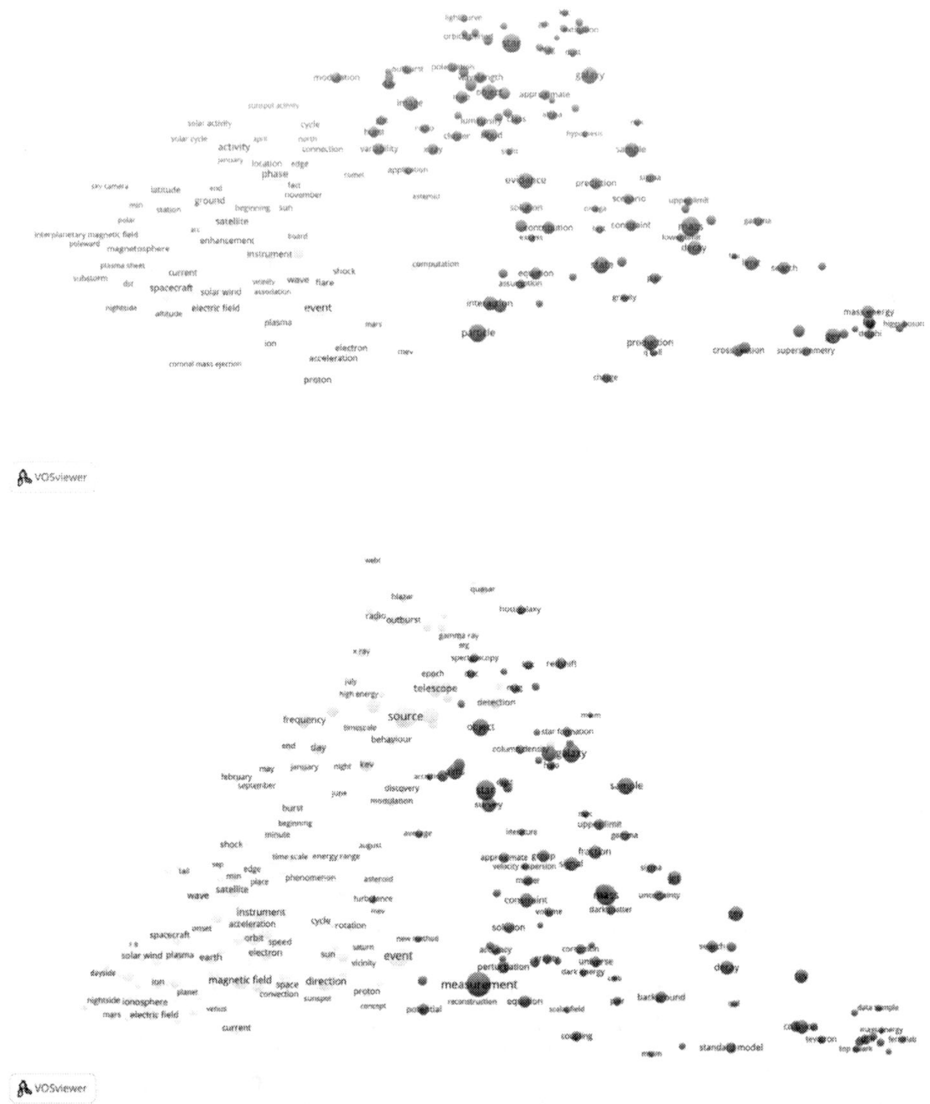

Figure 2. Term maps for years 2000-2003 (top) and 2004-2007 (bottom). More or less distinct clusters of astronomy, planetary sciences, and high energy physics are seen here. Astronomy can be seen in the top portions of the figures, planetary sciences on the left, and HEP on the right.

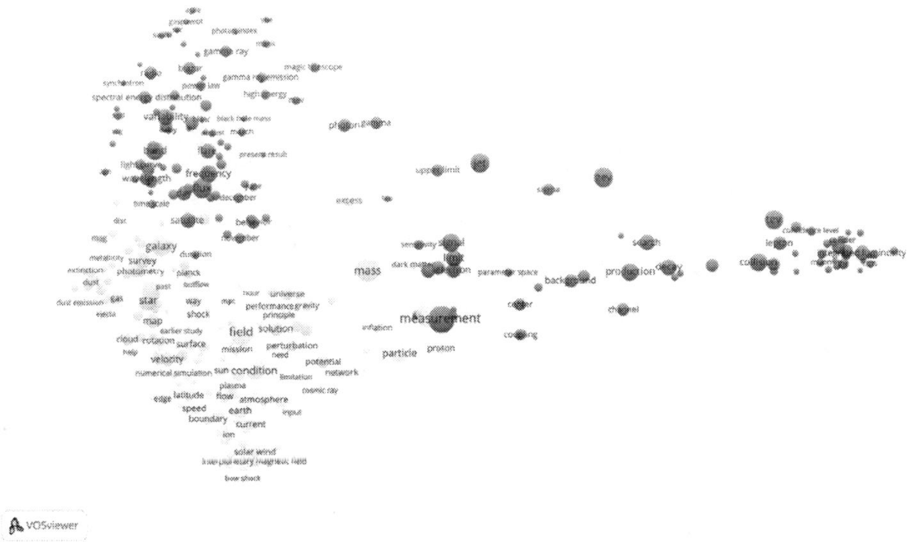

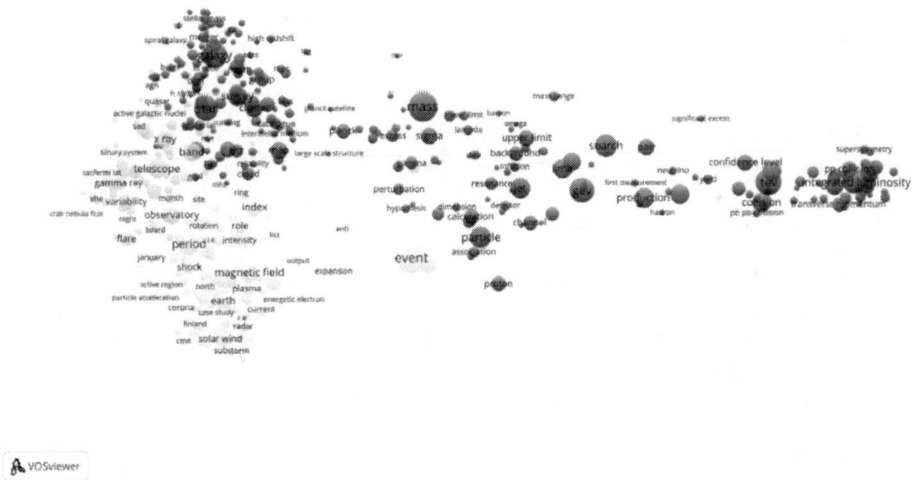

Figure 3. Term maps for years 2008-2011 (top) and 2011-2015 (bottom). Clusters of astronomy, planetary sciences, and high energy physics are clearly seen here. Of special interest is the apparent drifting of HEP away from astronomy.

The above results (CNCI and VOSviewer) show that the presence of two strong High Energy Physics (HEP) journals strongly affects the content and impact for the astronomy & astrophysics research category in the Web of Science. If we look at the worldwide results for this category, we see that *Physical Review D* tops the list. In Nordic countries, it is near the top, while the number one journal is still *Astronomy and Astrophysics*. This strong HEP presence means that astronomy evaluations are in fact astronomy and HEP evaluations when you use Web of Science categories.

Astronomy and HEP are obviously linked by subjects like dark matter and dark energy (see e.g., essay by Kolb [2]). But how to make a clear distinction between these two, and is it even possible? The dividing line is not linearly positioned between the fields. Astronomy has also been evolving towards large, HEP-style collaborations. That too will necessarily show up in bibliometric analysis.

In the period we looked at, experimental HEP publications have followed the pattern discussed by Pritychenko [3] for nuclear physics. There are fewer small facilities, and research is increasingly done with big instruments and by large collaborations connected by data-intensive networks.

The geosciences oriented journals suffer in comparison. One is tempted to ask whether you can rise above the world average if you are competing in the same category with LHC and the Higgs boson.

4 Scopus and SciVal: Better without HEP?

Scopus is a part of Elsevier and has established itself as the other big bibliometric service. In many respects, it is similar to Web of Science, with an analysis tool of its own, SciVal. There is a journal classification system that assigns one or more categories to each journal. The interesting thing is that SciVal has a more detailed classification than Scopus. On the other hand, the impact is fractionalized between categories, which tends to even out the differences.

For astronomy, there are two main categories: Astronomy & Astrophysics (A&A), and Planetary & Space Science (P&SS). Most astronomy journals belong to both, which means that these journals have to cut their results in half between both. For the top journals in the Nordic countries, the main differences for these two groups seemed to be a lone non-P&SS journal, *Journal of Astronomy and Astroparticle Physics*, and a number of geosciences oriented journals that were not included in the A&A group.

SciVal has a citation normalization indicator of its own, called Field Weighed Citation Impact (FWCI). Figure 4 shows the trends for Nordic countries. These results probably look good to most people. There are no HEP journals included, so administrators can see that their institution is above world average, and going upwards, no matter whether you look at the A&A or the P&SS category.

The halving of the results into two categories produces some notably distortions. Denmark is stronger in the A&A category, compared to the P&SS category. However, as astronomy journals tend to be found in both categories, it looks like the A&A results for Denmark are lower and the P&SS ones higher that they actually are.

5 Altmetrics: Mentions versus Impact

In the 2010s, we have seen the rise of altmetrics or alternative metrics that aims to catch the immediate, societal impact of research outputs by counting mentions from channels such

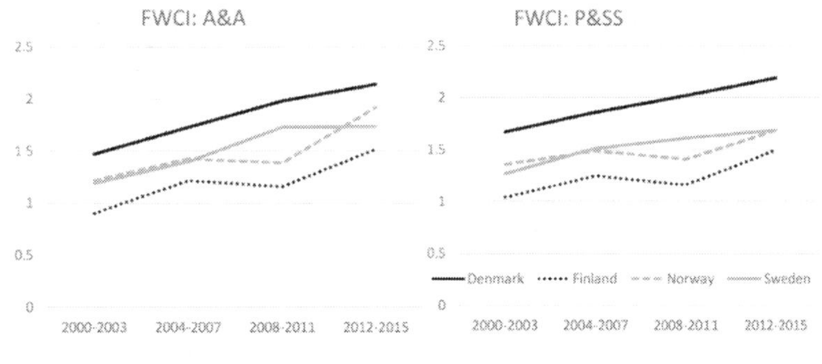

Figure 4. Field Weighed Citation Impact for Nordic Countries for SciVal categories Astronomy and Astrophysics (A&A) and Planetary & Space Sciences (P&SS).

as social media, blogs and news. Can we consider it as an additional tool in our evaluation toolbox?

What exactly are we measuring when we are counting mentions? What does a metric like mentions per output mean for a set of publications? To see how mentions are related to normalized impact, we took the Web of Science astronomy & astrophysics research area and chose the journal groups listed earlier in this paper. Instead of Nordic countries, we used the total worldwide output for these journals. Mentions were harvested from Altmetric Explorer for 2010–2016 (older results are not comprehensive enough). Impact was found from InCites for the same period.

The result is shown in Figure 5. The astronomy group of journals has the most mentions (4.4 per output) compared with particles and fields group of HEP journals (3.7) and geosciences journals (2.7). In particular, the astronomy journal group excels in wikipedia and news mentions. As for CNCI values, the geosciences journal group is clearly below world average with 0.6. The HEP journal group scores 1.3 and the astronomy group 1.2.

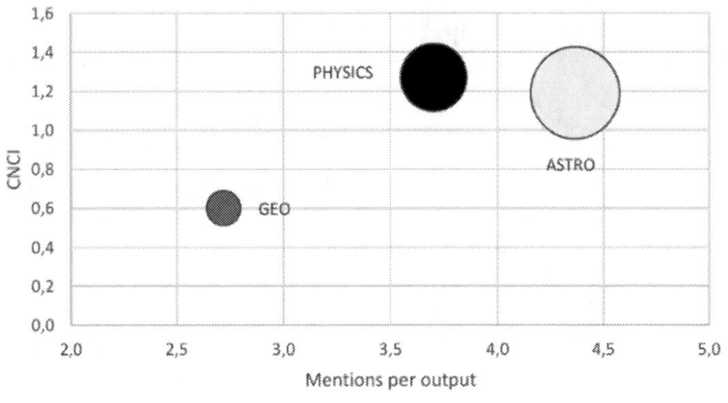

Figure 5. Category Normalized Citation Impact (CNCI) versus mentions per output for the astronomy, physics and geosciences journal subgroups of the Web of Science Astronomy & Astrophysics research category, 2010–2016. World average is set to 1.

What do the numbers mean? It is clear that the geosciences group of journals are getting less attention in terms of mentions. Are they any worse for that? These results tell us little about the mechanisms behind the numbers, and consequently, they are not suited for evaluation.

6 Journal levels

In the Nordic countries (except for Sweden), national journal ranking systems have been introduced mainly as a tool for funding university research.[2] In this system, a certain percent of the funding is distributed to the universities according to where and how much they publish. Journals and publishers are assigned levels determined by expert panels. Also those publication channels not found in Web of Science or Scopus (whether representing social sciences and humanities which have a low coverage in those databases, or published in national languages) are assigned a level.

Levels do not reflect the impact of research directly, as they do not measure the impact of individual papers. Also, they need to be balanced between disciplines, so that only a certain percentage of journals are assigned to a higher level for a given discipline. Levels often do not reflect the real quality of a journal.

Journal	Finland	Norway	Denmark
A&A	3	2	2
A&A Rev.	3	1	2
ApJ	2	2	2
ApJ Lett.	1	2	1
ApJS	1	2	1
JCAP	1	1	2
JGR: Space Phys.	3	1	2
MNRAS	1	1	2
Phys. Rev. D	2	1	1
Phys. Lett. B	2	1	1

Table 1. Journal levels (June 2017) are generally not identical for Finland, Norway and Denmark.

Journal levels are occasionally used for evaluation of units and individual researchers, against official recommendations. In Finland, the Ministry of Education compared both impact and publication points from journal levels for Finnish universities[4]. According to the analysis, astronomy publications by University of Helsinki had bigger impact compared to University of Turku. In publication points, the outcome was reversed for so-called productivity. In a situation like this, you can trust either Web of Science or journal levels. The latter tell very little about actual impact of research.

Table 1 lists some interesting differences between Nordic countries. The only journal all three systems agree on is *The Astrophysical Journal*. For other journals, there is a rough rule of thumb: the more you publish in a journal, the more likely your national panels are to choose a higher level. This seems to be true for Finland and Denmark. These different levels for the same journals could lead to some variation as to how astronomy output is rewarded in terms of national research funding.

[2]One can use the search tool at the Finnish portal for looking at journal levels in Nordic countries: https://www.tsv.fi/julkaisufoorumi/haku.php?lang=en

7 Conclusion

Research evaluations are usually built on comparisons of units. For an international effort like astronomy, the results are better when one participates in high profile international collaborations. Certain journal titles that publish such research will influence the expected impact of whole research areas. Evaluations that span several disciplines cannot use the Astrohysics Data System, as there are no normalization tools for the comparisons that are usually needed. The next best choice could be Web of Science with InCites, closely followed by Scopus with SciVal. Both of these have their own problems. Web of Science includes two extremely strong HEP journals in the Astronomy and Astrophysics Research Area. As a result, HEP results tend to overshadow some of the astronomy and astrophysics. There is also a geosciences (space physics, planetary science) related group of journals that suffer even more from the HEP presence.

In Scopus and SciVal, the Planetary Sciences have a category of their own, without any HEP journal complications. However, the fractionalizing of results between categories distorts the impact for most astronomy journals.

As for altmetrics, it is certain that astronomy is one of the disciplines with a significant number of mentions. There is a very long way from that to meaningful metrics. Instead of evaluations, altmetrics is better suited for promotion and visibility of research.

Of all approaches to evaluation, national journal ranking systems are the most problematic, as they were developed to serve as tools for university funding.

References

[1] F. N. Piro (ed.), *Comparing research at Nordic higher education institutions using bibliometric indicators: Covering the years 1999–2014* (NIFU, Oslo, 2017)

[2] R. Kolb, Reports on Progress in Physics, **70**, 1583 (2007)

[3] B. Pritychenko, Scientometrics **105**, 1781–1786 (2015)

[4] Summary. *Suomalaisen yliopistotutkimuksen tuottavuus ja vaikuttavuus. Productivity and impact of Finnish university research* (Ministry of Education and Culture. Helsinki, 2015), 7–8.

EPJ Web of Conferences **186**, 06003 (2018) https://doi.org/10.1051/epjconf/201818606003
LISA VIII

Evaluating High Impact Papers: Are We Missing Something?

Sherry Winkelman[1,★], *Arnold* Rots[1], and *Raffaele* D'Abrusco[1]

[1] *Harvard-Smithsonian Center for Astrophysics*

Abstract. The bibliographic science papers with high citations rates are often used as an indication of the science impact of an observatory. These high impact papers are presented as examples of the best science being done with an observatory's data. But, is the number of citations by itself a good indicator of the scientific impact of the paper, and is impact a good indicator of the scientific impact of the observatory? In this paper we will present results from a recent study of Chandra high impact papers and suggest some alternative methods for identifying such papers.
This work has been supported by NASA under contract NAS 8-03060 to the Smithsonian Astrophysical Observatory for operation of the Chandra X-ray Center.

1 Introduction

Observatory science papers with high citations rates are often used by observatories as an indication of the science impact of an observatory. Classically, the papers with the highest number of citations are deemed High Impact Papers (HIPs). In the case of the Chandra X-ray Observatory (CXO) we typically list the 50-100 papers with the highest citation counts (the top 1-2% of papers) and no further evaluation is done.

One reason for identifying HIPs is that High Impact Papers may tell us something about the most influential types of science coming from the observatory and therefore what sorts of observing programs will have long-lasting contributions to astronomy. Since it often takes many years for papers to reach the HIP category, we hope to identify objective measures which will allow us to identify High Impact Papers earlier to perhaps provide guidance in the request for proposals process.

2 What Makes a Paper HIP?

There are two components for defining a High Impact Paper: 1) evaluation of the science contribution from the observatory to the paper as a whole and 2) objective measures based on citation history and for which HIPs are outliers for the observatory. The first component is subjective, but the idea is that the science content from the observatory needs to be integral to the findings in the paper. Some reasons we have found for rejecting a paper as being a HIP are:

★e-mail: swinkelman@cfa.harvard.edu ORCID: 0000-0001-7354-6221

- the only connection to Chandra was to create a figure to confirm measurement of data from another observatory and the other observatory data was the focus of the paper
- cross-matching of sources in Chandra data with sources from another observatory but no further analysis using Chandra data is done
- X-ray data for only a few targets of a large sample of targets came from Chandra
- Chandra provided data showed a non-detection
- the Chandra analysis is in the supplementary materials

The objective measures are based on citation histories of papers. There are two caveats associated with the objective measures: 1) they may change as the observatory ages, so one must decide what to do about HIPs that no longer meet the current criteria and 2) they need to be determined in a programmatic way without requiring further (or minimal) analysis by curators.

3 Morphology of HIPs

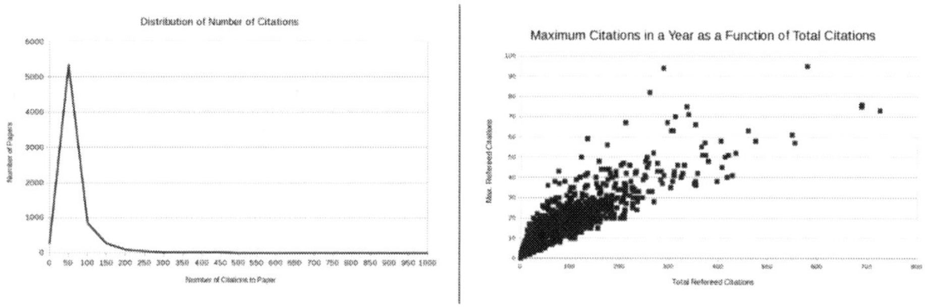

Figure 1. The plot on the left shows the distribution of number of citations in Chandra Science Papers. The long tail of the distribution starts at ~ 200 citations. The plot on the right shows the distribution of the total number of refereed citations to the maximum number of refereed citations in a single year. More than 95% of CSPs are clustered in the lower left of the figure with total citations < 200 (corresponding to the long tail papers in the left plot) and maximum annual citation rates < 30, suggesting that papers with a maximum annual citation rate > 30 may be high impact.

We looked at four objective criteria for identifying potential High Impact Papers: 1) the distribution of the number of citations in a paper: indicates papers with high citations compared to the full body of science papers from the observatory (shown in left panel of Figure 1); 2) the number of citations in a paper versus the highest citation in a year: highlights papers with unusually high citations of a paper in a single year (shown in right panel of Figure 1); 3) the number of papers with > X citations in a year: provides papers which have a moderate citation rate for several successive years (shown in left panel of Figure 2); and 4) the number of papers which previously published results from a particular paper: gives you papers whose results are frequently used for subsequent studies (shown in right panel of Figure 2).

Based on the four criteria which we examined, we chose four categories for identifying potential High Impact Papers and identified a fifth category which shows promise as a category but the data linking in the Chandra Bibliography is not complete enough to set any limits.

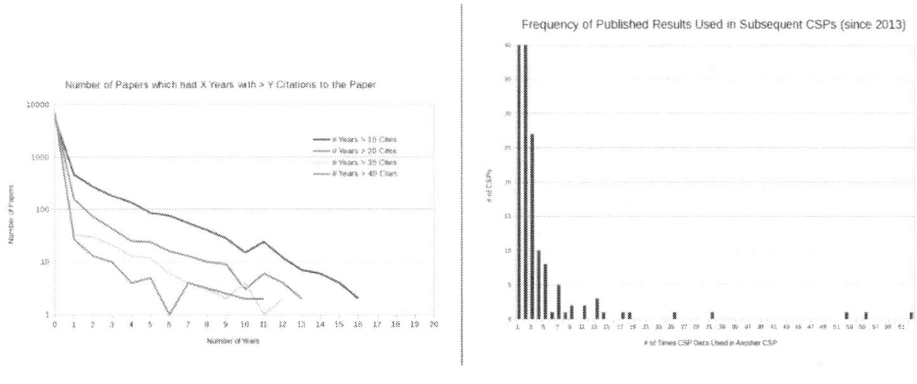

Figure 2. The left panel shows the number of papers which had X years with > *Y* citations to the paper. 5 and 10 years (~ 1/4 and ~ 1/2 the age of the Chandra archive) were chosen to as benchmark times. Categories C and D were somewhat arbitrarily chosen as the points were the curves fall below 10 papers to encompass those papers which steady growth rates in citations. The right panel shows the frequency that published results of the HIP are used in subsequent Chandra Science Papers. The data linking required for this plot are complete in the Chandra Bibliography for papers published after 2013.

- **Category A**: > 200 refereed citations is based on the left panel of Figure 1. The long tail of the distribution starts at ~ 200 citations.
- **Category B**: > 30 citations in a single year is based on the right panel of Figure 1. Papers with a peak citation > 30 is in the scatter portion of the distribution.
- **Category C**: > 10 years with > 10 citations in a year is based on the left panel of Figure 2 and corresponds to steady citation rate for ~ 1/2 of the mission.
- **Category D**: > 5 years with > 20 citations in a year is based on the left panel of Figure 2 and corresponds to a moderate citation rate for ~ 1/4 of the mission.
- **Category E**: > *X* Chandra Science Papers (CSPs) have used the results from the HIP for the analysis in the CSP is based on the right panel of Figure 2.

The left panel of Figure 3 shows the distribution of the various flavors of HIPs for Chandra Science Papers. Also shown in the figure are the numbers of papers in each category which fulfill the science relation criterion. While most HIPs fit into more than one category, each category identifies some papers that do not fall in another category. Note that Category E is not included in Figure 3. Preliminary results suggest that this is a promising category, but the indirect data linking required for this analysis is not complete in the Chandra Bibliography. Reference [1] gives a description of the new data linking being added to the Chandra Bibliography.

The hope is that the new categories of High Impact Papers will help us to identify HIPs earlier. This is clearly seen in the right panel of Figure 3. Category B in particular can be used to identify HIPs as early as 2-3 years after publication, while HIPs in Category C can be identified as early as 5-7 years after publication.

4 HIPs Compared to Full Bibliography

Other than citation histories, are there other differences between High Impact Papers and the complete bibliography of Chandra Science Papers? We looked at four characteristics of HIPs versus CSPs:

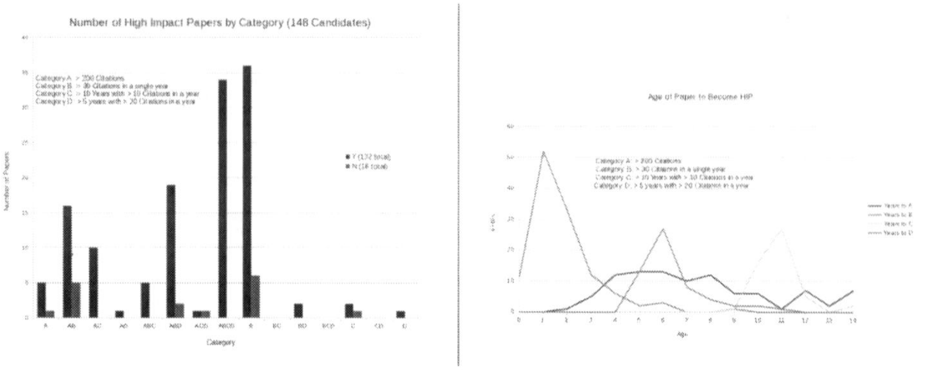

Figure 3. The left panel shows the number of HIPs by category combination. The blue bars indicate the number of HIPs where the Chandra analysis in the paper is integral to the findings in the paper. The green bars are potential HIPs for which the Chandra analysis is not significant to the paper. See Section 2 for examples of what falls within the green bars. The right panel is a plot of the age of a paper when it became HIP for a given category.

1. percentage of non-CSP citations in the paper
2. primary data usage in the paper
3. data from other wavebands also analyzed in paper
4. Chandra survey regions covered in paper

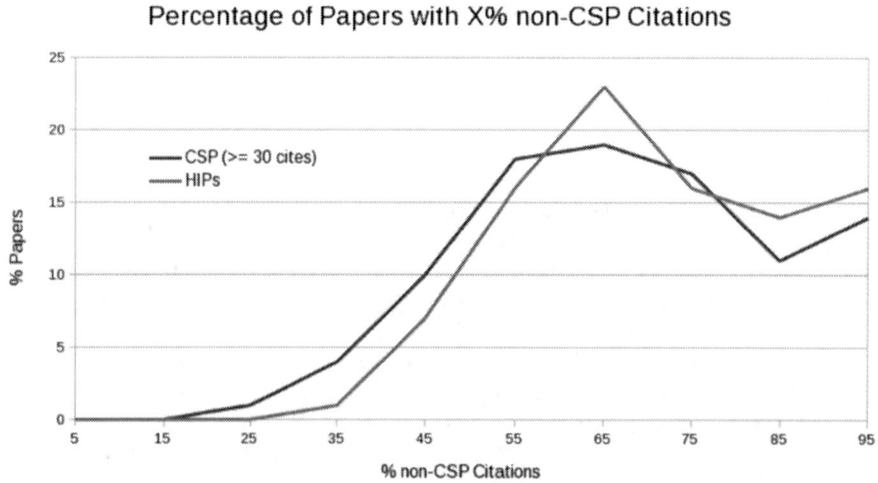

Figure 4. Plot of the percentage of papers with X% non-CSP citations. The blue line represents CSPs with >= 30 citations and the red line represents HIPs.

Data Comparison	CSPs	HIPs
Direct data analysis	79.5%	69.7%
Indirect data analysis	20.5%	30.3%
Multi-observatory	57.9%	52.8%
Theory/computation	10.4%	13.4%
Follow-up analysis	6.8%	6.3%

Table 1. Comparison of the primary Chandra data usage between CSPs published since 2013 and HIPs. Indirect data analysis is broken down into multi-observatory, theory/computation, and follow-up analysis.

Observatory Usage			Waveband Usage		
Observatory	HIPs	CSPs	Waveband	HIPs	CSPs
XMM	16%	22%	Visible	75%	30%
HST	13%	11%	X-ray	62%	46%
Spitzer	10%	5%	IR	23%	15%
VLA	9%	6%	Radio	19%	13%
VLT	7%	2%	UV	7%	5%
Swift	7%	8%	Submm	2%	1%
ROSAT	5%	2%	Gamma	1%	2%
Subaru	4%	1%	Microwave	1%	0%

Table 2. Other observatory data analyzed in CSPs published since 2013 and HIPs.

Figure 4 indicates that HIPs tend to be cited by non-CSPs more frequently than CSPs which could indicate that science in High Impact Papers is influencing non-Chandra science more than general CSPs.

Table 1 indicates that HIPs are more likely to be based on indirect analysis of Chandra data than CSPs as a whole. We flag three types of indirect analyses: multi-observatory analysis means previous Chandra results are used in a complementary fashion with data from another observatory; theory/computation means previous Chandra results are compared to models or used as input to theoretical computations; and follow-up analysis means previous Chandra results are as the starting point for further data analysis. For both CSPs and HIPs, multi-observatory analysis is the primary form of indirect analysis of Chandra data.

The Chandra bibliography also flags whether data from other observatories was analyzed in the paper. We directly flag a number of observatories and if an observatory is not in our list of observatories, we flag the waveband of the data used in the paper. We see in Table 2 that there are no significant differences between HIPs and CSPs which also have observatory flags when looking at the top cited observatories, but we do see a significant difference when looking at wavebands. HIPs combine analysis of Chandra data with data from visible and other X-ray observatories at a much high rate than CSPs as a whole.

Table 3 shows that HIPs are far more likely to be connected with a deep Chandra survey or catalog than CSPs which are connected to a survey. A secondary reason that these surveys are frequently used maybe the homogeneity of the sample. Not surprisingly, those surveys are also those areas of the sky which are covered by deep surveys in other wavebands.

Survey	HIPs	CSPs
Chandra Deep Field South	50%	24%
Chandra Deep Field North	43%	8%
C-COSMOS	12%	19%
Extended CDFS	7%	6%
Chandra Large Area Synoptic Survey	5%	0%
Chandra Source Catalog	3%	4%
Chandra Orion Ultradeep Project	3%	2%
AEGIS-X	3%	6%
X-Bootes	3%	2%

Table 3. Survey regions covered in CSPs published since 2013 and HIPs.

5 Requirements for Inclusion in Bibliography

To make HIPs an integral part of the Chandra bibliography, there will need to be additional fields in the database and code development for identifying potential HIPs. In the database, at minimum we will need to add a flag that the paper is considered to be high impact. To be more useful, we need to flag when a paper is a candidate for being a HIP, which HIP categories the paper qualifies for, and when the paper qualified for that category. Another useful piece of information to have associated with a HIP is a brief description about the paper to include with replies to requests for lists of HIPs.

Procedurally, when a potential HIP comes in, a curator will need to determine whether the paper should be considered a HIP or not based on an evaluation of the science contribution of Chandra data to the paper as a whole. The HIP flag should then be set to the appropriate value and additional metadata added to fully classify the HIP.

It should be noted that potential Category E HIPs will also require a determination by curators of whether the referencing papers are part of a series of papers from the authors rather than independent work. This may be a significant burden to the system and requires careful consideration.

Finally, one needs to deal with HIPs which no longer fit the current criteria. Some considerations are: once a HIP, always a HIP; a HIP with qualifications; or perhaps, HIP versions.

6 What Next?

During the course of this study we identified three areas which need deeper investigations before moving forward on adding HIPs to the Chandra bibliography:

- Perform a historical analysis of objective criteria at ages 5, 10, and 15 years of the Chandra mission to see how they may have changed and whether 'former' HIPs are still HIPs
- Determine if the age of median citations as a function of the age of the paper help inform the HIP determination
- Explore predictive traits based on citation history which could flag potential HIPs

7 Summary

In summary, we have found that there are two components in defining a High Impact Paper, a subjective evaluation of the science contribution from Chandra to the paper as a whole and an objective determination based on citation history. Based on citation distributions of CSPs we identified four categories for identifying potential HIPs and that two of those categories allow us to identify High Impact Papers much earlier than using the total count of citations.

When comparing HIPs to the full set of Chandra Science Papers, we find that: HIPs tend to be cited by non-CSP's more frequently than CSPs, perhaps because HIPs tend to use data from other wavebands more frequently than CSPs; HIPs are more likely to be based on indirect analysis of Chandra data than CSPs; and that HIPs are far more likely to be connected with a deep and/or homogeneous Chandra survey than CSPs connected with surveys.

Inclusion of HIPs into the Chandra bibliography will require expansion of the metadata connected with paper and code development. A deeper understanding of how the objective criterion change over the course of the mission is needed before determining the extent to which HIPs should be included in the bibliography.

This research was supported by NASA contract NAS8-03060 (CXC).

References

[1] S. Winkelman, A. Rots, R. D'Abrusco, *Growing a Bibliography*, These Proceedings (2018)

S. Guéhénneux and A. Wallis (Photo: M. Wenger)

Comparing People with Bibliometrics

Michael J. Kurtz[1,★]

[1] *Harvard-Smithsonian Center for Astrophysics*

Abstract. Bibliometric indicators, citation counts and/or download counts are increasingly being used to inform personnel decisions such as hiring or promotions. These statistics are very often misused. Here we provide a guide to the factors which should be considered when using these so-called quantitative measures to evaluate people. Rules of thumb are given for when begin to use bibliometric measures when comparing otherwise similar candidates.

1 Some things you should know when using bibliometrics to compare people

This is a serious business. Careers can be greatly affected by the decisions which flow from bibliometric evaluations. [1, 2]

It is important to know how accurate these measures are. Measures such as citation or download counts are not important in themselves, they are proxies for evaluating the (research) capabilities of an individual. [1, 3, 4]

While the relative completeness and accuracy of some sources of bibliometric information are frequently discussed, this is not particularly important to comparisons of people. The statistical scatter, even with "perfect" input data in these measures is quite large, and puts limits on any attempt to use the data to evaluate people. [1]

Before using citations or downloads to compare individuals these measures must be adjusted to account for the known systematic errors due to age, discipline, and co-authorship. [5]

The total number of citations a person's work receives increases approximately quadratically with the "age" of that individual, where "age" means time as an active researcher. [5, 6] The correction is that the square root of the number of citations, divided by "age" is an approximate constant for persons with uniform productivity. For younger researchers determining the effective "age" is problematic, and makes this correction uncertain. [1]

Different disciplines, and subdisciplines, have different citation cultures, citation counts cannot be directly compared without correcting for this. [5, 7]

★e-mail: kurtz@cfa.harvard.edu ORCID: 0000-0002-6949-0090

Multi-author papers are now the norm. Bibliometric measures must take the degree of co-authorship into account. [5, 8, 9]

For comparisons with individuals measured at different times one must account for the steady increase in the number of papers, which double about every 15 years [10, 11], and in the number of references per paper, which is experiencing a similar increase [12, 13].

The TORI statistic [5], which normalizes citations by both the number of authors on the cited document and the number of references in the citing document, is intended to remove the effects of co-authorship and citation culture. The RIQ statistic [5] is intended to remove the effects of age. Both are available on the ADS metrics pages.

Downloads, or reads, are also a useful bibliometric indicator [4, 9, 14]. Downloads by researchers have very similar properties to, and can predict citations [11, 14, 15]. Great care must be taken when using download information, however, as downloads by persons who are not researchers have very different properties, and can dominate the statistic [4, 7, 13].

The ADS metrics pages show the Read10 statistic [1, 9], which is the yearly sum of co-author number normalized reads by scientists via ADS of papers written in the preceding ten years. Because the use of internet based services has grown rapidly in the recent past care must be taken when comparing download counts from different time frames.

Using samples and measures designed to minimize the systematic errors in the measurements we can determine the intrinsic scatter in the use of bibliometric techniques to evaluate people. We do this by comparing different measures of exactly the same sets of papers by exactly the same people. Because the actual impact of these sets of papers is identical (because they are exactly the same) the differences in impact as measured by different techniques must be due to the intrinsic measurement error in the techniques. Complete details are in [1].

Because citations and reads of people form a log-normal distribution [1, 17] the error is in the logarithm of the measurements. For the comparisons using active mid-career astronomers the one standard deviation error is 0.17 dex, a multiplicative factor of 1.48. [1]

Citations (and downloads, by extension) are also used to predict performance; essentially this is why they are used in hiring decisions. For individuals five years past the PhD, about the time when assistant professor positions are granted, even using an a posteriori selected sample, substantially more prescient than any contemporaneously selected sample could be, the one standard deviation error is 0.395 dex, a multiplicative factor of 2.5. [1]

2 Conclusions

Bibliometric measures, citation and download counts, have been shown to be useful and accurate in a number of areas. When aggregated over large entities, such as universities [13, 18], countries [13, 19], or journals [20] bibliometric measures can provide accurate measures of the quantity and quality of research. Additionally, persons with very high citation counts are, almost always, very well regarded researchers. This means that citations can be accurately used to recognize these individuals, and predict high honors [3].

These successes have led some to use bibliometric measures to evaluate people. In most cases this constitutes a misuse of the statistics. Decision aids such as citation or download statistics may well be able to identify most of the top people in a field, but they have almost no power in making the more common 41st chair [21] type of personnel decision: deciding between a set of similarly qualified candidates.

Because in any real instance the various corrections for age, co-authorship, subfield, etc will not be as clear as in the near ideal case analyzed in [1] it is reasonable to require that, when comparing two individuals, the difference in their citation or download measures be greater than two standard deviations (95% confidence) before ascribing any difference to them on the basis of those counts. This would be about a factor of two when comparing the current status of mid-career scientists, and about a factor of five when predicting the future productivity of possible assistant professor hires. Outside of these bounds one would want to understand why the counts were so different.

References

[1] M.J. Kurtz, E.A. Henneken, Journal of the Association for Information Science and Technology **68**, 695–708 (2017)

[2] L. Bornmann, W. Marx, Zeitschrift fuer Evaluation **12**, 103-127 (2013)

[3] E. Garfield, M.V. Malin, *Can Nobel Prize Winners be Predicted?* Paper presented at the 135th Annual Meeting of the American Association for the Advancement of Science, (1968)

[4] M.J. Kurtz, J. Bollen, Annual Reviews of Information Science and Technology **44(1)**, 1-64 (2010)

[5] A. Pepe, M.J. Kurtz, *A Measure of Research Impact Independent of Time and Discipline*, PLoSONE e46428, (2012)

[6] J.E. Hirsch, *An index to quantify an individual's scientific research output* in *Proceedings of the National Academy of Science* (2005), Vol. 102(46), p. 16569–16572

[7] D. Nicholas, P. Huntington, T. Dobrowolski, I. Rowlands, M. Jamali, P. Polydoratou, P., Information Processing & Management **41**, 1441–1461 (2005)

[8] D.de S. Price, Journal of the American Society for Information Science **27 (5-6)**, 292-306 (1976)

[9] M.J. Kurtz, G. Eichhorn, A. Accomazzi, C.S. Grant, M. Demleitner, S.S. Murray, N. Martimbeau, B. Elwell, Journal of the American Society for Information Science and Technology **56(2)**, 111-128. (2005)

[10] D.de S. Price, *Little Science, Big Science*, New York: Colombia University Press, (1963), ISBN-10: 0231049560

[11] M.J. Kurtz, G. Eichhorn, A. Accomazzi, C.S. Grant, S.S. Murray, J.M. Watson, Astronomy and Astrophysics Supplement Series **143**, 41-59 (2000)

[12] E.A. Henneken, M.J. Kurtz, A. Accomazzi, C.S. Grant, D. Thompson, E. Bohlen, S.S. Murray, Journal of Informetrics **1**, 1-8 (2009)

[13] E.A. Henneken, M.J. Kurtz, *Usage Bibliometrics as a Tool to Measure Research Activity* to appear in: Handbook of Quantitative Science and Technology Research, Springer, ArXiv e-prints arXiv:1706.02153. (2017)

[14] S. Haustein, *Readership Metrics* in *Beyond Bibliometrics*, edited by B. Cronin and Cassidy R. DSugimoto, Cambridge:MIT Press (2014), 327-344
[15] H.F. Moed, G. Halevi, Journal of the Association for Information Science and Technology **67**, 412–431. (2016)
[16] T. Stankus, B.A. Rice, Collection Management **4(1/2)**, 95–110 (1982)
[17] W. Shockley, Proceedings of the Institute of Radio Engineers tectbf45(3), 79–290 (1957)
[18] Times Higher Education Ranking https://www.timeshighereducation.com retrieved (2017)
[19] National Science Board, "Science & Engineering Indicators" https://www.nsf.gov/statistics/2016/nsb20161/#/ retrieved (2017)
[20] E. Garfield, Science **178(4060)**, 471–479 (1972)
[21] R.K. Merton, Science **159**, 56 (1968)

Walking tour of the historical « Petite France » quarter (Photo: M.E. Jiménez)

 https://doi.org/10.1051/epjconf/201818607001

Institutional Repository

Know-how of a decade in Managing Digital Assets

Meera B.M.[1,*], *Krishnamurthy* S[1,**], and *Manjunath* Kaddipujar[1,***]

[1] *Raman Research Institute, Bengaluru, Karnataka, INDIA*

Abstract. The advent of the Internet and the mushrooming of information in various digital formats have resulted in looking at the library and its services with newer perspectives for conserving and disseminating scholarly publications. This paper discusses two mechanisms of digitization initiated at Raman Research Institute (RRI) in India, namely 1) the Digital Repository of RRI and 2) the Imprints Collection. The digital repository of RRI is the institutional repository built in 2006 to bring visibility to the scholarly publications of its founder Sir C V Raman. This paper discusses a decade long experience in building and sustaining an institutional repository. A value addition to this repository is the "Imprints Collection"- a bio-bibliographic database of retired scientists of RRI. The design and functionality of the Imprints Collection are also discussed in this paper.

1 Introduction

Clifford Lynch defined Institutional Repository (IR) in 2003 as – "A set of services that an institute offers to the members of its community for the management and dissemination of digital materials created by the institution and its community members. It is most essentially an organizational commitment to the stewardship of these digital materials, including long-term preservation where appropriate, as well as organization and access or distribution."

An institutional repository philosophically has an open access approach, and they aim to show case the scholarly output of an Institute. Notionally they are a digital library, quite similar to analogous functions of a conventional library, performing tasks such as collecting, housing, classifying, cataloging, curating, preserving, and providing access to digital content of scholarly publications of an Institute. IR is relatively a new phenomenon initiated in early 21st century. Open access initiatives and the development of various software to encourage Open Archives Initiative has acted as a spring board for the growth of IR's across the globe.

Raman Research Institute (RRI) started in 1948 by Sir C V Raman is a pioneering research institute in Physics in India. Being an Institute started by a Noble Laureate, RRI library has a

*e-mail: meera@rri.res.in ORCID: 0000-0002-8355-8776
**e-mail: skmurthy@rri.res.in
***e-mail: kaddipujar@rri.res.in

wealth of information resources such as scholarly publications of Raman and his students, and his correspondences with dignitaries from different walks of life, for example when India was undergoing socio-political turmoil. They represent the history of Indian science, and they have archival value. Most of the correspondences were hand written, and few were typed. Day by day and year after year, the papers holding these information were becoming fragile. RRI library also possessed many newspaper clippings reporting Raman's activities and his role in science policy soon after Indian independence in 1947. Handling newspaper clippings, which were in print format, was also becoming difficult. So, there was a need to preserve all of them for the future. The best option was to digitize, preserve and showcase Raman's publications and all the information resources associated with him through an IR. Subsequently, a decision was made to include scholarly publications of current research, photographs and audio/video files showcasing the diverse activities of the Institute.

2 Institutional Repository of RRI:

The institutional repository of the Raman Research Institute was initiated in 2004. The decision was made to use open repository software to build the IR. This was the time when many open source repository software was being developed and introduced in the market for free. RRI had three software options and they were:

1. Greenstone by the New Zealand Digital Library Project at the University of Waikato, and developed and distributed in cooperation with UNESCO and the Human Info NGO
2. Eprints developed at the University of Southampton School of Electronics and Computer Science and
3. DSpace developed by MIT and HP Labs.

The decision was taken to install 'Eprints' and 'DSpace, ' and a test bed was created to identify the most suitable software for RRI. The final choice to build the repository was - "DSpace". The repository was named the "Raman Research Institute Digital Repository (RRIDR)." A dedicated server was used to install DSpace version 1.0.1. Different communities and sub-communities were identified for the classified submission of a variety of information on the repository. Dublin Core metadata standards were adopted to build the repository. Copyright is an important aspect while making submissions to the repository. SHERPA/ROMEO Publisher copyright policies and self-archiving guidelines were followed to handle copyright issues. The decision was taken to provide full-text access wherever permissible, and the rest were tagged as 'Restricted access,' thus protecting from copyright infringement. Self-submission guide for authors was written. The home page was customized.

Initially, the following seven communities were identified for submission:

- Archives
- C. V. Raman and his works
- Science at the Raman Institute 1948-1970
- Astronomy and astrophysics
- Light and matter physics
- Soft condensed matter
- Theoretical physics

It was realised during the process of digitisation that most of the materials were not born digital. So, they had to be scanned, cleaned, curated, digitized, classified and finally submitted and approved. RRIDR became operational in 2006 having 1406 records. In 2008, the repository was upgraded to version 1.5.1. The customized layout of the Repository is as in figure 1.

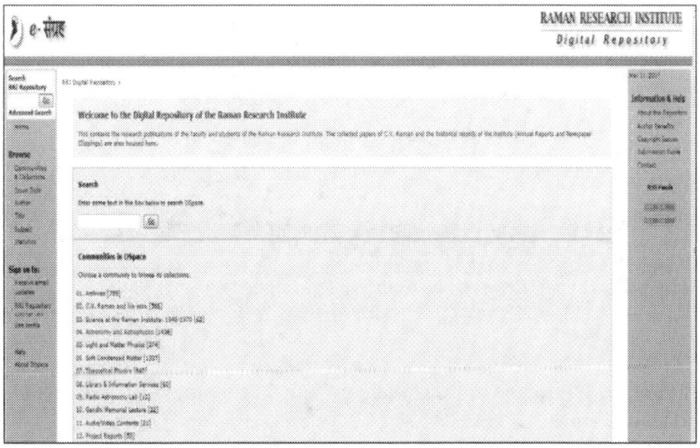

Figure 1. Raman Research Institute Digital Repository

The number of communities has steadily grown, and currently, there are 18 communities. The additional 11 communities are:

- Library and Information Services
- Electronics Engineering Group (EEG)
- Academic Events
- Gandhi Memorial Lecture
- RRI In-House Meeting
- Project Reports
- RRI Science News
- Audio/Video
- Photo Gallery
- Prof. Satish Dhawan's Collection
- Miscellaneous

Photographs, audio/video files were included as communities in 2011. Version 1.5.1 of DSpace did not have the feature of depicting thumb nail images of the photographs. So, searching a particular photograph was a difficult task. Video streaming was also not possible in the 1.5.1 version of DSpace. Video files had to be downloaded before viewing. This would test the patience of the users, mostly depending on the file size. Version 5.5 had additional helpful features to handle multimedia files. So in 2016, the repository was migrated to version 5.5 and subsequently to the current version 6.0 in 2017. The latest layout of RRIDR looks as it appears in figure 2.

Figure 2. New Layout of Raman Research Institute Digital Repository

RRIRDR has steadily grown, and the years growth of the uploaded records on the repository is graphically shown in Figure 3.

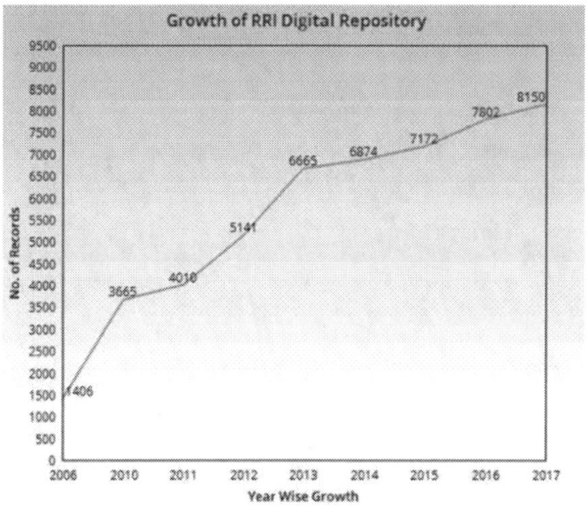

Figure 3. Growth of RRI Digital Repository

2.1 New features of DSpace version 6.0

The new version of the repository has certain additional helpful features and they are:

- High quality thumbnails available for image bitstreams and PDF bitstreams using Image Magick / Ghostscript.

- File downloads now can be tracked in Google Analytics
- Enhancements to DOI support is possible
- Video streaming option enabled
- Integrated with ORCID (Open Researcher and Contributor ID)
- Auto generation of PDF citation cover pages for all PDFs
- Sherpa/Romeo Guidelines (for copyright) is inbuilt and can be looked up during item submission
- Item visual indicators in browse and search results (Represented by an icon for Eg: Book chapter)

2.2 Challenges faced during Migration:

- As the data migration was from 1.5.1 to a much later version 5.5, the process was not smooth. Version 1.5.1 was installed in 2008 and Version 5.5 in 2016. In eight years there were many new developments in the software. So, special training was required along with lot of self study by two professionals (one from the Computer group and another from the Library group) for successful migration to the latest version.
- Customization was another area which created challenges. Version 1.5.1 used XML UI and in version 5.5, Bootstrap (JSPUI) is used for front end web framework. The new technology had to be learnt before implementation.
- Domain knowledge of both computer science and library science is essential for successful migration and implementation from version to version. So, this is a collaborative effort, which requires constant learning.

3 Imprints Collection – a Bio-bibliographic database

Imprints collection is the name of a bio-bibliographic database of all the retired scientists from the Raman Research Institute who have made one or more contributions to the world of scientific literature. This is designed to honor and archive the lifelong efforts of RRI scientists in science promotion and science communication. The Imprints collection, as a database, captures various activities of scientists which include their publications, research guidance, lectures delivered, education, scientific career, awards and honors received etc. Additionally, the database has a photo gallery, ResearcherID badge from Thomson Reuters, newspaper clippings and audio/video (if available) files in the well-designed profile page of each scientist.

3.1 Design of Imprints collection

This web based information retrieval platform for scholarly communication was designed in 2011 using Dreamweaver software. Base line for building the Imprints collection is the RRI Digital Repository. A profile page for each scientist is built giving primary information such as name of his/her parents, place of birth, primary education, higher education, careers, fellowship or association with any professional bodies, information about any awards and honors that are conferred during their career. Further, the profile page gets linked to their list of publications, list of research guidance that they have done, photographs of their association with RRI, any news items that appeared in newspapers, and their audio/video file. In the case of deceased scientist, an obituary column is also included. The figure below represents the different components of the Imprints collection.

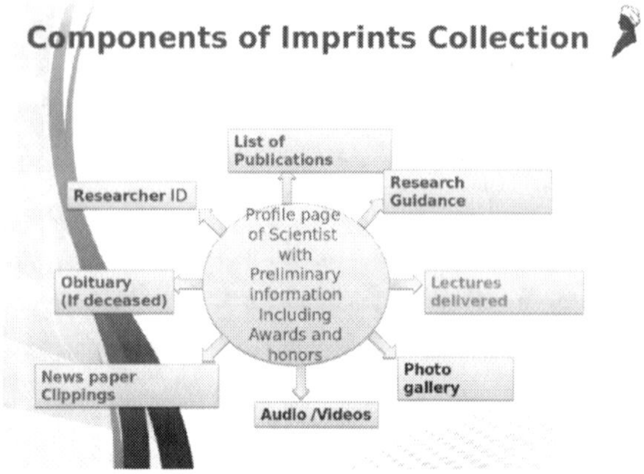

Figure 4. Components of Imprints Collection

The photo album is created using Jalbum.net software. The Imprints collection was initiated with the profiles of 18 scientists. Currently, the database has 32 profiles. It is accessible at http://www.rri.res.in/htmls/library/imprints_collection The main page looks as in figure - 5.

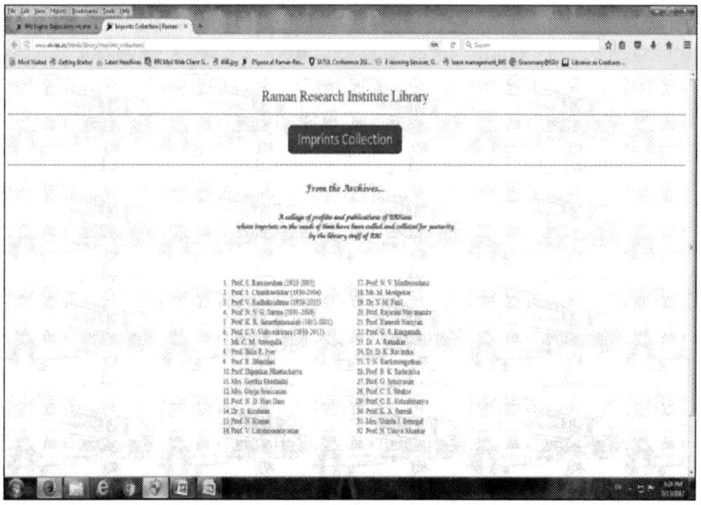

Figure 5. Imprints Collection Homepage

3.2 Navigation through imprints collection:

A click on any one of the names will take the user to the profile page of the respective scientist. Below is the profile page of Prof. N Kumar.

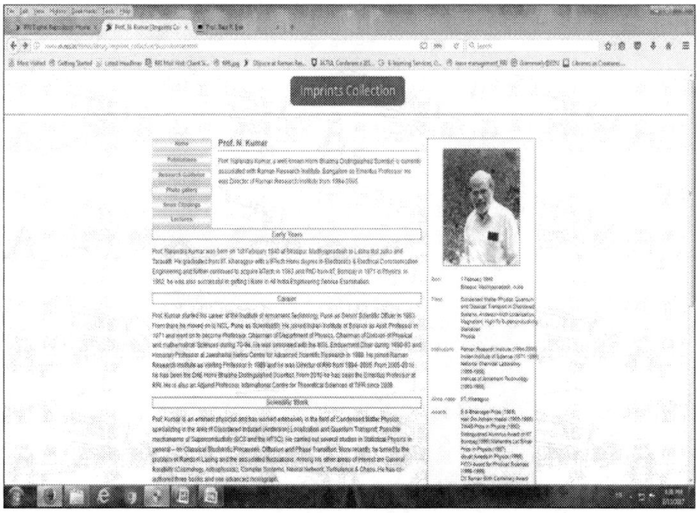

Figure 6. Profile page of a Scientist

Body of the profile page gives all the information about the scientist. Tabs on the left hand side top corner enables links to the scientists published papers, his research guidance, photo gallery etc. Right hand column provides quick information for a glance along with the ResearcherID badge. A click on publications tab will take the navigator to the list of publications in the repository. A collage of next levels of retrieval is given below in figure – 7.

Figure 7. Collage of Variety of Information in Imprints Collection

3.3 Advantages of Imprints collection:

- This is a new generation information retrieval tool assimilating the complete research output of a scientist as a one stop-shop, showcasing his/her life time research and accomplishments, spanning about 30-35 years.
- For one who is trying to collaborate with a scientist for a project or a scientific endeavor, Imprints collection could be a major source of information as it is complete in itself
- This is also a tool to showcase the areas of research of the Institute
- With the built in researcherID badge, this serves as a base for scientometric analysis for a library scientist.
- Above all, a bio-bibliographic database of this kind has incredible archival value.

4 Conclusion

Library services and their formats have been changing drastically ever since the Internet/WWW has crept into the library domain, facilitating librarians to design better, user friendly and also personalized services in addition to the conventional services. In the context of growing concern towards open access phenomena and publically funded research, libraries are expected to develop a platform to enhance the visibility of research carried out in an institute by showcasing them through institutional repositories. Raman Research Institute library has built an institutional repository with the intension of showcasing all the information associated with the founder, Sir C V Raman. Additionally, the repository also intends to bring visibility to the government funded current research carried out at the Institute. "Imprints collection", is an offshoot of Raman Research Institute Digital Repository, having a unique style of looking at the scholarship of an institute and presenting research activities in a novel way for better use. This is one of the best practices of RRI Library.

Acknowledgements

The authors acknowledge and thank Mr. Jacob Rajan for his technical support in installation and customization of DSpace.

References

[1] B.M. Meera, M. Manjunath, M. Kaddipujar, *Facets of digital data dissemination: value addition through "imprints collection"*, Library Hi Tech **31**, p308-322 (2013)

[2] B.M. Meera, *Best practices in scholarly communications: an open access phenomenon*, In: Open access: free flow of science Edited by Shivalingaiah, D et al, p75-84 (2016)

[3] G. Sirnivasan, Y.M. Patil, J. Rajan, *Scholarly communication in a digital world: the role of the digital repository at the Raman Research Institute*, ICSD-2007, Edited by ARD Prasad and Devika P. Madalli , p285-294 (2007)

[4] G. Srinivasan, Y.M. Patil, M. Manjunath et al., *Five Years of the RRI Digital Repository: Some Lessons and Reflections* in *21st Century Astronomy Librarianship, From New Ideas to Action*, edited by Eva Isaksson, Jill Lagerstrom, András Holl, and Nirupama Bawdekar (2010), Vol. 433 of Astronomical Society of the Pacific Conference Series, p. 308–322

[5] Wikipedia, *Open Access*, https://en.wikipedia.org/wiki/Open_accessl (accessed on 28th Jun 2017)

[6] Wikipedia, *Institutional Repository*, https://en.wikipedia.org/wiki/Institutional_repository (accessed on 28th Jun 2017)
[7] DSpace @MIT, , http://libraries.mit.edu/dspace-mit/build/policies/format.html (Accessed on 3rd July 2017)
[8] Eprints, , http://en.wikipedia.org/wiki/EPrints (Accessed on 3rd July 2017)
[9] Green Stone Digital Library Software, , http://www.greenstone.org/ (Accessed on 3rd July 2017)
[10] Jalbum, , http://en.wikipedia.org/wiki/Jalbum (Accessed on 5th July 2017)

EPJ Web of Conferences **186**, 07002 (2018) https://doi.org/10.1051/epjconf/201818607002
LISA VIII

Polvere di stelle

The Italian platform for data-sharing and data-preserving of modern and ancient astronomical resources

Mauro Gargano[1,*], *Antonella* Gasperini[2,**], *Luisa* Schiavone[3,***], and *Francesca* Brunetti[2,****]

[1] *INAF Astronomical Observatory of Capodimonte, Salita Moiariello 16, I-80131 Napoli, Italia*
[2] *INAF Arcetri Astrophysical Observatory, Largo Enrico Fermi 5, I-50125 Firenze, Italia*
[3] *INAF Turin Astrophysical Observatory, Via Osservatorio 20, I-10025 Pino Torinese, Italia*

Abstract. *Polvere di stelle* (Stardust) is the web portal of the National Institute for Astrophysics (INAF) dedicated to the Italian astronomical libraries, archives and museums. It offers different tools and databases created to support astronomical research and increase the value of one of the richest astronomical heritages in the world.

In a single virtual space one can find useful tools for sharing digital resources and other services for current research. Besides the OPAC, consisting of bibliographic data of ancient and modern books and serials, the portal offers to astronomers, scholars, students, amateur astronomers and historians of science the possibility to search at the same time other databases: manuscripts, instruments, archival documents and biographies of astronomers. Furthermore, a digital showcase of rare books plays a relevant role in the portal.

This paper illustrates the ongoing developments and perspectives of *Polvere di stelle*.

1 The project

The project *Polvere di stelle* (Stardust) was conceived for the International Year of Astronomy 2009. It originated from the growing interest in historical archives that has engaged the archivists and librarians of Italian observatories in the last 30 years. It can also be considered the final and public result of the *Specola 2000* project [1]. *Polvere di Stelle* has increased the knowledge of the historical legacy of Italian astronomy, presenting in a complete and homogeneous way the treasures hidden in the folders of the Italian astronomical archives.

The first release of this portal was presented at the LISA VI conference held in Pune in 2010 [2]. *Polvere di Stelle* was the stargate to access twelve historical archives, to consult a first nucleus of data stored in a homemade database, and to browse a wide gallery of the most important documents. In the same year INAF published an elegant booklet to present and describe its fonds, series and documents [3]. In addition, a video presentation was published on the web and on social media, showing the

*e-mail: mauro.gargano@inaf.it ORCID: 0000-0002-4582-704X
**e-mail: antonella.gasperini@inaf.it
***e-mail: luisa.schiavone@inaf.it
****e-mail: francesca.brunetti@inaf.it

Stardust pearls held by the historical archives of Italian observatories and sometimes hidden in old folders or inside ordinary documents[1].

1.1 The databases

The project had a turning point in 2013, when INAF decided to create a web platform to host the cultural heritage held by some of the oldest scientific institutes in Italy, as the astronomical observatories are. INAF in 2010 had already acquired two professional databases for all the twelve observatories.

Bibliowin is the web-based interface to catalogue books and journals, both modern and old, in accordance with the International Standards. The application program, ASP based, allows the shared and derived cataloguing, and is provided with tools for library circulation and management.

SicapWeb is the web interface of a MS SQL Server and MSDE database. It includes six cataloguing modules for different kinds of historical holdings. The module for manuscripts follows the standards approved by the Standing Committees of the IFLA Section on Cataloguing and the IFLA Section on Rare Books and Manuscripts. The inventory module for the archival documents complies with the ISAD(G) General International Standard Archival Description and ISAAR (CPF) International Standard Archival Authority Record for Corporate Bodies, Persons and Families. The modules for the scientific and technological instruments, the artworks such as the painting of father Piazzi (see fig. 5), the astronomical plates and photographs are defined in accordance with the rules of the Italian Minister of Culture. Finally, the last module is structured to record the biographies of Italian astronomers.

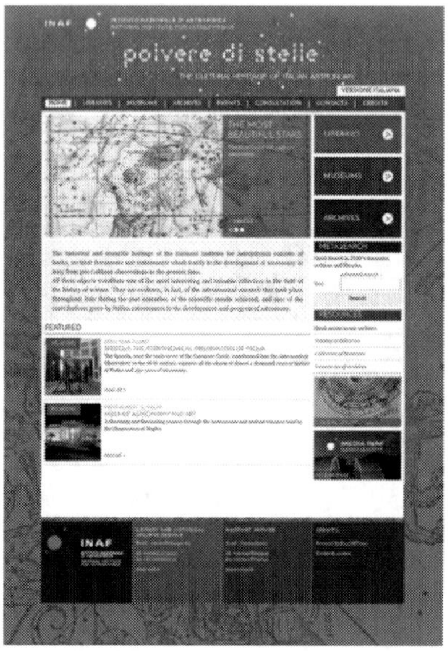

Fig. 1. The home page of *Polvere di stelle*.

A new release of the web portal was realized in 2014 and was presented at the LISA VII conference held in Naples (see fig.1). *Polvere di stelle* is reachable at the web address www.beniculturali.inaf.it.

2 The rare and valuable heritage of Italian astronomy

The historical and scientific heritage of Italian astronomy consists of manuscripts, ancient books, scientific instruments, archival documents, artworks, astronomical plates and photographs which prove the important developments of astronomy in Italy, from the pre-Galilean observations to the present time. Held in twelve astronomical observatories, all these objects are one of the most outstanding and valuable collections for the history of science worldwide.

2.1 Rare books

The Italian observatories hold more than 7000 rare books (from 1470 to 1830), including 19 incunabola and 551 cinquecentine, together with 30 manuscripts. Some of these books represent true cultural milestones: the works of Galileo, Copernicus, Ptolemy, Kepler and Newton (often first editions), are considered the symbols of the scientific revolution and have cleared the way for modern science.

[1]The video was published on different web sites, i.e. www.youtube.com/watch?v=4MCleV7PxwA.

Besides their importance for the history of Western culture, these books are also interesting for their fine and precious illustrations. The amazing star atlases of Hevelius, Doppelmayer, Flamsteed and Bode, or the cometographies and selenographies of Northern European astronomers, published in the 16th and 17th centuries, reveal a variety of details, illustrated with painstaking care in works of rare beauty that merge art, mythology and science. The oldest document hold by the National Institute of Astrophysics is a manuscript preserved in the Library of the Copernican Museum in Roma, that dates back to the end of 1300 (see fig. 2). It is a collection of medieval astronomical texts, probably the most popular essays of that period.

2.2 Museum collections

The ancient astronomical instruments collection of the Italian observatories counts over 1200 items, from the 11th century to the first half of 1900. Exhibited in 13 museums, it mainly consists of quadrants, telescopes, theodolites, clocks, stopwatches, globes (see fig. 3), mathematical and meteorological instruments. Among the most precious pieces, the museum collections hold an Arabic astrolabe dated back to the year 1090, one of the biggest telescope mirrors made by William Herschel, the telescope used by Schiaparelli to describe the surface of Mars, and the instruments manufactured and used by the Italian astronomers to realize a "new science", astrophysics.

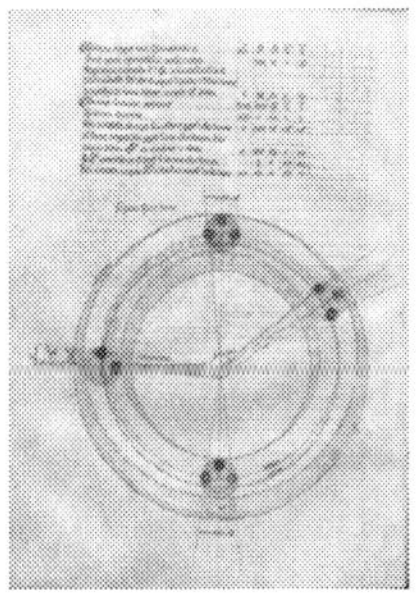

Fig. 2. *Theorica planetarum* by J. de Sacrobosco (14th century).

Fig. 3. The celestial globe by G. Roll and J. Reinhold, 1589.

2.3 Historical archives

The historical archives of the astronomical observatories preserve the documents of some of the oldest scientific institutions in our country.

Our archives keep records from countless nights spent at the telescope, beautiful sketches of solar phenomena (see fig. 4), nebulae, comets and planets, drawn during the observations under an astronomical dome, letters revealing the history some scientific discoveries, journey logbooks, maps, bills and meteorological observations as a continuum line throughout the centuries. A cultural legacy consisting of over 3 million documents and including 122 series of archival fonds of various astronomers especially.

2.4 Astronomy and astronomers in artworks, biographies, and photographs

Besides being the oldest scientific institutes in Italy, the observatories have also had cultural leadership at large. Many of them were installed in antique buildings, such as the medieval tower of Ezzelino da Romano in Padova, the royal Norman Palace in Palermo or the Brera Palace in Milano [1]. Therefore the Italian astronomical observatories preserve valuable artworks as frescoes, sculptures, paintings and honor medals. A catalogue of these kinds of objects still has to be completed: some hundred pieces of art can be roughly counted at a first estimate.

For what concerns astronomical plates and photographs, the Italian observatories own solar and lunar eclipses photos, such as the plates realized by father Angelo Secchi during some astronomical expeditions in the second half of 19th century, and images of historical and social events, like the pictures of the participants at the First International Astronomical Union Congress in 1922 [4]. So far, two collections of drawings and plates of solar observations have been recorded and scanned: the 2516 sketches of the Sun realized from 1865 to 1925 at the Catania Observatory, and the 5250 CaIIK and 6941 Hα spectroheliograms acquired during 5042 observing days at the Arcetri solar tower from 1926 to 1974 [5]. In next future these records and their digital copies will be uploaded in *SicapWeb*.

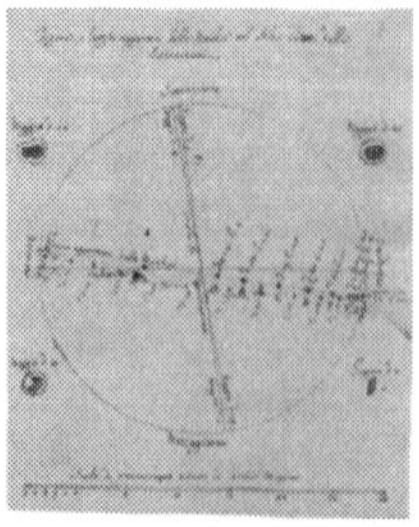

Fig. 4. *Figures and sites of sunspots deduced from observations*, 1703.

Fig. 5. *The astronomer Piazzi* by C. Angelini, abt. 1825.

In 1997 the Italian Astronomical Society published a book with 277 biography records of Italian astronomers from the Unity (1861) to the present [6]. The astronomers who worked in the Italian observatories were clearly many more, so the next goal will be adding the biographic profiles of Italian astronomers in *SicapWeb*.

The holdings of Italian astronomical observatories catalogued and/or recorded in INAF databases are listed in Table 1.

3 The web portal framework

The contents and the services offered by *Polvere di Stelle* are available to every user: astronomers, students, historical researchers, amateurs. A general introduction presents the three different thematic areas constituting the web platform. It describes activities and projects, offers access to catalogues and tools, and provides tips.

The actual structure of the portal is the result of a long-time cooperation among Italian astronomical observatories in the library, archives and museum areas.

From 1995 till 2010 every astronomical library had a local catalogue and used the CUBAI[2], a multiOPAC which made a simultaneous search on all databases [7]. In 2010 was acquired *Bibliowin*, where all the various catalogues have merged in an unified database and OPAC for all INAF libraries. Now INAF librarians are working to normalize records and authority files.

Table 1. Items recorded in the databases until June 2017.

Records	#
Manuscripts	19
Ancient books	6674
Modern books	75574
Serials	206
Journals	2570
Museum instruments	545
Artworks	66
Archival records	6334
Biographies of astronomers	58
Authors and craftsmen indexes	974

In 2001 astronomical libraries decided to participate to ACNP (Italian Union Catalogue of Serials) [8] and to create a national catalogue of astronomical journals (Astronomical ACNP) [3].

[2]CUBAI is the acronym of *Catalogo Unico delle Biblioteche Astronomiche Italiane* (Unified Catalogue of Italian Astronomical Libraries).

[3]The catalogue can be searched at the following address: acnpsearch.unibo.it/custom/astro

A similar cooperation path was followed by the historical archivists working in the astronomical observatories, who realised *Specola 2000*. The aim of this project was the preservation, reorder, and inventory of the ancient astronomical documentation [9]. The rearrangement activities and the digital inventories, realized with this project, allowed the retrieval of outstanding documental sources, crucial to the understanding of the evolution of scientific thought and the reconstruction of the Italian "scientific politics" starting from the 18th century.

The project "Astrum 2009", a temporary exhibition realized at the Vatican Museums during the International Year of Astronomy 2009, was the first result of a national collaboration among observatories in the museum areas. This circumstance highlighted the local astronomical collections and how they could be considered as a part of a unique national museum of instruments illustrating the developing of Italian astronomy [10].

Now *Polvere di stelle* is the web portal of:

- 13 libraries

  of the twelve Italian astronomical Observatories and the library of the *Fundación Galileo Galilei*, which operates the TNG (*Telescopio Nazionale Galilei*) at Los Roches de los Muchachos in La Palma island. Each single library section presents the institute, staff, opening hours, local services, rules, and historical notes;

- 12 historical archives

  kept in the observatories, including the one owned by the Department of Physics and Astronomy of Bologna *Alma Mater Studiorum*[4]. Besides information on staff and schedules, every archives describes how it was formed, its structure, and the sorting and inventory status of documents. Moreover, one can find the complete tree structure of every archives. In some cases, such as in Torino and Milano, more than a single archives are kept and presented;

- 10 museums

  or scientific collections of nine Italian observatories and "La Specola" Museum owned by Bologna University. To complete the framework, still lack the pages of the instrument collections of Cagliari and Trieste observatories. In addition to general information, opening hours, contact and staff, some museums' web pages show peculiar aspects of exhibits such as historical notes, a virtual tour, and detailed descriptions of domes or gardens.

3.1 The queries and the records

Beyond presenting library catalogs, archives inventories, and museum collections, and providing historical descriptions for different kinds of items, *Polvere di Stelle* offers a double way to search the databases: a specific form and a simultaneous form. The specific form has different types of queries: simple, advanced, by list, by collapsing (in archives case). The simultaneous search performs a query in the two databases and in all kinds of materials: books, archival documents, manuscripts, artworks, scientific instruments, photographs, and astronomers' biographies records.

Some records, such as those of old books, show not only the bibliographic description of materials but also the specimen notes and the biographical notes of authors and printers.

The astronomers records contain links that launch queries on *Bibliowin* catalogue, ADS, and *Sicapweb*, thus producing a complete bibliography and a list of the cultural materials related to that person.

[4]An agreement signed in 2014 by INAF and Bologna University permits a complete collaboration to study and promote the ancient astronomical heritage of both institutions.

Other records, in particular those referring to instruments, documents, and astronomers, are structured so that one can launch a query by clicking on an icon. The query will show all records in the database having the same metadata in the same field (i.e. the institute preserving the item, persons cited in the document, category, authors and craftsmen, as shown in fig. 6).

All these useful interconnections permit to build some specific aspects of the history of astronomy, and to realize thematic pathways of important astronomical events, discoveries and instrument employments.

4 Leafing through rare books

A digital showcase plays a relevant role in *Polvere di Stelle*. With the same technology used to store and share astronomical data, the Italian datacenter for Astronomical Archives (IA2), housed in the Astronomical Observatory of Trieste, has set up a platform to create a national repository for the digital copies of the ancient books [11]. IA2 is a national infrastructure based on the cloud paradigm, hosts data from the main ground-based Italian Telescopes and offers proprietary and public data access through user portals and Virtual Observatory (VO) services. All software tools are based on Open Source application and are developed in C++, Java and PHP.

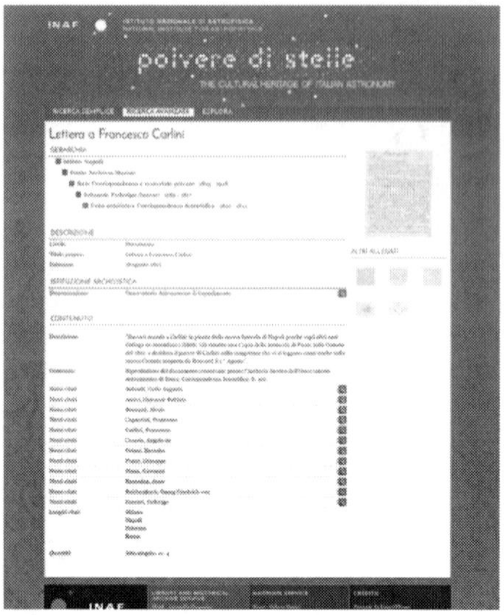

Fig. 6. An archival record on the OPAC.

Fig. 7. The main page of the Digital Showcase.

Using this infrastructure, and in particular the data cloud storage, we have uploaded over 110,000 files of 324 ancient books, in different resolutions: the images at 150 DPI are used to publish the books on the web with a watermark on each page, while the 300 and 600 DPI images are used to keep a digital copy of each book [12].

At present, the digital showcase enables the reading, in a colour digital form and in very good resolution, of the rarest and valuable editions of the old book collection (in fig. 7 its main page). It presents some highlighted volumes and four tabs listing Stellar atlases, Incunabula, Cinquecentine and Seicentine.

Moreover the timeline allows users to surf all the digital copies uploaded in the repository and catalogued in Bibliowin. Until July 2014 there are digital volumes published from 1326 to 1842. In addition, a search field enables to select a specific digitized edition.

People can leaf through some wonderful volumes like the *Almagestum* by Claudius Tolomeus published in 1528, the *Cosmographiae vniuersalis* by Sebastian

Munster published in March 1572, and the *Tabule astronomice* by Alfonso X, King of Castile and Leon, printed in 1521. Some amazing stellar atlases have a special scientific and artistic relevance: the *Vranometria* by Johann Bayer, published in 1661, includes 51 engravings of detailed astronomical plates, and the *Atlas novus coelestis* by Johann Gabriel Doppelmaier, printed in 1742, presents 30 hand-coloured stellar maps and a marvelous "antiporta". The digital showcase, from where you can leaf the books, can be reached at www.beniculturali.inaf.it/eng/digital-showcase.

We have also defined a policy to upload the documents in the national repository of the digital copies of ancient cultural heritage. Each Observatory can create a main folder to preserve the high resolution copies and various sub-folders to upload the digital copies of each archival record (see fig. 6). A pilot project is in progress, and 31 folders have been uploaded in the repository at present.

In the last years, more than a half of the astronomical archives of Bologna University have also been digitized, and the digital copies will be soon available on the web. This work, combined with a reordering and rearrangements of both archival files and single sheets, allows archivists to recover many others series of documents and astronomical logbooks of observations, among the oldest observations ever realized in an observatory [13].

5 A place to offer utilities and to promote events

Polvere di stelle has a specific section, a launch window, to promote the events realized to valorize the cultural heritage and disseminate the astronomical knowledge for the general public and schools, even in a multidisciplinary approach.

This section is used to inform patrons about updates and new features supporting the scientific research: astronomers can find which journals, serials and digital libraries are available online, how to access to these resources with a remote connection, and how to use NILDE, the library services for document delivery.

A long series of cultural activities have been realized in the last years in the observatory libraries, archives and museums. To name but a few, we can cite the concerts organized at Arcetri Observatory where the "Albert Einstein piano" is placed in the library hall[5], or the stage shows like *Settemillimetri di Universo* (Sevenmillimeters of Universe) conceived for the *Starlight* project[6].

A special mention goes to a particular event that took place in January 2015. On the 70th anniversary from the liberation of Auschwitz concentration camp, some astronomy librarians, archivists, and historians organized the conference *Under the same sky?* in collaboration with the University of Bologna, the Jewish Community of Bologna, the Italian Astronomical Society and the "Primo Levi" University [14]. It was the first official commemoration of the six Jewish Italian astronomers Azeglio and Giulio Bemporad, Luigi Jacchia, Guido Horn d'Arturo, Tullio Levi-Civita and Bruno Rossi, who at the end of 1938 were dismissed from their jobs because of the racial laws [7].

Finally, the portal offers the *Bibliography of Italian astronomical books for children*, a periodically updated selection of Italian astronomical books for children, dedicated to teachers, parents, public librarians, science communicators. This kind of bibliography is a unique tool ever made in the Italian scientific libraries [15].

[5] From 2016, the central hall of Arcetri library houses the piano that Albert Einstein gave to his sister Maja during her stay in Firenze. The piano has been the "star" of many of Arcetri library events.

[6] *Starlight* was a shared exhibition realized in five observatories in 2016 (Palermo, Napoli, Roma, Firenze and Padova). See the virtual tour at the following address: www.starlight.inaf.it

[7] In September-December 1938 the Fascist Government issued a series of provisions regulating the exclusion of foreign and Italian Jews from the school, the academia, politics, finances, professional world, and all sectors of public and private life.

6 Towards new challenges

Polvere di stelle is a platform to support current research and, at the same time, a web portal to enhance the Italian astronomical heritage. An interlinked web tool that looks to the present and includes the past, in the name of the historical and scientific knowledge considered as an "unicum". After three years since its first release, we are now working to implement new contents, to realize new projects and to publish a new layout for *Polvere di stelle*. The digitization of old books will also continue, and a new set of rare plates and great pages of astronomy will be soon available in the showcase.

INAF is also interested in new technologies. Through Turin Astrophysical Observatory, we are participating in a pilot project on Linked Open Data which is fully detailed in a poster presented at LISA VIII conference (see the related paper [16]).

Polvere di stelle is and will be even more a focal point supporting astronomical research with its information services, and preserving the great cultural heritage of astronomy which is "the most beautiful monument of the human mind" (Laplace).

References

[1] A. Mandrino, D. Randazzo, L. Schiavone, *Specola 2000: A Project for the Preservation of the Historical Archives of the Twelve Italian Astronomical Observatories*, in *Common Challenges, Uncommon Solutions*, edited by S. Ricketts, C. Birdie, E. Isaksson (2007), Vol. 377 of *Astronomical Society of the Pacific Conference Series*, p. 361

[2] M. Gargano, A. Gasperini, A. Mandrino, *Polvere di Stelle — Stardust: The National Historical Archive for Astronomy on the Web*, in *21st Century Astronomy Librarianship, From New Ideas to Action*, edited by E. Isaksson, J. Lagerstrom, A. Holl, N. Bawdekar (2010), Vol. 433 of *Astronomical Society of the Pacific Conference Series*, p. 244

[3] M. Gargano, A. Gasperini, A. Mandrino, eds., *Polvere di stelle* (Istituto Nazionale di Astrofisica, Roma, 2010)

[4] A. Gasperini, Giornale di Astronomia **38, n. 2**, 47 (2012)

[5] E. Marchei, I. Ermolli, M. Centrone, F. Giorgi, C. Perna, Memorie della Società Astronomica Italiana Supplementi **9**, 51 (2006)

[6] G. Foderà Serio, D. Randazzo, *Astronomi italiani dall'Unità d'Italia ai nostri giorni* (Società Astronomica Italiana, Firenze, 1997)

[7] L. Abrami, A. Balestra, M. Ferrucci, A. Gasperini, Vistas in Astronomy **39**, 265 (1995)

[8] F. Brunetti, O. Bonora, G. Filippucci, *ACNP and NILDE: Essential Tools for Access to Scientific Research*, in *Open Science at the Frontiers of Librarianship*, edited by A. Holl, S. Lesteven, D. Dietrich, A. Gasperini (2015), Vol. 492 of *Astronomical Society of the Pacific Conference Series*, p. 275

[9] I. Chinnici, A. Mandrino, F. Bònoli, Journal of Astronomical History and Heritage **9**, 200 (2006)

[10] O. Gingerich, Journal for the History of Astronomy **42**, 116 (2011)

[11] E. Olostro Cirella, M. Gargano, A. Gasperini, A. Mandrino, D. Randazzo, V. Zanini, *The AstroBID: Searching through the Italian Astronomical Heritage*, in *Open Science at the Frontiers of Librarianship*, edited by A. Holl, S. Lesteven, D. Dietrich, A. Gasperini (2015), Vol. 492 of *Astronomical Society of the Pacific Conference Series*, p. 143

[12] M. Gargano, A. Gasperini, E. Olostro Cirella, R. Smareglia, V. Zanini, Communications in Computer and Information Science **612**, 71 (2016)

[13] F. Bònoli, E. Cenacchi, A. Mandrino, R. Stasi, D. Zuccato, in *Physics, Astronomy and Engineering. Critical Problems in the History of Science* (2013), Vol. 492, p. 147

[14] F. Bònoli, A. Mandrino, eds., *Sotto lo stesso cielo? Le leggi razziali e gli astronomi in Italia*, Vol. 41, n. 2 of *Giornale di Astronomia* (2015)

[15] F. Brunetti, A. Gasperini, *Astronomical Libraries Make the Future Happen: Support to Public Communication of Science as Part of the Library Mission*, in *21st Century Astronomy Librarianship, From New Ideas to Action*, edited by E. Isaksson, J. Lagerstrom, A. Holl, N. Bawdekar (2010), Vol. 433 of *Astronomical Society of the Pacific Conference Series*, p. 33

[16] L. Schiavone, F. Morando, The CoBis Communication Working Group, *The CoBiS Linked Open Data Project and Portal*, in *These Proceedings* (2018)

V. Stoll, F. Saconnet and A. Tisseau des Escotais (Photo: M. Wenger)

EPJ Web of Conferences **186**, 07003 (2018) https://doi.org/10.1051/epjconf/201818607003
LISA VIII

Project PHaEDRA: Preserving Harvard's Early Data and Research in Astronomy

Daina Bouquin[1,*], *Katie* Frey[1], *Maria* McEachern[1], *James* Damon[1], *Daniel* Guarracino[1], *Alex* McGrath[1], *Edwin* Henneken[2], and *Lindsay* Smith Zrull[3]

[1] *Wolbach Library, Harvard-Smithsonian Center for Astrophysics, USA.*
[2] *SAO/NASA Astrophysics Data System, USA.*
[3] *Harvard College Observatory, Astronomical Photographic Plate Collection, USA.*

Abstract. The staff of Wolbach Library, in collaboration with partners at both the Smithsonian Institution and Harvard University, has begun a complex digitization and transcription effort aimed at making a large collection of historical astronomy research more findable, accessible, interoperable, and reusable (FAIR). This collection of material was originally produced from the mid-18th century through the early 20th century by researchers at the Harvard College Observatory and was recently re-discovered in the HCO Plate Stacks holdings. The team of professionals supporting the effort to make this century and a half old science FAIR have developed a novel, distributed workflow to ensure that people can engage critically with this material to the fullest extent possible. The project's workflow is guided by the collections as data imperative conceptual frameworks and is now being referred to as Project PHaEDRA, or Preserving Harvard's Early Data and Research in Astronomy.

1 Introduction

A prodigious collection of historical astronomy research produced from the mid-18th century through the early 20th century by researchers at the Harvard College Observatory (HCO) was recently rediscovered after decades of being overlooked in an off-site storage facility owned by Harvard University, the Harvard Depository. This unique collection helps document the history of women's contributions to astronomy, the history of HCO, and also comprises remarkable examples of primary source material showing the evolution of observation methods and early astronomy as a whole. The staff of Wolbach Library at the Harvard-Smithsonian Center for Astrophysics (CfA), recognizing the need to share this collection with the broadest possible community, built a multiskilled team and comprehensive workflow to ensure that the necessary infrastructure was in place to digitize the collection and enrich its metadata to the fullest extent possible using transcription. Throughout the process of defining the workflow, Wolbach was guided by the collections as data imperative conceptual frameworks and worked to meet goals originally developed to support modern open data practices: The FAIR Data Principles (Table 1).

*e-mail: daina.bouquin@cfa.harvard.edu ORCID: 0000-0003-2626-3688

1.1 Project PHaEDRA Goals

The Project PHaEDRA team is working to achieve the following:

1. Digitization of the entire PHaEDRA collection
2. Indexing of the collection in HOLLIS and OASIS (Harvard's library catalog systems) and the NASA/SAO Astrophysics Data System (ADS) [1]
3. Transcription of as much of the collection as possible using the Smithsonian Transcription Center [2] citizen science platform
    - Define a markup procedure to tag instances of sketches, calculations, and data tables
    - Subset the sketch collection – Astronomers and scholars in the History of Science can then help contextualize sketches and assign alt text to the images for increased accessibility
    - Subset calculations and data tables for further metadata enrichment and potential future projects
4. Incorporation of transcriptions into machine-readable metadata to enable full text search capability on ADS

1.2 The PHaEDRA Collection

The materials that make up the PHaEDRA collection include 2,518 journals and logbooks, as well as numerous tip-ins, and loose papers. These items are spread across 118 paige boxes. The earliest material the Project PHaEDRA team has encountered in the collection thus far is from 1750. Due to size of the collection and initially sparse metadata, the team has not yet fully determined the more recent end of the date range covered by the collection. It is important to note, however, that a sizable portion of the collection is hand written work of the "Harvard Computers".

The Harvard Computers were women who were predominantly hired by Edward C. Pickering (director of HCO, 1877–1919) to process astronomical data. During their time at HCO, the Computers studied over 130 years of the night sky preserved on glass plate photographs. These women cataloged stars, identified variables, interpreted stellar spectra, counted galaxies, and measured distances in space. Much of the Computers' work was done using methods they developed themselves and that are still used in modern astronomy and astrophysics. Several of the women even made their own discoveries [3].

In addition to the work of the Harvard Computers, the PHaEDRA collection also contains the early work of William Cranch Bond and his son George Phillips Bond, the first two directors of HCO as well as many other notable astronomers who visited Cambridge in the mid 1800s. Moreover, the collection's volumes are significant in that they contain many examples of hand drawn observations that predate photography of any sort. Observations done by hand are both exceedingly rare and have the potential to inform modern time domain astronomical studies, thus they are of interest to historians of science and contemporary scientists as well. Artists may also be interested in examining the collection for its aesthetic qualities and ties to the art community. For example, the collection contains previously unknown early sketches by the artist and astronomer E.L. Truvelot [4].

This diverse range of potential applications of knowledge gleaned from the PHaEDRA collection emphasizes that the collection is no less valuable than it was when its items were cutting edge science; rather the possibilities associated with the collection's use have grown more complex and varied.

https://doi.org/10.1051/epjconf/201818607003

Above: Pickering and Harvard Computers. 13 May 1913. Back row (L to R): Margaret Harwood (far left), Mollie O'Reilly, Edward C. Pickering, Edith Gill, Annie Jump Cannon, Evelyn Leland (behind Cannon), Florence Cushman, Marion Whyte (behind Cushman), Grace Brooks. Front row: Arville Walker, unknown (possibly Johanna Mackie), Alta Carpenter, Mabel Gill, Ida Woods. Image courtesy of the Astronomical Photographic Plate Collection, Harvard College Observatory.

1.3 FAIR Data Principles and the Collections as Data Imperative

The Project PHaEDRA team aims to achieve its goals of ensuring that the PHaEDRA collection is as broadly useful as possible by working toward meeting the standards outlined in the FAIR Data Principles. These principles came about in 2016, when representatives from academia, industry, funding agencies, and scholarly publishers shared them as general guidelines for enhancing people's ability to engage with research data by making them more findable, accessible, interoperable, and reusable by people, as well as by machines [5]. Although the principles were developed in the context of modern scientific research data needs, they are of vital importance to ongoing conversations within the broader library community concerning library collections in regards to what Thomas Padilla calls, "a data imperative" [6]. According to Padilla, collections as data imperative "entails developing the means to help all members of society, across all classes and backgrounds, working within the academy and outside of it to engage critically with the traces of human activity we collect in the fullest manner possible" [6]. The Library of Congress' National Digital Initiatives team, in response to the collections as data conversation, facilitated the development of three "conceptual frames" [6] to help guide thinking through how libraries can participate in a collections as data imperative:

- Generativity – to increase meaning making capacity
- Legibility – to document and convey provenance and possibility

 https://doi.org/10.1051/epjconf/201818607003

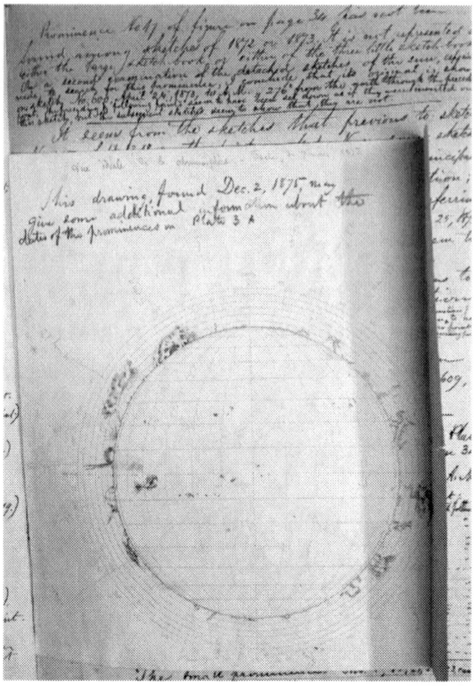

Above: PHaEDRA collection material – previously unknown sketch by E.L. Truvelot, 7 March, 1873

• Creativity – to empower experimentation

The conceptual frames provided by the Library of Congress supplement the FAIR Data Principles in such a way as to allow the Project PHaEDRA digitization and transcription effort the opportunity to more fully foster the conditions for "meaning making" [4] with this historic collection.

2 Project PHaEDRA Workflow

The Project PHaEDRA workflow is distributed across multiple institutions and systems and requires collection materials to be tracked at both the item and box level. To achieve this the PHaEDRA team is developing a database and custom front end to allow all members of the team to interface with both the collection metadata and perform item tracking. The database incorporates all steps of the process outlined in Figure 1. These steps ensure that the FAIR Principles are enacted and are guided by the collections as data imperative conceptual frameworks of "Generativity" and "Legibility". This is to say that the motivation behind the defined steps is to create the cyberinfrastructure necessary to help diverse communities navigate the possibilities that arise during the collection's use, and that as much information as possible about the collection and the project is accessible to them [6].

2.1 Project Partners

The partnerships engendered by Project PHaEDRA collaboration are proving to be an exceptional example of how the collections as data imperative conceptual framework of "Creativity" is also foundational to the successful implementation of the FAIR Principles. "Pursuing a collections as data

Table 1. Table 1: FAIR Guiding Principles

To be Findable	F1. (meta)data are assigned a globally unique and persistent identifier F2. data are described with rich metadata (defined by R1 below) F3. metadata clearly and explicitly include the identifier of the data it describes F4. (meta)data are registered or indexed in a searchable resource
To be Accessible	A1. (meta)data are retrievable by their identifier using a standardized communications protocol A1.1 the protocol is open, free, and universally implementable A1.2 the protocol allows for an authentication and authorization procedure, where necessary A2. metadata are accessible, even when the data are no longer available
To be Interoperable	I1. (meta)data use a formal, accessible, shared, and broadly applicable language for knowledge representation. I2. (meta)data use vocabularies that follow FAIR principles I3. (meta)data include qualified references to other (meta)data
To be Reusable	R1. meta(data) are richly described with a plurality of accurate and relevant attributes R1.1. (meta)data are released with a clear and accessible data usage license R1.2. (meta)data are associated with detailed provenance R1.3. (meta)data meet domain-relevant community standards

imperative requires creative thinking" and for creative thinking to occur, it is necessary to create a team that "administratively and programmatically... encourages a wide range of experimentation" [6]. The team brought together in PHaEDRA's workflow have been empowered to create a protocol that builds on the benefits afforded to them through the collaboration between the Smithsonian Institution and Harvard University that makes up the Harvard-Smithsonian Center for Astrophysics (CfA). The CfA collaboration was formalized in 1973 to coordinate the related research activities of the two observatories (the Harvard College Observatory and the Smithsonian Astrophysical Observatory) and to share resources to the two institutions' mutual benefit. Therefore, the partners on the project already have access to a set of administrative and programmatic infrastructures to support their collaboration and have been able to develop completely new ways to take advantage of it.

The partners and their roles:

1. Wolbach Library, Harvard-Smithsonian Center for Astrophysics
    - Coordinating all project partners and efforts
    - Defining procedures for tracking both physical and digital objects
    - Determining metadata procedures and standards
    - Prioritizing batches of material for digitization and transcription
    - Engaging with the public in transcription efforts
    - Planning for long term curation

2. Harvard College Observatory Plate Stacks
    - Providing historical metadata and references
    - Supporting outreach and transcription workflows
3. NASA/SAO Astrophysics Data System
    - Hosting of digitized images of collection
    - Helping define metadata and file structure needs
    - Creating indexed records containing all available metadata (e.g. text resulting from transcription enabling full-text search)
    - Minting DOIs and bibcodes for collection objects
4. Harvard Library Preservation and Imaging Services
    - Providing collections care and preparation for scanning
    - Managing all elements of imaging workflow
    - Creating files meeting needed metadata standards and file structures for ADS image hosting ingest procedures
5. Smithsonian Transcription Center
    - Providing a citizen science platform and interface for the public to engage with the digitized material
    - Promoting the PHaEDRA project
    - Providing structured metadata with transcription files
    - Allowing Wolbach to define markup procedure for transcription volunteers ("volunpeers")
6. Harvard University Archives
    - Providing access to and support for ArchivesSpace – the collection's finding aid platform
    - Offering guidance on archival metadata practices (EAD)

2.2 Workflow Narrative

The general procedure for PHaEDRA workflow is as follows:

Wolbach Library recalls a small batch of material (e.g. three boxes) from off-site storage. Wolbach examines the material and adds as much metadata as possible from the items using the Library's tracking database. The metadata will subsequently be added to the item records in the collection finding aid on OASIS and in ADS records. Next, the boxes are sent to Preservation and Imaging Services where they are prepped and safely scanned. Imaging Services sends the boxes back to Wolbach with an accompanying physical hard drive containing the image files in a predefined structure. Wolbach receives the boxes and hard drive and returns the boxes to off-site storage. Wolbach then gives the hard drive to the NASA/SAO Astrophysics Data System along with access to a spreadsheet containing the collection's item-level metadata. ADS creates records for each item in the batch and hosts the collection image files on their own servers. ADS mints appropriate unique identifiers and creates persistent links. ADS shares the resulting records with Wolbach, which are added to the OASIS record and finding aid. Wolbach then generates a list of URIs for the images that ADS is now hosting. Wolbach provides that list of URIs to the Smithsonian Transcription Center, which uses them to populate "projects" (one for each collection item) on its platform. The public can then transcribe and review transcriptions of the images on the Transcription Center site even though the image files are being hosted by ADS. Once a "project" is transcribed completely (e.g. one notebook) Wolbach can download the resulting transcription file to be ingested by ADS. ADS can then use their existing procedures to enable full-text search of the item via their public facing search interface.

3 Metadata

The PHaEDRA workflow relies on multiple metadata structures to ensure that the collection's digital objects meet the standards outlined in the FAIR Principles. No proprietary file formats are used and the pipelines rely on open source languages and protocols. More specifically, the workflow employs Python, SQL, CSV, TIF, and EAD at various points in the process. The PHaEDRA finding aid is managed using ArchivesSpace.

3.1 Initial Catalog and Finding Aid

Initially, the Project PHaEDRA team prepared to start from scratch in documenting the collection contents as exceptionally little metadata actually existed in a machine-readable format. The information that did exist was generally at the box level and was abruptly truncated, likely in the process of adding records to the earliest library computer systems at Harvard. The lack of metadata was also likely a product of the collection being relocated on a number of occasions and eventually being left to sit in the Harvard Depository. Luckily, a hand-written item-level catalog documenting the material was serendipitously discovered at the CfA. The hand-written catalog was produced in 1973 by a person named Joseph Timko and was later typed up in 1975. The type-written catalog was discovered soon after the initial catalog was found. Much to the surprise and relief of the PHaEDRA team, a little more searching unearthed a spreadsheet that was created in 1999 based on the Timko catalog. The discovery of the spreadsheet sped up the project exponentially and became the basis on Wolbach Library's work to create a collection finding aid [7].

Wolbach librarians, in consultation with the Harvard Archives, cleaned up the spreadsheet and normalized all dates, names, and existing metadata to the greatest extent possible so that the spreadsheet could be converted into EAD. Python scripts were used to create the EAD and the resulting file was ingested into ArchivesSpace to create the collection finding aid. This data was also shared with the ADS team who used this same information to create item records and mint DOIs and bibcodes for the digitized materials.

Wolbach assigned the following metadata elements to each digital object whenever that information was available:

- id – unique ID defined by Wolbach Library (e.g. phaedra0001)
- kg call number – call number that was likely assigned to items when they were housed in the New England Deposit Library [8] (possibly in the 1950s)
- secondID – a sequence identifier that was applied by the researchers themselves
- titleIdentifier – A non-unique title to describe each item
- yearStart – Earliest date available (year)
- monthStart – Earliest date available (month)
- dayStart – Earliest date available (day)
- yearEnd – Last date available (year)

- monthEnd – Last date available (month)
- dayEnd – Last date available (day)
- boxNumber – Number for the paige box the item is housed in
- firstAuthor – A primary author is assigned whenever one can be determined
- secondAuthors – Any other contributors identified in the material
- notes – Content notes gathered in initial examination of the material
- date certainty – Dates are sometimes speculative

The PHaEDRA team will be continuing to flesh out the collection's metadata as it is transcribed and explored more fully.

3.2 Transcription Metadata

The metadata provided through the Smithsonian Transcription Center (STC) emphasizes Project PHaEDRA's collections as data imperative particularly well in that the STC "seeks to engage the public in making our collections more accessible" by working "hand-in-hand with digital volunteers to transcribe historic documents and collection records to facilitate research and excite the learning in everyone" [2]. The STC's focus on engaging actively with the public in meaning-making and the creation of digital objects is essential to the Project PHaEDRA mission to make the collection more FAIR. The STC gives the PHaEDRA team the opportunity to enrich the collection metadata to the point where full-text search is possible. It will also be possible to assign additional meaning to the digitized material thanks to the markup being provided by volunteers. Wolbach worked with the STC to augment their standard instructions for transcribers to ensure that sketches, equations, and data tables were transcribed uniformly, which will allow for them to be more flexibly used as the project progresses [9].

4 Obstacles and Future Work

Despite the Project PHaEDRA team's successful implementation of the project's general workflow, a number of obstacles place limitations on the speed of project implementation and the comprehensiveness of the collection's metadata. Wolbach has begun a social media campaign to adjust to these challenges, but barriers still exist.

4.1 Obstacles

1. The transcription component of the current workflow relies on "volunpeers", making outreach and engagement paramount. This engagement defines the project timeline.
2. Some logbooks were initially scanned as part of a separate digitization effort [10] at the CfA and the resulting images were hosted and transcribed separately. These files now need to be incorporated into the new comprehensive workflow.
3. Multi-institution and multi-platform workflow. Communication and documentation are essential and changes to the workflow would be exceedingly challenging
4. Wolbach is receiving only a small amount of funding from the Smithsonian Astrophysical Observatory to scan the material. Funding for physical conservation of materials and support from other potential partners will likely need to be secured externally.

5. Workflows still need to be defined for the sub-setting of images for annotation

6. Approaches must be developed to enable the machine-readability of transcribed data tables

4.2 Potential Future work

The Project PHaEDRA team is confident that they can achieve their initial goals and hope to continue increasing the usefulness of the digital objects that they are creating. One potential avenue of support for that objective would be to link digitized journals to their corresponding DASCH plates. The DASCH project, or Digital Access to a Sky Century @ Harvard [10], is an effort to digitize the entire Harvard College Observatory Glass Plate Collection. The plates themselves are numbered and the PHaEDRA transcriptions will contain those same numbers. Additionally, it may be possible to employ multidimensional image processing techniques to identify sketches and tables without the need for volunteers to tag them, which would dramatically speed up the plan to subset those sketches so that alt text can be added to them. To ensure that the alt text is accurate, the PHaEDRA team is also hoping to develop and fund a post-doc fellowship to contribute to the evolving project and contextualize the materials in the history of astronomy as a whole. Another idea that the team has is to tag the PHaEDRA collection using terms from the Unified Astronomy Thesaurus [11] to ensure that the collection is as findable as possible.

5 Thank you!

Wolbach Library extends a grateful thank you to our partners at the SAO/NASA Astrophysics Data System, the Smithsonian Transcription Center, and Harvard University Library's Imaging Services, Archives, and Preservation Services groups. The team is especially grateful to the many members of the public who have contributed to the transcription of the PHaEDRA material and who will continue to support the project moving forward.

References

[1] *ADS Search*, `https://ui.adsabs.harvard.edu/`

[2] *Smithsonian Digital Volunteers*, `https://transcription.si.edu/`

[3] *Women Computers*, `https://platestacks.cfa.harvard.edu/women-computers`

[4] *The Trouvelot Astronomical Drawings (1882)*, `https://publicdomainreview.org/collections/the-trouvelot-astronomical-drawings-1882/`

[5] M.D. Wilkinson, M. Dumontier, I.J. Aalbersberg, G. Appleton, M. Axton, A. Baak, N. Blomberg, J.W. Boiten, L.B.d.S. Santos, P.E. Bourne et al., *The FAIR Guiding Principles for scientific data management and stewardship* (2016), `https://www.nature.com/articles/sdata201618`

[6] T. Padilla, *On a collections as data imperative*, `http://www.digitalpreservation.gov/meetings/dcs16/tpadilla_OnaCollectionsasDataImperative_final.pdf`

[7] *Harvard College Observatory observations, logs, instrument readings, and calculations : an inventory*, `http://oasis.lib.harvard.edu/oasis/deliver/deepLink?_collection=oasis&uniqueId=wol00001`

[8] K.D. Metcalf, *The New England Deposit Library* (1942), `http://www.jstor.org/stable/4302994`

[9] *Women Computers at the Harvard College Observatory | Smithsonian Digital Volunteers*, https://transcription.si.edu/instructions-women-computers-workbooks
[10] *DASCH Home Page*, http://dasch.rc.fas.harvard.edu/project.php
[11] *Unified Astronomy Thesaurus*, http://astrothesaurus.org/

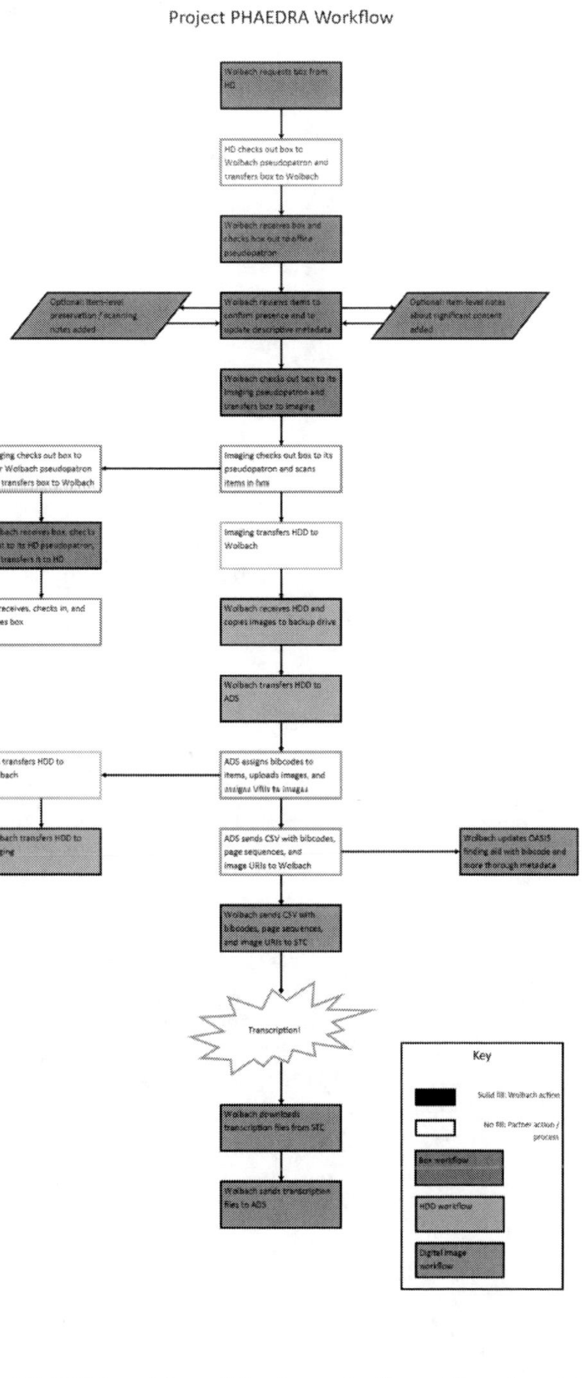

Figure 1. Full Project PHaEDRA workflow. HD refers to the Harvard Depository (off-site storage). HDD represents the physical hard drive used to transport the digitized image files. Diagram by Daniel Guarracino

EPJ Web of Conferences **186**, 08001 (2018) https://doi.org/10.1051/epjconf/201818608001
LISA VIII

New ADS Functionality for the Curator

Alberto Accomazzi[1,*], *Michael J.* Kurtz[1], *Edwin A.* Henneken[1], *Carolyn S.* Grant[1], *Donna M.* Thompson[1], *Roman* Chyla[1], *Steven* McDonald[1], *Taylor J.* Shaulis[1], *Sergi* Blanco-Cuaresma[1], *Golnaz* Shapurian[1], *Timothy W.* Hostetler[1], and *Matthew R.* Templeton[1]

[1]*Harvard-Smithsonian Center for Astrophysics, 60 Garden St., Cambridge, MA, USA*

Abstract. In this paper we provide an update concerning the operations of the NASA Astrophysics Data System (ADS), its services and user interface, and the content currently indexed in its database. As the primary information system used by researchers in Astronomy, the ADS aims to provide a comprehensive index of all scholarly resources appearing in the literature. With the current effort in our community to support data and software citations, we discuss what steps the ADS is taking to provide the needed infrastructure in collaboration with publishers and data providers. A new API provides access to the ADS search interface, metrics, and libraries allowing users to programmatically automate discovery and curation tasks. The new ADS interface supports a greater integration of content and services with a variety of partners, including ORCID claiming, indexing of SIMBAD objects, and article graphics from a variety of publishers. Finally, we highlight how librarians can facilitate the ingest of gray literature that they curate into our system.

1 Introduction

The NASA Astrophysics Data System plays a central role in the ecosystem of information providers in Astronomy. As the main discovery platform for the scientific literature used by the community of research astronomers, ADS attempts to provide a timely and complete record of scholarly artifacts for our discipline. One of the features that makes ADS such a compelling platform for a scientist is not just its disciplinary completeness but also the richness of the information that it collects. Rather than simply serving as a library catalog, ADS enriches its bibliographic records via the integration of content from a variety of data providers, including links to datasets hosted by NASA and other archives and bibliographies maintained by collaborating curators and librarians.

As an information system, ADS has become a platform where decisions about the inclusion and classification of both conventional and non-traditional content has become a community-driven effort. The integration of community-curated content and the need to support the needs of librarians which maintain such content is so important that both these activities are explicitly mentioned in ADS's official mission statement. Our participation in conferences such as LISA and in the conversations surrounding these activities is a testament to the importance of the partnership between ADS and the library community.

*e-mail: aaccomazzi@cfa.harvard.edu ORCID: 0000-0002-4110-3511

2 Content

ADS aggregates content from a variety of scholarly information providers, and adds value to such content by enriching and linking its records with a wealth of internal and external resources. The policies and decisions which inform what content should be indexed in the system are generally dictated by the needs of the research community in Astronomy. As these needs and expectations change, so do our policies.

An example of such change is the way in which arXiv eprints have been ingested and used in ADS. Originally these publications were indexed as a separate collection of records available for search on demand, and lived alongside the traditional literature indexed in ADS. At a later time, ADS started integrating references published in eprints in its citation database, so that citations from eprints would contribute to the total (non-refereed) citations accrued by published papers. Finally, once we started cross-identifying eprints with the corresponding published records, citations made by and to eprints were automatically attributed to the corresponding published papers. Today, ADS's new search engine and user interface simply consider eprints as an additional source of publications which may or may not overlap with the refereed literature.

Similarly, we are now seeing an evolution in the way a variety of non-traditional content is considered not only relevant but rather essential to the research process. Transparency, reproducibility, and credit are the three major reasons behind current publishing trends which encourage data and software citation [1, 2]. While not traditional articles, all of these resources have begun appearing in citation lists of papers published in Astronomy, and as our community is moving to support data and software citations [3], it becomes crucial for ADS to support these efforts. By making sure that whatever citable scholarly astronomy resource is indexed in our system, ADS makes these resources easily discoverable, while at the same time giving them the proper citation credit. Such content expansion does not come without additional curation challenges, imposing new requirement on the workflows managed by the project. We are currently working with the AAS and Zenodo on a Sloan-funded effort to implement software citation indexing in ADS.

We are also seeing a rapid change in the way individuals and organizations "publish" content. As more options become available to scientists and librarians alike to make a variety of scholarly content available on a variety of digital platforms, the question remains of how discoverable and citable this information may be. As our community looks to ADS as the primary search engine for scholarly astronomy content, it becomes essential that new scholarly publication venues be included in what we index. We are currently seeing the rise of general-purpose repositories such as Zenodo as publishing platforms for gray literature, software, and data products. We have started ingesting such material to the extent that we can determine its relevancy and reputation. Our partnership with librarians in vetting specific communities (such the Zenodo Astronomy Thesis Collection) and creating specific event venues (such as presentations and posters given at a particular conference) provides the appropriate level of trust in this material which would be otherwise onerous to evaluate. If you are supporting the publication and archiving of such content please contact us to ensure that it gets properly ingested in ADS.

Finally, we continue to collaborate with a number of different archives and institutions in the maintenance of bibliographic groups which represent collections of papers belonging to either an institutional or so-called "telescope" bibliographies [4]. Such collections, once indexed in ADS, provide a very effective and powerful way to select papers containing observations from a particular telescope or mission (e.g. HST), or published by scientists at a particular institution (e.g. CfA). Their existence makes it easy to generate useful statistics about missions, projects and research centers, understand the science being enabled or pursued by them, and even locate the datasets which were generated or analyzed by them.

3 Services

The ADS is built around a database, and access to its contents is available through a number of services. These services are used by a number of different applications, some of which act on behalf of individual users (e.g. somebody interacting with the system through a web browser), others on behalf of curators (e.g. a librarian using a tool to maintain a bibliography), scholarly information providers (e.g. a publisher validating bibliographic links), or even search engines (e.g. the Googlebot crawler). The technical challenge in maintaining these services is providing the proper level of responsiveness and functionality to all our constituents. In this respect, prioritization of tasks and features is not always a simple matter: while a bug or the lack of a feature may only affect a few librarians, its impact may be great overall, if it were to affect the curation of an entire data center bibliography or the access to its data holdings.

Managing a system such as the ADS also presents administrative and legal challenges such as protecting the privacy of its users and safeguarding its content, while at the same time providing useful services to the community. Since data is given to the ADS under specific terms of usage from each publisher, we need to make sure that our search services and interfaces protect this content from abusive behaviour such as the indiscriminate download of most of our records or full-text papers by any one individual. From a technical prospective, this has led us to the implementation of rate limits for all of our new services: each (possibly anonymous) user is allowed only N amount of queries for any given period of time. There is often a fine line between securing the system against misuse and providing users with the access they need to conduct research studies based on ADS content. In general, we aim to have N be large enough that no individual researcher interacting with our system on a daily basis will ever run out of available queries, but that applications which systematically query and download records from ADS might. If you find yourself in need to have these limits raised for legitimate reasons, feel free to contact us with details of what you are trying to accomplish and we will work with you to make sure you can get your work done.

3.1 Application Programming Interface

Access to the ADS services is delivered through a so-called RESTful Application Programming Interface (API), which provides a single, top-level entry point to all of its services. The API supports self-discovery of all its services and corresponding endpoints via a versioned public URL. Internally, the API connects to a number of different microservices which implement parts of the ADS system. Among them one can find interfaces to search, metrics, user libraries, ORCID, visualizations, export, and recommender services. We expect this list to grow over time.

Access to the API and the underlying services is best accomplished through libraries and applications. For those curators interested in programmatic access to the API, we provide online documentation and recommend the use of existing libraries such as the ads python module, which is available for direct install from PyPI.

3.2 Search Engine

At the core of ADS's system is a search engine which provides access to the entirety of the corpus indexed by it. The current search engine is an ADS-modified version of the open-source project Apache SOLR, which is an enterprise level search solution used by a large number of academic institutions and commercial companies. The adoption of a robust, open source system such as SOLR has allowed us to index for the first time not only the main article metadata, but also the full-text content of the articles in our possession. This has increased the amount of data indexed in our system by one order of magnitude compared to the prior generation search.

While most users limit their searches to the basic article metadata (author, title, keywords and abstract), the implications of being able to perform full-text searches is a game-changer for curators who have a need to find mentions of facilities, telescopes, instruments or even specific observations in the literature. ADS-specific enhancements to SOLR give users easy access to the citation and usage data indexed in the system via so-called second-order operators. As an example, the *citations(query)* operator allows retrieval of all articles citing the results returned by the inner search *query*. Additional operators allow retrieval of trending, useful, and review articles on a particular topic.

3.3 User Interface

In 2015, the ADS project launched a new User Interface, which was code-named ADS Bumblebee. Bumblebee represents the latest product in the evolution of our UI and attempts to integrate the best practices from the fields of usability and graphic design. Our goal was to create an interface that is intuitive and forgiving while simultaneously providing access to the full range of ADS's sophisticated query and analytical capabilities.

Bumblebee is entirely a javascript application and as such runs on a user's web browser, requesting content from the ADS services via the API described above. In that sense, it is nothing more than an application built on top of the public ADS infrastructure, and as such could be modified and improved by anybody. The interface integrates and highlights a number of useful features offered by the API services, among which are the availability of facets (filters) for query refinement, scalable metrics for any set of query results, and detailed metrics for collections of records. A number of visualizations have been implemented that provide additional insights into the search results. Visualizations of author networks and co-citation networks provide a hierarchical view of research groups and research topics, respectively. Displays of paper analytics provide plots of the basic article metrics (citations, reads, and age). All visualizations are interactive and provide ways to further refine search results. The detailed view of an individual article now includes a preview of the images and plots included in the paper thanks to our collaboration with publishers such as the AAS, EDP Sciences, and Elsevier.

Of particular interest to librarians and curators is the inclusion in any search results of features provided by the SOLR search engine and our new API. These include the highlighting of search terms in context, a histogram of the publication dates for any set of results, the distribution of citation counts, and a breakdown of records by collection, bibliographic group, or keywords.

3.4 ORCID

The new ADS interface and search engine provide integration with the ORCID (Open Researcher and Contributor ID) system. ORCID provides a system to uniquely identify scholars across all disciplines, thus disambiguating people with similar names in the literature. While everybody tends to agree that there is a compelling need for author disambiguation and that ORCID is currently providing the best platform for it, accomplishing this goal is no easy feat. For ORCID to deliver on its promise, it is necessary that the entire ecosystem of authors, publishers, and information providers work together to further its use.

The ADS and many publishers have been supporters of ORCID from its inception and have been instrumental in promoting its adoption. Currently, we are participating in this effort in two significant ways. First, ADS is facilitating the process of claiming papers in the ORCID system via the implementation of an "ORCID mode" in the new Bumblebee interface which allows authenticated users to send their claims to the ORCID API, thus populating their bibliography list using ADS as the reference database. Second, ADS is leveraging ORCID data in our search system by indexing ORCID identifiers provided to us in manuscripts or claimed via the user interface. Using these two features, a scientist interested in uniquely identifying their publications can use ADS to create a bibliography and associate it with their ORCID identifier, thus making it possible for other people to discover it in ADS.

While the current initiatives by publishers and systems such as ADS can go a long way in promoting the use of ORCID, librarians can play a significant role as well. Several observatories and research institutions currently maintain a bibliography of their staff's publications. Some of them (such as the CfA) have started collecting ORCID identifiers for their staff, thus being able to create a mapping between bibliographic records in their collection with the ORCIDs of staff members at their institution. These ORCID claims, curated by a trusted party, represents a source of additional data which ADS can ingest in its system to increase its ORCID coverage, without requiring any additional action on the part of the authors. We plan to start working on the integration of ORCIDs collected by the CfA library in the fall of 2017 and to expand these efforts further in 2018.

3.5 Future Plans

The effort to recast the ADS technology platform began a decade ago, when it became clear that the current infrastructure was neither scalable nor maintainable, and that the development of a new system would require a significant increase in the resources available to the project. Over these past ten years we have been able to grow the ADS team, test new technologies, gather community feedback, and plan the complex system architecture which we are now implementing.

As is often the case, the road to this new system has presented more challenges than originally anticipated, and the transition of ADS operations to the new system has been postponed as a result of this. We are now planning to start this transition in 2018, when use of the current ADS Classic search system will be deprecated in favor of the new ADS API and the Bumblebee user interface. We are fully aware of the impact this change will have on our community and we are therefore planning a staged migration of our user base to minimize disruption. Initial feedback about modifying the search habits to use the new interface and search syntax will influence the timing and specifics of a broader roll-out.

After achieving a significant migration of our users to the new system during 2018 we will focus on migrating API usage of ADS Classic, followed by access by search engines, which end up accounting for approximately half of total ADS usage. The current plan calls for discontinuing the ADS Classic services in 2019, pending a successful transition of all of our constituencies to the new platform. While this is a very aggressive schedule, we believe that making the transition sooner rather than later is in the interest of all parties involved: our users, our collaborators, our team of developers and curators, and our sponsors. Our main focus at this point is therefore meeting this goal as early as possible while minimizing the impact of this change to our constituents.

There is no doubt that our transition to a new platform may cause significant disruption to both scientists and librarians. However, the introduction of new technology and new capabilities also offers an opportunity for updating the ADS-related tools, applications, and workflows which have been developed over the years. One such example, mentioned in section 3.4, is the integration of ORCID claiming into the new UI and the subsequent indexing of ORCID data in the search engine, which became possible thanks for the refactoring of the system functionality as a set of separate services talking to each other. Another example is the development of a tool based on Google Sheets which integrates with the ADS API to facilitate the management of bibliographies by librarians. The "ADS bibliography tool," described in [5], requires no programming experience and can be installed as an add-on from the Google Sheets interface or directly from its GitHub repository.

4 Conclusions

We live in a time of rapid technological change, with the expectation that this change will continue to make things steadily better for all of its consumers. This certainly includes the field of scholarly research and publishing, where use of information technology is ubiquitous from the moment of data capture at the telescope to that of publication of scientific results. As information providers, it is our responsibility to leverage such advances and continue to improve our services to the astronomical community in order to maximize the efficiency of the research process as well as its impact on it and society at large.

At the same time the role of librarians in research institutions has been changing, and now includes activities such as digital collections curation and management, data analytics, and publishing advice. Much of this work can benefit from the use of a system such as ADS in a number of ways: content discovery, text mining, document classification, and bibliometric analysis are all services which we make available through our new API. Similarly, several curation activities performed by librarians can greatly benefit ADS and its user community: from article-data linking to bibliography management to ORCID enrichment. Having your curated data indexed in ADS extends its usefulness greatly beyond the boundaries of your local institution. In this respect, sharing your data with ADS is the first step in embracing the FAIR principles.

Over the next several years expect to see significant changes within the ADS project. From an access point of view, the major updates described earlier will cause both discomfort and delight as users will struggle to navigate a new user interface and learn a new search syntax before realizing the benefits of the new system and its capabilities. A significant amount of effort will go into ensuring that all relevant content is properly represented in our system, including non-conventional items such as scientific software and high-level data products. Existing and new initiatives in the use of persistent identifiers for people, institutions and projects will make it easier to reliably identify the contributions of both individuals and collaborations. We believe that librarians can make significant contributions in all these areas and look forward to partnering with them. If you have suggestions or would like to get involved please let us know by contacting us directly at ads@cfa.harvard.edu.

The NASA Astrophysics Data System is operated by the Smithsonian Astrophysical Observatory under NASA Cooperative Agreement NNX16AC86A

References

[1] A.M. Smith, D.S. Katz, K.E. Niemeyer, PeerJ Computer Science **2**, e86 (2016)

[2] D.C.S. Group", *"joint declaration of data citation principles"* (2014), `https://www.force11.org/group/joint-declaration-data-citation-principles-final`

[3] P. Teuben, A. Allen, G.B. Berriman, K. DuPrie, J. Mink, T. Robitaille, K. Shortridge, M. Taylor, R. Warmels, ArXiv e-prints (2016), `arXiv:1611.06232`

[4] A. Accomazzi, E. Henneken, C. Erdmann, A. Rots, *Telescope bibliographies: an essential component of archival data management and operations*, in *Observatory Operations: Strategies, Processes, and Systems IV* (2012), Vol. 8448 of *SPIE Proceedings*, p. 84480K, `arXiv:1206.6352`

[5] J. Damon, E. Henneken, A. Accomazzi, These Proceedings (2018)

EPJ Web of Conferences **186**, 08002 (2018)
LISA VIII
https://doi.org/10.1051/epjconf/201818608002

Toward Binding Database Interfaces with Scientific Papers

Laurent Michel[1,⋆], *Sinan* Acar[2], *Gilles* Landais[1], and *André* Schaaff[1]

[1]*Observatoire astronomique de Strasbourg, Université de Strasbourg, CNRS, UMR 7550, 11 rue de l'Université, F-67000 Strasbourg, France*
[2]*UTBM, Rue de Leupe, 90400 Sevenans, France*

Abstract. Despite a large variety of facilities helping to either select or manipulate data from Web interfaces, it remains difficult to provide users with relevant scientific or technical annotations for those data. Introducing such content by hand into a Web interface is a tedious job with a risk of providing incomplete or inadequate content. To overcome this difficulty, we are exploring the possibility of using the names of exposed quantities to index a text corpus. This index can be used to show the most relevant text snippets in a given context. The full text can be displayed by user request and automatically scroll down to that snippet. Our approach is based on the conversion of PDF papers into machine readable files that are indexed by a search engine. Index entries are reported as PDF annotations that are used to control the display. This workflow has been tested on the IVOA standard corpus as a proof of concept. It has then been applied to the XMM-Newton user guides for our catalog interface. Finally, it has been adapted to find resources within portals exposing a lot of various data collections.

1 Introduction

The idea for this project comes from our experience with the XMM-Newton [1] database interface which we are operating in Strasbourg. XMM-Newton is an X-ray space observatory orbiting since 2000 and working in pointed observation mode. Observations are requested by guest observers who get proprietary access to the data over one year. After this period, the data are made public. Every year or so, all data are compiled in a catalog [2] merging all source detections and linking them with other scientific products. The whole catalog is accessible through different interfaces, among those is the XCatDB [3] deployed in Strasbourg by the XMM-Newton Science Survey Consortium. In addition to source parameters, XCatDB users can take advantage of accessing technical data such as filter responses, energy band definitions and so on and so forth. This kind of information is spread out over various documents such as the XMM user's documentation or Web pages. Although these documents are publicly available, they would be more valuable if they were automatically connected with a database user interface.

2 Providing Users with a Rich Contextual Help

Scientific data products provided with the XMM-Newton catalog do not contain any significant information about the instruments or the data reduction process. This information is available in separate

⋆e-mail: laurent.michel@astro.unistra.fr ORCID: 0000-0001-5702-0019

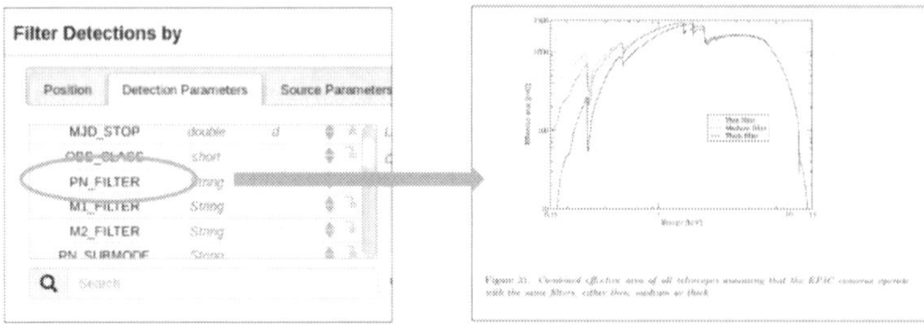

Figure 1. Implementation concept: the left scrolling list is a partial screen shot of the actual XCatDB interface; it contains all catalog columns available for filtering data. The search engine is invoked after some user action on a list item, to look for information related to that keyword within the documentation.

technical documents, Web pages or scientific papers. Our purpose is to make this ancillary data available on demand from the XCatDB interface. For instance, a user selecting sources detected with a given filter might be interested in knowing more about the response curve of that filter (fig. 1).

The most basic way to proceed would be to put links to those documents somewhere. However this wouldn't help the user get information focused on what he/she is doing. A more advanced possibility would be to extract by-hand information of interest and to feed tooltips [1]. This is an interesting solution, but with one major drawback: compiling by-hand such a data corpus is a big and expert job which is to big for our team. The solution we are exploring here is based on the use of a search engine, à la Google, running on a limited text corpus. The user could trigger search queries from different locations on the database user interface. Queries would be automatically setup according to the current users action. What would be returned is a selection of relevant text fragments displayed in their original context, while keeping access to the whole document. Thus, the user could start browsing the text corpus from a section related to what he/she is doing. It is important to note that unlike Web search engines we are not looking for documents but for fragments of them.

In other terms, we are considering merging in a single interface both data and documentation interfaces.

3 Component Breakdown

This project, achieved in the frame of a student internship, is a pathfinder project. It is split into independent modules, listed below, and intended to be put together after validation:

- Gathering a corpus of relevant texts
- Building an indexer enable to process the whole corpus
- Building a rendering tool providing entry points in the corpus from the text fragments found by the search engine
- Connecting these modules to the database interface.

This last step has not been achieved during this work.

[1]A small "hover box" with information about the item being hovered over (Wikipedia)

4 Text Corpus Creation

This is a critical point. All interesting entities must be described in various aspects (definition, calibration description, ...), but the documents must be close enough to the main topic in order not to pollute the results with irrelevant data. On the other hand it must be rich enough to provide non-trivial information. We started with indexing different XMM-Newton user guides (PDF), the catalog Web pages (HTML) and the reference paper [2] (PDF). The indexation process being rather short, it remains easy to modify the corpus and to rebuild the database.

5 Text Corpus Indexation

5.1 The Search Engine

A text corpus limited to a few tens of documents does not require high computing performance neither for the index storage nor for the query processing. Assuming we do not need specific features, we choose Elasticsearch [4], one of the most popular NOSQL database engine. The installation is easy and the tool comes with a convenient Web interface helpful to check both data ingestion and indexation. It also offers a powerful query language. Basically, Elasticsearch ingests text fragments formatted in JSON. Queries return those messages, still on JSON, matching at best the query. Elasticsearch is schema-free; there is no constraint on the JSON keys present in the ingested messages. Indexes are created for all keys of all ingested messages. Specific types (numerical, text, enumeration, ...) can however be attached to specific keys.

5.2 The Indexation Process

The text is first split in to short fragments. The fragments are then located within the original text and tagged with their natures (title, section, caption ...). The annotated fragments are packed in JSON files ready to be ingested by Elasticsearch (Fig 2). The search results must remain independent from the format of the searched documents. That means that the structure of the fragments must be the same for all text formats. Although the indexation process is enabled to ingest the whole corpus, it implements specific fragmentation routines for all supported text formats.

5.2.1 PDF Fragmentation

PDF is a printer format that maintains the same layout (paging, text location, font, colors, ...) for any device (screen, mobile, printer ...) but ignores the text structure. In fact, extracting text elements from a PDF document is akin to a data mining task more than a parsing task. We were using Grobid [5] to achieve this. Grobid is a machine learning tool enable to convert PDF documents into TEI files [6]. Grobid algorithm relies on knowledge bases built from large document samples. This tool is at the heart of the Science Miner project [7]. We had to patch the Grobid code to extract data of interest before the TEI generation. Our Grobid version returns a list of text fragments with their natures (section, title, ...) and their exact locations (page number and coordinate within the page). These data are packed in JSON messages ready for Elasticsearch.

We have to keep in mind that the text analysis with Grobid is not 100% reliable as any other machine learning tool.

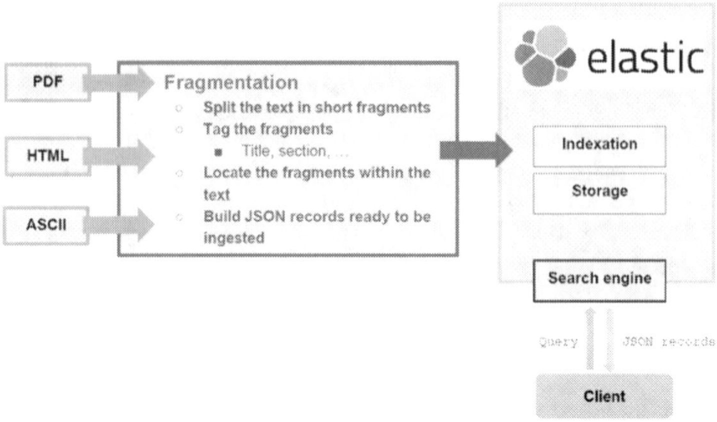

Figure 2. Indexation workflow: fragmentation tasks are specific for each text format but the indexation process is the same for all text fragments.

5.2.2 HTML Fragmentation

Unlike PDF, HTML documents are split into several independent pages spread out over different Web sites. We index one Web site at a time. A so called Web site is a set of HTML pages referenced by a master (home) page and confined in one domain. It is a sub-part of the corpus and each page is a document. Our indexer works on a local copy of the site downloaded with `wget` and analysed with `Craw4J` [8]. Using a local copy prevent issues with dynamically generated content. In addition, this allows the annotation of the content, for instance marking the fragment positions. The nature of the fragments is easy to retrieve with the HTML tags, but their locations are more difficult to get since the layout is very flexible and strongly dependent on the display device. The indexer extracts HTML fragments from the documents with an arbitrary length. These fragments are stripped from HTML tags and indexed by Elasticsearch. At the same time an empty tag with an incremental identifier is inserted into the HTML file. The fragment is then located by the text URL suffixed with `#Identifier`. As this feature is not working properly yet, it is replaced with a search facility similar to those implemented by all browsers.

6 Rendering Tool

Before connecting the search engine to the database interface, we built a Web application to validate the queries and to test different layouts for displaying searched results (Fig 3). Searched keywords are typed by the user on the top left search bar. A few buttons are available to setup both query and display. The fragments returned by Elasticsearch are listed below the search bar either ranked by relevance or grouped by documents. On the top of each fragment is an anchor running the display, on the right panel, of the whole text scrolled down to the fragment location. PDF files are displayed by the Mozilla Javascript library PDF.js [9]. The Javascript code has been patched to open the text at the requested position and to highlight all searched keywords. HTML files are displayed in a simple frame. The fragment location is achieved by a simple search based on JQuery selectors. This is a workaround while the tagging at indexation time does not allow the proper location of fragments.

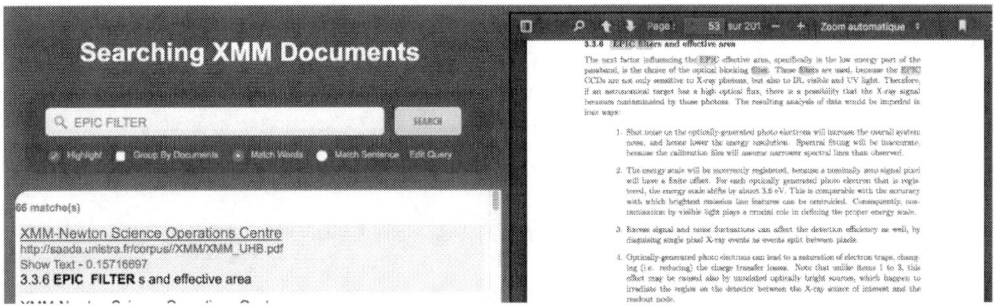

Figure 3. Rendering Tool developed to test the search engine.

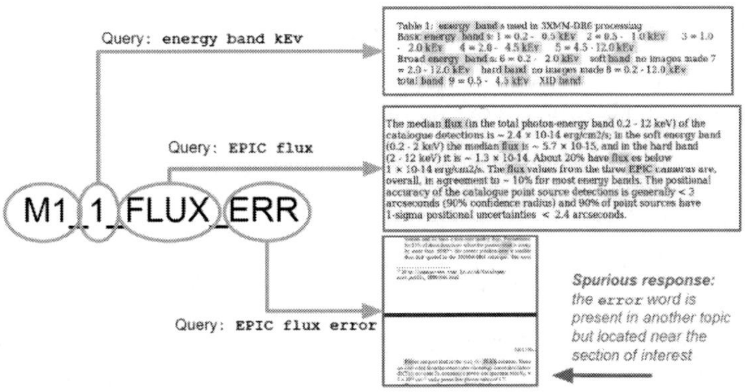

Figure 4. Compound keyword breakdown: We get relevant results by splitting compound keywords in individual components and then submitting their definitions to the search engine.

7 Database Integration

7.1 Rough Test

The initial scope of the project, connecting any meta data displayed on the screen with the search engine, has been limited to the 332 XMM catalog keywords. The reason for this restriction is that the catalog is the main resource exposed by the XCatDB. Catalog quantities are used to both filter scientific queries and to provide additional data to the searched results. We expect that users will be first interested in knowing more about these quantities. To get an idea of the system efficiency, we first submitted all column names to the search engine. The success rate is given in the table below.

Although not satisfactory, these results are not that bad. We skip the cases returning either confusing responses or just text fragments mentioning the keyword without further explanation. These cases are rare enough to be considered as noise.

Most of the successful keywords are related to one specific quantity (e.g., DR3SRCID is the identifier of the same source in the third release of the catalog) whereas most of the failing keywords (No responses) refer to compound quantities.

Good	None	Confusing	Just mentioned
77	240	6	9

Table 1. Search engine success score for the the catalogue keywords: Good) helpful response, None) no response at all, Confusing) irrelevant response, Just mentioned) the keyword is cited without further explanations

XMM is a complex observatory which runs, among others instruments, 3 cameras (MOS1, MOS2 and PN) working in 9 different energy bands (numbered from 1 to 9). A given quantity (e.g., a flux) can be used in 27 different keywords, or even 54 if we consider the errors attached to that quantity. The documentation describes all cameras, all energy bands and all measurements but it doesn't give a specific description for each set of camera/band/measurement. If we want to get information about a compound keyword, we have to first separate each component and to submit one of them to the search engine. The choice of that component depends on what the user is expecting. Furthermore, components cannot be directly submitted, they must be translated in more comprehensive expressions. Proceeding this way works for most of the keywords. Fig 4 shows an real example.

7.2 Baselines for an Effective Integration

These rough tests show that compound keywords cannot be used to query the search engine. The actual queries must be inferred from the individual components. The list below summarizes such a preprocessing.

1. **Detecting** whether the keyword is a compound keyword. If not, it can be used without further processing
2. **Extracting** the components. This can be helped by the syntax rules used to build the keywords
3. **Translating** each component in one or more words is comprehensive for the search engine. For instance, the component 1, which refers to the energy band 1, is absolutely not discriminating for Elasticsearch, whereas the expression `energy band` will return the precise definition of all energy bands
4. **Asking** to the user with the component he/she wants to submit (e.g., camera M1, energy band, flux or flux error)
5. **Submitting** the translation of the selected component

Fig 5 illustrates such a workflow. This processing has not been implemented yet. We just had a quick look at ontologies which are the natural framework to handle quantities described by plain text and bound with semantic links. This is, however, ahead of the current work.

8 Conclusion

Our pathfinder project validates the concept of a service that allows access to documentation from a database interface through a search engine. However, it showed the necessity of having an interface between the database metadata and the search engine. This will be the last step before any deployment on a production system. In addition to the XMM catalog, we operated it on the 13,000 Vizier [10] catalogue descriptors and on the IVOA [11] standard corpus with interesting results. This shows that our system could be transposed in other contexts than the XMM catalogue.

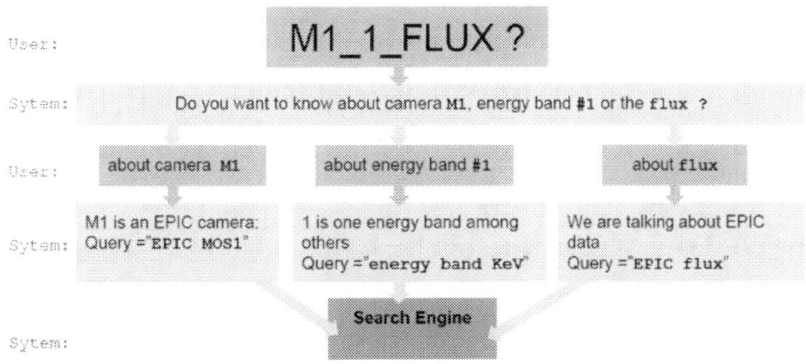

Figure 5. Getting information about compound keywords is not straightforward. The user must first say on which component they are interested. The component must be translated in comprehensive words which can be submitted to the search engine.

References

[1] ESA, *Esa science & technology: Xmm-newton*, [Online; accessed 19-July-2017], `http://sci.esa.int/xmm-newton/`

[2] S.R. Rosen, N.A. Webb, M.G. Watson, J. Ballet, D. Barret, V. Braito, F.J. Carrera, M.T. Ceballos, M. Coriat, R. Della Ceca et al., **590**, A1 (2016), `1504.07051`

[3] L. Michel, F. Grisé, C. Motch, A.N. Gomez-Moran, *The XCatDB, a Rich 3XMM Catalogue Interface*, in *Astronomical Data Analysis Software an Systems XXIV (ADASS XXIV)*, edited by A.R. Taylor, E. Rosolowsky (2015), Vol. 495 of *Astronomical Society of the Pacific Conference Series*, p. 173

[4] Elasticsearch, *Elasticsearch — Wikipedia, the free encyclopedia* (2017), [Online; accessed 19-July-2017], `https://en.wikipedia.org/wiki/Elasticsearch`

[5] Patrice Lopez and individual contributors, *kermitt2/grobid: A machine learning software for extracting information from scholarly documents*, [Online; accessed 19-July-2017], `https://github.com/kermitt2/grobid`

[6] Wikipedia, *Text encoding initiative — Wikipedia, the free encyclopedia* (2017), [Online; accessed 19-July-2017], `https://fr.wikipedia.org/wiki/Text_Encoding_Initiative`

[7] P. Lopez, *Semantic Scholar*, `https://www.semanticscholar.org/`, [Online; accessed 19-July-2017]

[8] Yasser Ganjisaffar and individual contributors, *yasserg/crawler4j: Open source web crawler for java*, [Online; accessed 19-July-2017], `https://github.com/yasserg/crawler4j`

[9] Mozilla and individual contributors, *Pdf.js*, [Online; accessed 19-July-2017], `https://mozilla.github.io/pdf.js/`

[10] CDS - Strasbourg, *http://vizier.u-strasbg.fr/*, [Online; accessed 19-July-2017], `http://vizier.u-strasbg.fr/`

[11] IVOA, *www.ivoa.net*, [Online; accessed 19-July-2017], `http://www.ivoa.net/`

EPJ Web of Conferences **186**, 09001 (2018) https://doi.org/10.1051/epjconf/201818609001
LISA VIII

Who is LISA?

Donna Thompson[1,*]
[1]*Harvard-Smithsonian Center for Astrophysics, 60 Garden St., Cambridge, MA, USA*

Abstract. The field of library science is always changing. Even professional library organizations have conducted studies about changing the name of the profession. All of the attendees of the LISA meeting are involved in Library and Information Sciences in Astronomy and we represent a number of different organizations and professions. In crafting a white paper about the profession, an ethnographic study of the field was pro posed with the research reported on at the LISA meeting. The research methods include an online survey and in person interviews with various members of the community. We hope to identify trends, differences and a road map for the future.

1 Introduction

At the 2015 International Astronomical Union General Assembly (IAU) meeting, and specifically at the focus group meeting for *Scholarly Publication in Astronomy: Evolution or Revolution?*, librarians, scientists, editors and others involved in the distribution of astronomy literature recommended that a white paper documenting the roles and responsibilities of astronomy libraries be written. The LISA meetings involve a number of astronomy librarians. In this paper I am going to present some information about the history and attendance at the LISA meetings. This information will be helpful in the preparation of the white paper for the IAU.

2 Who are astronomy librarians?

Astronomy librarians work at observatories, colleges, universities, government agencies and other research institutions. Frequently, astronomy librarians have responsibilities for other topics in addition to astronomy and there are few librarians who focus solely on astronomy. This being said there are at least five professional groups that represent astronomy librarians. Each organization is different but there can be overlap in membership as some astronomy librarians may belong to more than one group.

The International Astronomical Union (IAU) has a working group on Information Professionals. The working group was previously named the working group on libraries and was officially recognized by the IAU in 1991. This group meets every 3 years at the IAU General Assembly. The aim of this group is to foster communication between scientists, librarians and other members of the astronomical community. The membership of the working group reflects the diversity of the parties involved. One of the strengths of this group is that the membership is international.

* e-mail: dthompson@cfa.harvard.edu ORCID: 0000-0001-6870-2365

The Physics-Astronomy-Mathematics (PAM) Division of the Special Libraries Association is an active group of librarians who meet yearly.[4] Membership is typically about 150 to 200 librarians representing the three major fields of the Division. The PAM Division maintains an active website and listserv that are great resources for the astronomy library community. Membership of the PAM Division is mostly North American although it is open to international participation.

The Forum for Research Sharing in Astronomy and Astrophysics (FORSA) is an active group of astronomy librarians in India. The focus of this group is for sharing and streamlining resources for the many universities, observatories and research institutions in India.

The Astronomy Librarians of Hawaii Association (ALOHA) serves the community of Hawaii with its large number of telescopes. The librarians in this group meet to foster communication about events in their libraries and institutions.

3 History of the LISA meetings

The first LISA meeting was held in July 1988 in Washington D.C. just before the XXth General Assembly of the IAU. [1] The IAU provided some financial support and also served as a catalyst to bring the scientific and library communities together. (See Corbin and Grothkopf, 2006, for more complete information.[2]) To date there have been eight LISA meetings in total. The aim is to have a meeting every three to four years in a different location so that interested individuals from different parts of the world can participate. The average attendance for the eight meetings has been 106. (See figure 1.) At the conclusion of LISA VIII it was announced that a number of libraries have expressed interest in hosting LISA IX and hopefully this will happen.

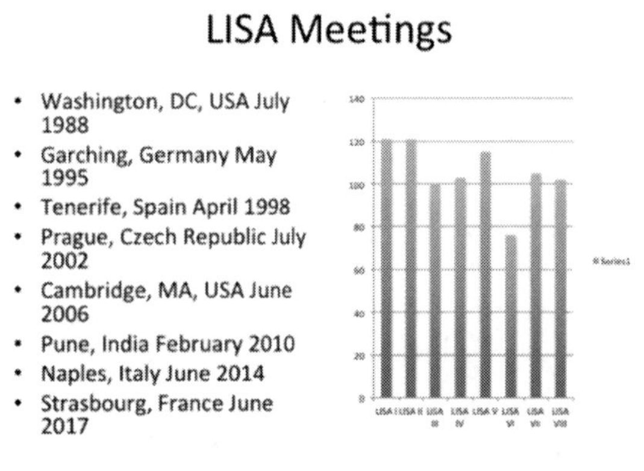

Figure 1. Location and registrant numbers by conference

4 LISA attendees

To investigate the attendance at the LISA conferences, I compiled spreadsheets listing the names and the country affiliation for each attendee. I also developed a spreadsheet for each conference,

Argentina (5)
Australia (5)
Austria (3)
Bangladesh (1)
Belgium (3)
Brazil (1)
Canada (8)
Chile (5)
China (8)
Czech Republic (9)
Czechoslovakia (2)
Democratic Republic of Congo (1)
Denmark (4)
Estonia (1)
Finland (3)
France (66)
Georgia (2)
Germany (41)
Greece (1)
Hungary (5)
India (47)
Indonesia (2)
Ireland (1)
Israel (1)
Italy (32)
Japan (6)
Korea (2)
Malaysia (2)
Mexico (6)
Morocco (1)
Netherlands (10)
Nigeria (1)
Norway (2)
Pakistan (1)
Poland (6)
Russia (5)
South Africa (7)
Spain (15)
Sweden (3)
Switzerland (6)
Ukraine (6)
United Arab Emirates (1)
United Kingdom (27)
USA (162)
USSR (2)
Vatican City (1)

Figure 2. Alphabetical list of participants by country: number of participants from each country listed in parantheses.

followed by a master list of all conference attendees, which I annotated to include the number of conferences each person had attended and specifically which conferences each person had attended. All of the LISA conference proceedings are online and can be accessed from the LISA website *http://www.eso.org/sci/libraries/lisa.html* maintained by Uta Grothkopf. When available I used the printed participant list from each volume of the conference Proceedings; if there was not a participant list, I used the list available on the website of the meeting.

In all there have been 843 people registered for the LISA conferences. 530 individuals have attended one or more LISA meetings. Not one person has attended all eight LISA meetings. Six individuals have attended LISA II through LISA VIII–Monique Gomez (Spain); Uta Grothkopf (Germany); Jane Holmquist (USA); Maria Elena Jimenez (Mexico); Soizick Lesteven (France) and Francesca Martines (Italy).

Attendees of the various LISA meetings have come from 44 countries from around the world. (See figure 2 for an alphabetical list by country and figure 3 for a map depicting the various countries from which LISA participants have come.)

Since one of the main goals of the LISA conferences is to have representation from astronomical communities around the world, the map shows that the conferences have indeed been truly international. The fact that the six individuals who have attended seven of the eight LISA conferences came from six different countries is also an indication of the international scope of LISA. Not surprisingly there is increased local attendance at each conference.

5 PAM Survey

As a member of the PAM Division, I am able to access a membership listing for the division. As a first step in surveying astronomy librarians, a list of 10 questions was developed and sent to PAM Division members who self-identified as astronomy librarians. The 10 questions were:

 https://doi.org/10.1051/epjconf/201818609001

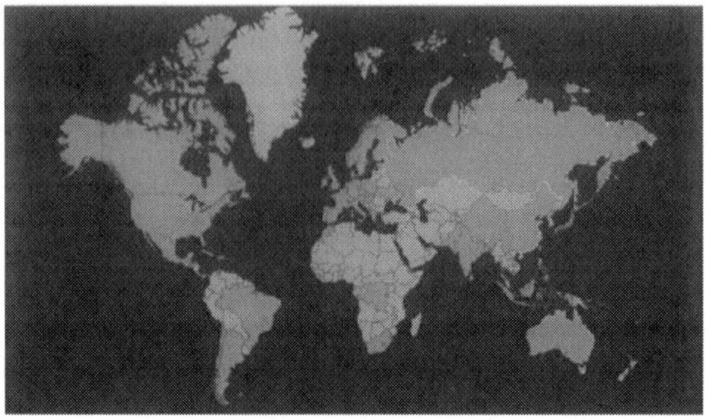

Figure 3. Blue shaded areas of map depict countries represented at LISA conferences.

1. What is your job title?
2. Regardless of your official job title, do you consider yourself an "astronomy librarian?" If not, what other title would you give yourself?
3. Please list any additional subject areas that you are responsible for:
4. What is the highest level of education you have completed?
5. What was your major in college?
6. What type of organization do you work in?
7. What are your major responsibilities in your position?
8. How do you see your position evolving over the next three years?
9. Are you aware of the Library and Information Services in Astronomy (LISA) or the IAU Working Group of Information Professionals meetings?
10. One of the goals of my research is to develop a listing of astronomy librarians. If you would like to be included in this list, please provide the information below:

Question 7 had a list of topics from which the respondent could choose answers and a section to add more responsibilities. Question 9 had a list of answers which would indicated whether or not the respondent was aware of the meetings and if they had attended. The other questions were open-ended. Thirty-one people out of the fifty-three people who were sent the survey completed it.

While this was not a large sample, some of my initial findings support some assumptions that I had made about astronomy librarians. Some general findings about astronomy librarians from this short survey:

- Very few librarians have the job title of astronomy librarian

- Very few librarians cover only astronomy: 22% also covered physics; 13% math; 9% chemistry; 6% computer science; 5% engineering; 4% earth and planetary sciences and about 30 other disciplines including nursing, nutrition, robotics and general science
- 67% have a master's degree in library science as his/her highest degree
- 58% have a bachelor's degree in the sciences; 16% in history, 16% in literature
- Major responsibilities include reference work, teaching/training, researching and bibliography maintenance
- expected changes upcoming in the next three years
  - digital assets management
  - building digital archives
  - helping faculty demonstrate funding requirements
  - scholarly communication
  - bibliometrics
  - disappearing library space
  - digital librarianship
  - STEM data
- 75% were aware of the LISA meetings
- 29% were aware of the IAU Working Group on Information Professionals

These preliminary results lead me to conclude that the topics being covered in the LISA conferences are relevant to the future of astronomy librarianship and that the astronomy librarian community could do a better job of informing its members of groups and conferences that might be helpful to individuals. While the results of this short survey showed that *traditional* library tasks such as cataloging and bibliography and reference work will continue to be part of the profession, there is also awareness that some additional duties have not always been *traditional* librarian skills.

6 Future study

In furthering the research necessary to write the white paper for the IAU, I will continue to research groups such as PAM Division members and LISA participants for insights into the profession. In the future I may ask LISA and IAU Working Group meeting participants to identify themselves as scientist, librarian, editor, etc. so that we can find out if the collaborations that we are hoping to develop are already taking place or if we need to do more outreach.

If you have any interest in pursuing the topic of astronomy librarianship in the future, I would be happy to have collaborators. I welcome any feedback or contributions for the white paper.

Long live LISA!

Acknowledgments

I would like to thank Jill Lagerstrom, former Chief Librarian at the Space Telescope Science Institute, for her ideas which led to this paper and for her input into the questionnaire.

References

[1] http://www.eso.org/sci/libraries/lisa.html

[2] B.G. Corbin, U. Grothkopf (2006): *LISA ? The Library and Information Services in Astronomy conferences*, in *Organizations and strategies in astronomy* (OSA), vol. 7, A. Heck (ed.), Springer, Dordrecht, ISBN 1-4020-5300-2, p. 285-306

[3] U. Grothkopf, F. Murtagh, C. Birdie, M. Bishop, L. Bobis, D.J. Coletti, B.G. Corbin, M. Gomez, H. Naimova, *Division XII / Commission 5 / Working Group Libraries*, in *Transactions of the International Astronomical Union, Series A* (2009), Vol. 27, p. 363-365, doi:*10.1017/S174392130802588X*

[4] http://pam.sla.org

The most faithfuls to these conferences :
Uta Grothkopf, Soizick Lesteven, Monique Gomez, Jane Holmquist, Maria Elena Jimenez
(Photo: M. Wenger)

EPJ Web of Conferences **186**, 09002 (2018) https://doi.org/10.1051/epjconf/201818609002
LISA VIII

Developing an e-Strategy for the South African Astronomical Observatory Library and Information Service: A case study

Samantha Muriëll Bennett[1,*] and *Theresa* de Young[1]

[1]*South African Astronomical Observatory*

Abstract. The South African Astronomical Observatory (SAAO) Library and Information Service (LIS) forms part of the SAAO - which is a business unit of the National Research Foundation (NRF) in South Africa. Until recently each business unit had a library service that operated independently. With the formation of the NRF Knowledge Center (KC) this is set to change. Previously it was possible for a library service to conduct its processes and procedures in isolation as long as the LIS aligned with the business unit. This paper covers the study that investigated the needs and possible guidelines for an e-Strategy for the SAAO LIS that is both aligned to the broader NRF strategy but which also incorporates recommendations to upgrade the library products and services offering. The pool of Information and Communication Technology (ICT) -based products and services is growing. 'Upgrading' therefore requires that it is acknowledged that ICT provides essential infrastructure for the relevant the library. Library focused trends show continuous development. Bringing the SAAO LIS up to standard will require that these developments are continuously monitored for relevance and incorporation. This paper provides information about the SAAO LIS and its history. Both 'SWOT' and 'PESTEL' analyses were used to direct the discussion on the internal and external influences on the SAAO LIS. The analyses also provided insight into the challenges the SAAO LIS faces, the strengths that could be capitalized and the opportunities that could be exploited. The paper recommends that the SAAO LIS, before engaging in a strategic change, addresses the challenges of sustainable funding and accessibility to core resources. It is also recommends that the SAAO LIS uses its e-Strategy to 'fast track' the process of bringing it up to the standard required of world-class astronomy libraries.

1 Introduction

The South African Astronomical Observatory (SAAO), a National Research Foundation (NRF) facility, is the South African national resource for optical/infrared astronomy. Its primary function is to conduct fundamental research in astronomy and astrophysics. It not only provides a world-class facility that promotes astronomy and astrophysics in Southern Africa, but it also connects South African astronomical research to the rest of the world. The Library and Information Service (LIS) of the SAAO is seen as an integral part of the facility as it provides the research staff with access to essential information and tools that allow them to function as informed professionals within their community.

*e-mail: sbennett@hct.ac.ae

Because the SAAO is part of the larger NRF-facility family the LIS of the SAAO also aligned its strategic actions with those of the NRF. Any strategic actions adopted by the library need to contribute also to the NRF's strategic direction and should assist in achieving the NRF's corporate goals.

This e-Strategy case study was conducted in 2014 and NRF's 2012-2015 annual performance plan ([1])was applicable, stating its vision as: "ensuring research excellence within a transformed research workforce, and a sustainable environment, not only for the benefit of the present generation but also for future generations of South Africans." In short, this vision seeks to create:

- World-class research
- A transformed society
- A sustainable environment

The plan also indicated the NRF's mission as the following: To contribute to the knowledge economy in South Africa by attaining at least one percent of global research and development (R&D) output by 2015.

In support of achieving its objectives the NRF adopted the following set of values:

- Passion for excellence
- World class service
- Ethics and integrity
- Respect
- People centered
- Accountability

These values highlight a passion for excellence and world class research and service through the organisation's sound ethical practices and accountability. The NRF also values its staff.

The SAAO is one of 6 reporting facilities to the NRF. It appears though that the LIS of the SAAO has not kept pace with the growth within the rest of the SAAO. It is therefore essential that the LIS re-looks at its reason for existence and that it fully and explicitly aligns with the NRF values as well as with developments within the LIS sector. It especially appears that the LIS of the SAAO has not stayed aligned with the technological developments in either the library sector or with the changes that the rest of the SAAO has experienced. It is therefore anticipated that the opportunities, brought about by changes in technology, may need to be investigated in considerable detail.

1.1 Overview of the SAAO historic background

The SAAO was formed in 1972 by the amalgamation of two older institutions, the Royal Observatory, Cape of Good Hope, established during 1820 and the Republic Observatory, Johannesburg, established during 1905. ([2]) "The LIS at the SAAO dates its existence to the arrival of the first Royal Astronomer in the 1820s. The SAAO LIS is now the national library for astronomy and space science. The larger collection is located in the Main Building at SAAO, Cape Town, with a smaller collection at the observing site in Sutherland. The library has a unique richness of pre-twentieth century journals with the core astronomical journals starting at volume 1." ([2]) The Observatory buildings, including the main building which hosts the library, are old and due for restoration and refurbishment. Restoration of the Observatory would be beneficial for the public and incline preservation of Astronomical heritage of South Africa. "Friends of the Cape Town Observatory" have supplied advice and are encouraging the South African Heritage Resource Agency (SAHRA) to declare the Observatory as a National Heritage Site. ([3])

1.2 LIS holdings

The LIS online catalogue was started in 1993 and since 2004 the library personnel have been engaged in retrospective online cataloguing. The online catalogue currently gives access to 9,521 records and 3,454 staff publications (compared to the card catalogue which holds just over 30,000 records). The online catalogue is currently only available from the SAAO intranet.

The library currently uses DB Text software powered by Mindex / InMagic as a library system. This system provides the necessary customisation and personalisation to fulfil the library's needs. The holdings of the library are partly catalogued. The libraries' holdings consists of journals (electronic and print), books (electronic and print), databases, atlases and catalogues, brochures, maps, photographs, plates, posters, staff publications, slides, media collections, CD's / DVD's and many historic documents. Not all of these holdings are catalogued on the library system or are in any organised form. Most of the holdings are scattered throughout the observatory buildings.

The library maintains a database of staff publications and papers produced by observers using data obtained at the SAAO. These publications are available on the library catalogue from 1998–current.

The NRF invested in DSpace, an open source repository software package used for the creation of open access repositories for scholarly and/or published digital content. DSpace could also be used as a digital archives access system when focusing on long-term storage, access and preservation of digital content. DSpace offers a mobile interface for easy accessibility as an important needed feature. The SAAO library has rights to upload its collections on DSpace to be available. However, the astronomy community's needs are very peculiar and distinct.

1.3 Explaining the research

The study intended to at least catalogue the library holdings at collection level. It also evaluated the library system and the value of the library content by guiding the organisation in evaluating the collection(s) hosted by the library. Furthermore, it assisted in identifying which collections were worth digitising and which collections needed to be preserved in the longer term.

The main focus of any library is to avail all collections, i.e. to be accessible as Buckland (1992: 14) indicated. Accessibility for the SAAO library is twofold:

- The collections are widely spread throughout the SAAO buildings. Many of these collections are not reflected in the card catalogues and searching for the collections is a huge task
- The online catalogue is equally outdated and research staff only have access to the electronic catalogue when they are on site.

2 Research question and sub-questions

Although the physical library may have defects it is anticipated that, going forward, the focus should be on a hybrid library with predominantly electronic content. This case study therefore focused on developing an e-strategy for the SAAO library. The research question is therefore: Given its unique context, what would be an appropriate e-strategy for the SAAO library to follow? In support of the research question above the following sub-questions have been developed:

- What, according to literature, are the most important factors to consider when developing an e-strategy for a Special Library?
- What could be learned from other astronomical libraries?
- What is the current status of the SAAO library – also within the broader NRF family?

- What are the explicit information needs of the SAAO researchers?
- What would the implementation priorities at SAAO be (with regards to technologies, collections and staffing)?

3 Scope and limitations of the research

This research only focused on the SAAO library collection and current library system prior to 2015 as per the status perceived by the researcher and benchmarked with senior staff at the SAAO. Astronomical and science library needs are distinct and leave librarians with challenges as Gomez ([**?**]) highlighted. However, in the scope of these challenges a comparison with world class astronomical libraries was done for benchmarking purposes. Furthermore, the study did not focus on the different user groups of researcher and astronomers, but generalised its focus to that of the SAAO. The study also provided a technology map of the current systems used prior to 2015 within the library, and made no provision to investigate new systems.

4 Research Approach

A qualitative approach was used for analysing and interpreting the content of documents, articles and web pages relating to the SAAO and the e-strategies of other similar institutions. A literature review was done to determine what the modern library should be. This review also highlighted best practices used in other libraries as a means of comparison. The third part of the data collection was an online survey administered to staff and students to determine what is needed to successfully implement accessibility, be it online or in hard copy.

5 Value of the study

This study provided useful information to the SAAO (and also the NRF) regarding the proposed 'updated' library and its collections or holdings. It aims to assist the librarian to create focus in terms of priorities, and to create order and access to the valuable and distinct information it could host online. It also provides an explicit 'key' or map as to where information and collections are stored or kept.

This is the first study for the SAAO library that focuses on the accessibility of information and the preservation of valuable information collections. Since the SAAO is the sole owner of South African astronomical data and information, giving guidance regarding the digitisation of the most valuable collections is a specific valuable aspect of this study.

This study adds value to the world of astronomy libraries as many of them ([**?**]) face the same challenges such as paper based information, which is unorganised or is accumulated over decades or even centuries without having kept pace with the digital revolution and the creation of new spaces. By developing the e-strategy, it will be possible to provide focus and guidance and it will assist the SAAO library to make informed decisions regarding its online future.

6 Obstacles during the study

During the study a few obstacles were encountered, which need to be noted. The focus group selection was opened to be inclusive of all divisions using the library. In the end the group only consisted of those staff members who had an interest in the library and it excluded administrative staff and representation from instrumentation department.

Even though the survey was sent to a sample of friends and colleagues to test for errors, it still forced people to answer questions that were not relevant to them. This was discovered and dealt with during the data analysis. Astronomy library literature is limited and left a gap in finding information relevant to the subject field. General library information needed to be used as guideline. The internal organisational culture was another hurdle and forced the researcher to use softer approaches in getting information.

7 Recommendations

ICT has become an important driver to the library. Due to the ever changing environment the library finds itself in, it has become clear that the pool of IT services and products is growing. For the library to best utilise these systems they have to ensure that the support supplied by these systems share the goals of the strategic plan. To best interpret the services required the library needed an e-strategy which was provided as an end result of the research.

8 Update on the current status

The recommendations of the study have been adopted, although funding has been the main concern in implementing them. These recommendations have been categorized into five main groups for ease of reporting on the current status:

8.1 Expand access to information

Retrospective online cataloguing, to ensure that most of the items in the card catalogue are captured on the library management system, is making good progress although it is slow and time consuming. It takes a lot of effort and is not a dedicated project but rather done whenever possible. Accessibility to catalogue records and ultimately actual library items is crucial for improving the library and information service at the SAAO.

Collections are being efficiently organised and archived by very enthusiastic "Friends of the Cape Town Observatory". Many of the items that were stashed away in different locations at the observatory have been worked through, weeded, categorised and preserved where necessary. During this clean-up exercise, many valuable items were found, among them two rare books. Access to these collections are of value to historical researchers, many of whom approach the SAAO library for information.

In order to provide access to cutting edge journal papers within astronomy, the SAAO needs to become involved in consortia deals that would be a benefit, providing wider access to information. The SAAO is a member of the SCOAP3 partnership that negotiates lower subscription fees on journals. SCOAP3 journals are open for any scientist to publish in without the worry of costs.

The National Research Foundation (NRF) is currently embarking on an initiative to preserve the heritage collection at the SAAO and has purchased major digitization equipment to accomplish the scanning of these items and then the preservation thereof. Digitisation of collections is scheduled to be complete by 2019. Funding has been secured to revamp the library by building a mezzanine floor in order to alleviate some of the space limitations in terms of shelving of library material that is being used. All of this is in preparation for the bicentennial of the establishment of the observatory at the Cape of Good Hope, to be held on 20 October 2020. This will also enhance accessibility to the collections that have been scattered in buildings around the observatory and that are not easily available to staff and the public.

8.2 A centralised library system

The study recommends the standardisation of library and information services within the NRF. There is a need to acquire a library management system that is more affordable and appropriate for the SAAO as well as other NRF facilities. The study also proposes a centralised library system within the NRF.

Conversations have started in this direction keeping in mind the benefits for the NRF in terms of cost and sharing of and access to all collections. Important for centralisation is the roll of NRF head office staff in taking the lead. Major changes with the staffing at the NRF appears to be the main obstacle to this; the lead can only be taken once structures are stable.

Shared e-services have commenced in 2016 with the centralised procurement of journals for all NRF facilities via a multi-year tender contract. This has proved to be a burden on the SAAO library budget as it had become more expensive than envisioned.

8.3 Develop ICT infrastructure

ICT infrastructure development hinges very much on availability of funding, whilst procedures for standardising processes, systems and tools are within reach as long as there is strong championing of these initiatives. Collaboration efforts together with various library partners within the SAAO will be considered and through collective efforts a stable infrastructure could be maintained. However, the standardisation of ICT within the NRF as a whole is a lot more tricky and very much dependent, once again, on head office leadership. The central point of access to one website that is recommended by the study as well as the unified use of cloud based services and device neutral platforms demands tight cooperation across the NRF including both head office and the facilities.

8.4 Provide research support

Research support, one of the key performance areas of the library and information service at the Observatory is recognised at the SAAO, with the librarian reporting directly to the Head of Astronomy. Along with research support comes the digital curation of intellectual work at the observatory. The NRF has taken the lead in this, adopting DSpace-CRIS open source module as the platform for managing data within the organisation including the setting up and maintaining of repositories at every one of its institutions, as well as making metrics (altmetrics) accessible, all of which are of interest in terms of research support. The NRF has also issued a 'Statement on Open Access to Research Publications from the National Research Foundation (NRF)-Funded Research' in January 2015, recognising the importance of open access to scientific research and instructing its researchers to deposit their final peer-reviewed accepted papers into the institutional repository with an embargo period attached.

8.5 Collaborate with astronomy partners

There is a dire need for more interaction and collaboration with astronomy library services partners across South Africa and Africa. Librarians from this continent need to be more engaged in astronomy internationally, making use of the networking opportunities to learn and share information, thus improving services to researchers. The SAAO, together with the Hartebeesport Radio Astronomy Observatory, will endeavour to encourage astronomy librarians across Africa to participate in meetings such as those organised through global initiatives such as the LISA conference.

9 Conclusion

The process of change may take some time as the SAAO LIS needs to first address the challenges of funding as well as accessibility limitations. It is however essential that priorities are set and a strategic plan is developed so that the process of LIS rejuvenation can commence.

References

[1] NRF: Annual performance plan 2012–2015, (2014), http://www.nrf.ac.za/document/annual-performance-plan-2012-2015

[2] SAAO: History of the South African Astronomical Observatory, (2014), http://www.saao.ac.za/about/history/

[3] Friends of the Observatory, (2014), https://friends.saao.ac.za/

[4] M. Gomez, *IAC library: Some Challenges.* in *21st Century Astronomy Librarianship, From New Ideas to Action*, edited by Eva Isaksson, Jill Lagerstrom, András Holl, and Nirupama Bawdekar (2010), Vol. 433 of Astronomical Society of the Pacific Conference Series, p. 47–54. Library and Information Services in Astronomy VI. ASP Conference Series, p. 47–54.

D. Bordelon, R. Cortés, L. Malaver and D. Depootere (Photo: M. Wenger)

 https://doi.org/10.1051/epjconf/201818609003

Managing Contributions to the Unified Astronomy Thesaurus

Katie Frey[1,*], *Sarah* Weissman[2], *Barbara* Kern[3], *Jill* Lagerstrom[2], *Josh* Peek[2], and *Alberto* Accomazzi[4]

[1]*Wolbach Library, Harvard-Smithsonian Center for Astrophysics, USA*
[2]*Space Telescope Science Institute, USA*
[3]*Science Libraries, University of Chicago, USA*
[4]*SAO/NASA Astrophysics Data System, USA*

Abstract. The Unified Astronomy Thesaurus (UAT) project managers have long defined the UAT as "an open, interoperable, and community-supported thesaurus." How do we solicit the detailed, comprehensive, and consistent community feedback that is required to keep the UAT relevant? The Steering Committee for the UAT has developed a visual organizational tool that lets reviewers suggest new concepts and restructure the existing hierarchy. Researchers and librarians can use this "Sorting Tool" to submit contributions and feedback to the UAT. The UAT Curator adds feedback to the UAT's GitHub Issues, which allows for tracking, searching, and referencing the suggestions. The UAT Curator implements the accepted suggestions, preparing to include them in a future release of the Unified Astronomy Thesaurus. This process of continual improvement ensures that the UAT project remains community supported.

1 A Formal Vocabulary for Astronomy

One of the guiding principles for the Unified Astronomy Thesaurus (UAT) is the understanding that labels are important. If something does not have a label, you cannot talk about it. To make matters even more complicated, different perspectives on a topic can lead to very different labels and therefore a different understanding. It is equally important to realize that the organization and relationships between labels also drive how a subject is understood and discussed. The field of Astronomy has a highly publicized example of the impact that a change in understanding can have on a field of research. As scientists' understanding of the word "planet" became more refined and specific, the International Astronomical Union (IAU) realized it had to reclassify Pluto as a "dwarf planet." This new label for Pluto has fundamentally altered the way people everywhere, not just scientists, understand the Solar System. This only serves to emphasize that in order to discuss a topic cogently, there has to be a mutual basis of understanding, a common ground from which to grow.

A thesaurus provides a consistent set of keywords that classify, describe, and map the understanding of a particular topic or subject. It builds a structure that can be mapped and navigated, leading from one concept to another in a logical way. It organizes the concepts into similar groups, adding additional context to help clarify the meaning of any particular concept.

*e-mail: kfrey@cfa.harvard.edu ORCID: 0000-0001-9891-4465

Many service providers and stakeholders in astronomy have expressed interest in embedding the UAT into a variety of applications, which means that the vocabulary will act as a bridge between these systems. We anticipate use of the UAT increasing and believe that the UAT will underpin and inspire a new range of cross-system data-sharing applications by creating a formalized source of vocabulary words. Using the same set of concepts to describe datasets, instruments, and astronomical objects, it enables an increased level of connectivity between data and content, an ability to enhance literature searching and indexing, and a huge improvement to discoverability.

2 Beyond the Astronomical Subject Keywords

The Astronomical Subject Keywords initiative ([1]) is one of the most recent efforts to build a unified list of keywords across publishers and publications in astronomy. In fact, this list has been approved by editors from many of the premier astronomy journals, who all saw the inherent value in sharing keywords in order to enhance findability. However, this keyword list is currently stagnant, having not received an update since 2013. Similarly, other previous efforts to create comprehensive keyword lists have stalled: the IAU Thesaurus ([2]) was last updated in 1995, the IVOA Thesaurus ([3]) was last updated in 2009, and the Physics and Astronomy Classification Scheme ([4]) was officially retired in 2010.

The UAT began as a direct result of efforts by the Institute of Physics and the American Institute of Physics. Independently, these two publishers recognized that current keyword lists, in addition to being woefully out of date, were also not in a format suitable for new linked data applications. By coincidence, each publisher began working with Access Innovations, a content and knowledge management business, to update, modernize, and improve their keyword lists. When each learned of the other's efforts, the two publishers decided to work together, merging their projects and donating the results to the American Astronomical Society with the intention that the thesaurus be an open and community supported resource. This thesaurus, referred to as the beta version of the UAT, included 1910 concepts, 15 top-level concepts, and a maximum depth of 12 levels.

A more detailed accounting of the provenance and history of efforts that lead up to the Unified Astronomy Thesaurus is available in a paper by Alberto Accomazzi, et all from 2014 ([5]).

3 Major Updates to the Unified Astronomy Thesaurus

Although the beta version of the UAT was a huge improvement over the older and no longer supported vocabularies, it had many flaws that needed to be addressed. The original concepts were serviceable, but not organized in the way that made logical sense to researchers in the field of astronomy and astrophysics. Top-level categories in the original beta version included "Astronomical objects," "Lunar physics," and "Positional astronomy," none of which are major fields of study today.

One of the first major undertakings was to align the top-level categories of the UAT with the Divisions of the International Astronomical Union, which are widely recognized as sub-disciplines within astronomy and astrophysics. This process involved a small committee of librarians and researchers who evaluated the structure of the UAT on a broad and large-scale basis, merging related top-level categories, promoting subsections to top-level categories, and enacted other overarching changes. After the bones of the reorganization were in place, and the Steering Committee felt comfortable with the overall structure, we quickly realized that the UAT needed detailed analysis on a concept-by-concept basis.

Since the UAT still had over 1900 concepts at this time, we worked with researchers to divide the UAT into logical and discrete subsections, which could be easily examined in detail. We ended up

Figure 1. Partial image of a hand drawn hierarchy, describing the relationship between concepts under "Compact objects."

with 40 subsections across the 11 top-level categories, each of which had about 50 to 100 concepts. Librarians participating in the UAT project from many institutions across the United States volunteered to take on the task of working with researchers to evaluate each subsection. They examined the subsection one concept at a time, exploring the validity and usefulness of each concept, along with the inter-relationships between the concepts in their subsection.

The feedback we received from this exercise was extremely varied. Some returned with a table listing each concept, its validity, comments regarding how it would connect to other concepts in the section, and sometimes a short list of new concepts that should be included on the topic. Others returned with hand drawn thought diagrams (see Fig. 1), explicitly describing the relationships between concepts in a field of study. Still others used an early version of the Sorting Tool to return a structured, but barebones, list of suggested changes. Finally, we also received suggestions in prose form, consisting of written paragraphs describing how a section of the UAT should be modified to more accurately describe its field of astronomy.

Based on this feedback, major changes were made to the substructure of each top-level category, but we were happy to see that our high-level reorganization remained intact. By the end of this huge undertaking, we removed about 300 concepts and added approximately 200 new concepts. We also learned a lot about the kinds of feedback we would receive and the kinds of tools our users preferred to use.

The result of this entire process was version 1.0 of the Unified Astronomy Thesaurus, released on December 23, 2015. It included 1836 concepts, with 11 top concepts, a depth of 10 levels, and 329 related concept links. True to our intent, the UAT was released in RDF/XML, using format standards compatible with linked data so that it could be ingested and used by applications that also followed those standards.

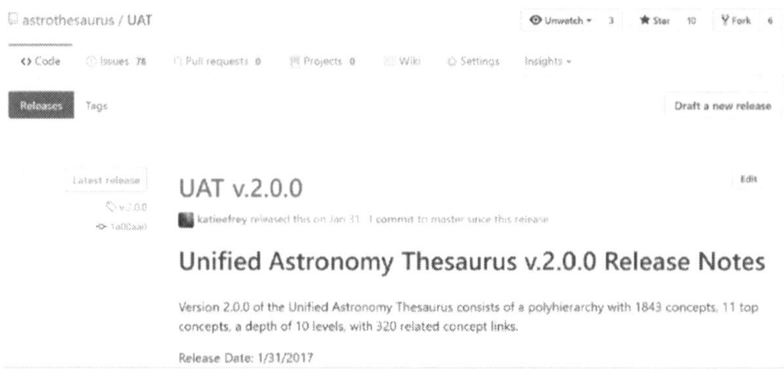

Figure 2. The current version of the Unified Astronomy Thesaurus is available on GitHub.

We decided to experiment with descriptive identifiers and released a minor version update (version 1.1.0) about six months later that also re-standardized the Uniform Resource Identifiers (URI). It was argued that descriptive identifiers would help to make the UAT more human-readable; i.e. a user would know just by looking at the URI which concept was being referred to. However, we found descriptive identifiers to be too rigid for the amount of changes and updates still being made to the UAT. We reverted to numeric identifiers in the version 2.0.0 release on January 31, 2017.

The version 2.0.0 release also included an attempt to add context, clarity, and consistency to a large number of concept labels. For example, instead of just using a label such as "A dwarf," we extended the label to "A dwarf stars," likewise "Double-mode Cepheids" became "Double-mode Cepheid variable stars," and so forth. These changes added additional information to each concept so that they could stand without the context of the UAT hierarchy, allowing them to be more useful as keyword tags. During this update we also discovered and removed 9 duplicate concepts, added 16 new concepts, and moved a few concepts from one parent to another. As of this writing, the version 2.0.0 of the Unified Astronomy Thesaurus is the current version and it includes 1843 concepts, with 11 top concepts, a depth of 10 levels, and 320 related concept links.

4 Contributing to the Unified Astronomy Thesaurus

As described above, the librarians and researchers often used a web-based visualization tool to explore the hierarchy of the UAT. It was a popular tool for analyzing the structure of the UAT since it easily displays the relationships between concepts, allowing researchers to understand how each concept fit into the greater context of the Thesaurus.

During the course of updating the UAT from beta to version 2.0.0, we made extensive updates to this web-based visualization tool. Instead of simply being a way to browse through the Thesaurus, it can now be used to directly submit suggestions and feedback. Upon launching the Sorting Tool ([6]), as we now call it, the user selects a specific branch of the UAT to work on. Then, as with the original visualization tool, they may click open a concept node to view its child concepts. Those concepts can also be expanded to show their children, and so on. The entire hierarchy of a branch can be displayed this way, within the limits of the display device. The extended capabilities now allow a user to click and drag a concept, moving it from one node to another. Changes are immediately visible on the screen, so the user can get a sense for the full context and implications of their suggestions.

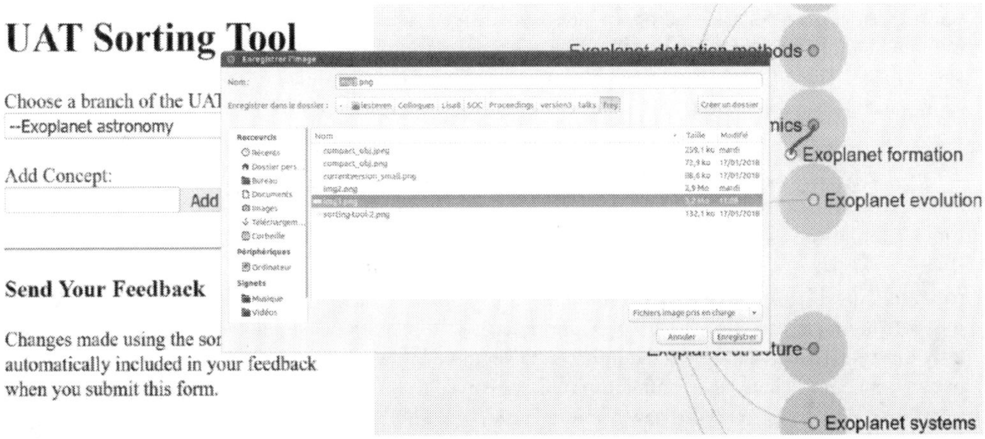

Figure 3. Example of the UAT Sorting Tool in use, concept "Exoplanet formation" is being moved to "Exoplanet dynamics."

Should a concept be deemed unnecessary, it can be moved to the "orphans" node, indicating that the user believes that this concept does not belong in this particular section of the UAT and should be considered for deprecation. Finally, by using the "Add Concept" input box, the new Sorting Tool lets users suggest new concepts. The new concepts will appear on the "orphans" node, and can then be dragged wherever in the hierarchy a user thinks they may belong. (see Fig. 3)

Once a user has finished re-arranging the concepts and has made all of the suggestions that they believe are necessary, they are asked to include some personal information, such as name, email address, and institution. We ask for this information so that we may verify the validity of the suggestions and reach out for questions if needed. A radical suggestion sent from a respected specialist in the field would garner more weight than a similar suggestion from an unknown user. Lastly, there is also a space provided for additional comments, should the user wish to make note of anything in particular or give an explanation for their suggestions. At that point they simply need to click the "submit" button and an email is sent to the Curator for the Unified Astronomy Thesaurus.

5 Tracking Suggestions to the Unified Astronomy Thesaurus

Throughout each revision of the UAT, we have been accepting and incorporating feedback, most of which was received via email. In addition to feedback generated by the Sorting Tool, we would also receive unstructured emails, often containing multiple suggestions. Although many of these suggestions were clear and easily incorporated, some suggestions were complicated, requiring further thought and consideration. Tracking which suggestions from which emails had been enacted and which were still pending quickly became very difficult. Additionally, by holding the suggestions in the personal email inbox of the UAT Curator, we noticed that we were essentially creating a "dark archive" of feedback that only one person could see, act upon, or reference.

Since we had already been using GitHub to publish release versions of the UAT, we decided to use GitHub Issues, the associated platform for managing suggestions for software updates, as our web-based platform for tracking feedback. We opened the first Issue on GitHub on November 18, 2016 by copying the contents of emailed suggestion into the web-based platform. One benefit of using GitHub was that each suggestion now has an open discussion forum. Not only can users search Issues to see if their feedback has already been submitted, they can create an account to add to the discussion. GitHub also has functionality to collect Issues together into a "milestone," which can then be linked to a release. This helps to create and maintain a log of changes, showing the provenance of the UAT as new versions are released.

6 The Future of the Unified Astronomy Thesaurus

In addition to the pending feedback that has already been received, the Curator for the Unified Astronomy Thesaurus has identified concrete areas that need improvement in order to move the UAT forward. Specifically, the UAT has some obvious gaps in content surrounding the topics of publishing, astronomical modeling, and software. Each of these subjects stands out as something that will only become more important in the future, and yet none are well covered in the UAT. Work has already begun on a comprehensive examination of these topics to draw out relevant concepts that can be used as a starting point for building new sections within the UAT.

To make full use of its linked data capabilities, we need to build connections between the UAT and other existing vocabularies. These would include general linked data sources such as Wikidata, specific vocabularies such as the Space Object Behavior Sciences taxonomy, helmed by the University of Arizona, and a potential taxonomy of astronomical instrumentation.

Another goal for improving the UAT is to include definitions and examples for concepts that are new to the field of astronomy or somehow ambiguous.

As for soliciting feedback in the future, GitHub Issues has created a fantastic springboard for launching conversations. Many of the suggestions are not trivial, meaning that to accept the change as stated might have broader implications over the UAT as a whole. These things require conversation, discussion, and contextualization. As we have learned from our work on the Unified Astronomy Thesaurus over the last few years, the most comprehensive feedback we have received came out of these types of discussions. To that end, we will be piloting focus groups with the aim of hashing out some of the thornier suggestions for the UAT. It is our expectation that these discussions will yield not only reasonable solutions, but also additional suggestions and feedback which will need to loop back into future focus groups.

In order to remain relevant as new discoveries are made in astronomy, the UAT must be considered a continual work in progress.

References

[1] *Astronomical Subject Keywords* (2013), `http://journals.aas.org/authors/keywords2013.html`

[2] *IAU Thesaurus* (1995), `http://www.mso.anu.edu.au/library/thesaurus`

[3] *IVOA Thesaurus* (2009), `http://www.astro.physik.uni-goettingen.de/~hessman/rdf/IVOAT/index.html`

[4] *Physics Astronomy Classification Scheme* (2010), `http://journals.aps.org/PACS`

[5] A. Accomazzi, N. Gray, C. Erdmann, C. Biemesderfer, K. Frey, J. Soles, *The Unified Astronomy Thesaurus.* (2014), Vol. 485 of *ASP Conference Series, vol. 485*

[6] *The UAT Sorting Tool* (2017), `http://uat.altbibl.io/`

EPJ Web of Conferences **186**, 09004 (2018) https://doi.org/10.1051/epjconf/201818609004
LISA VIII

A technologist view of the next decade evolution of the librarian work environment

André Schaaff[1,*] and *Marc* Wenger[1]

[1] *Université de Strasbourg, CNRS, Observatoire astronomique de Strasbourg, UMR 7550, F-67000 Strasbourg, France*

Abstract. The work environment has deeply evolved in recent decades with the generalisation of IT in terms of hardware, online resources and software. Librarians do not escape this movement and their working environment is becoming essentially digital (databases, online publications, Wikis, specialised software, etc.). With the Big Data era, new tools will be available, implementing artificial intelligence, text mining, machine learning, etc. Most of these technologies already exist but they will become widespread and strongly impact our ways of working. The development of social networks that are "business" oriented will also have an increasing influence. In this context, it is interesting to reflect on how the work environment of librarians will evolve. Maintaining interest in the daily work is fundamental and over-automation is not desirable. It is imperative to keep the human-driven factor. We draw on state of the art new technologies which impact their work, and initiate a discussion about how to integrate them while preserving their expertise.

1 Introduction

Anticipating what the future will be when it comes to the professional development of people is not easy. One must convince people who are in disbelief and will tell you that this is not possible. It is therefore necessary to argue, and above all not to forget past events, putting them forward to show that what you expect is not always a utopia. To begin with, we draw on an inventory of the main evolutions, in terms of volume and capacity, for data storage, calculation power, and communication and exchanges that have been boosted by the advent of the Internet and the Web. These technological evolutions have gradually built up an extremely elaborate working environment, which has been shaken by the new possibilities offered by increasingly intelligent tools based on machine/deep learning. There is a strong question about the gradual replacement of many trades with intelligent tools that can adapt and produce results that are identical to those of humans, but without them.

*e-mail: andre.schaaff@astro.unistra.fr ORCID: 0000-0002-8790-7048

2 Foreword

2.1 Why a decade ?

A decade is a reasonable duration. It is possible to go further on a purely technical level, but for aspects involving the human factor (acceptance of sometime radical evolutions) it is more complex and going beyond becomes random.

2.2 A technologist view ?

A technologist is often seen as having the following considerations:

- A geek loving science fiction
- A supporter of extreme automation
- The human aspect is not the most important
- Technology always first

We assume to have an asymptomatic approach against these considerations.

3 Librarians and documentalists in astronomy

According to our daily experience of working with librarians and documentalists, these jobs appear to us as having the following main skills:

- An intellectual job with a high level of general knowledge
- A sharp knowledge of the scientific field
- An ease with computer tools including scripting
- A strong interaction with scientists and computer engineers [1]
- A strong adaptability to change

4 Past decades

It is essential to review the main developments of the last decades before giving any indications of evolution. Computer science is a relatively young and constantly evolving field. This evolution has for a long time been almost purely technological and closely linked to that of physics and electronics. This has made it possible to miniaturize the components and consequently to increase the performances both in terms of calculation power and storage.

4.1 Storage revolution

In 2017 it seems quite normal to have, in the pocket, a smartphone with 128GB of memory, to store on it many photographs, to listen to hundreds of hours of music and many other things. It is also possible to store 1 TB of data on a USB key that weighs a few grams. At the time of the perforated cards (example: IBM punched card [2] capacity was 80 columns * 12 rows), the same volume of storage required an entire hangar. Figure 1 gives an example of an astronomical catalogue available on microfiche in 1977. Nowadays the online VizieR service distributes more than 16,000 catalogues.

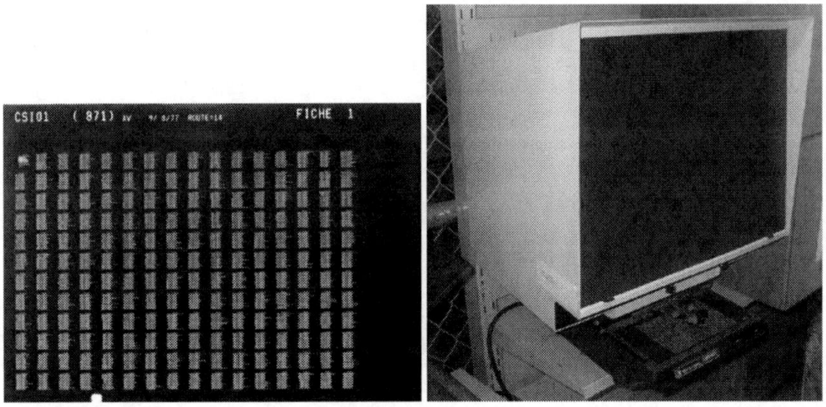

Figure 1. CSI catalogue on microfiche and a screen to read it

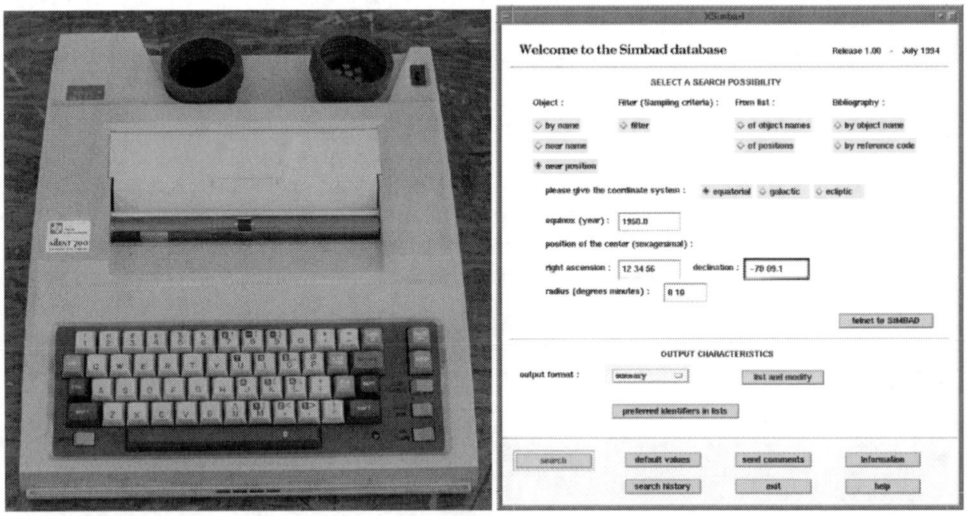

Figure 2. Accessing Simbad in 1981 and through the Web in 1994

4.2 Calculation revolution

The calculation has followed the same curve as the storage. UNIVAC 1 (1951) weighed 7.3 tons and had a calculation power of 0.001905 MIPS (Million Instructions Per Second). A mid-range Smartphone has a computing power of about 100,000 MIPS, and weighs 150 grams.

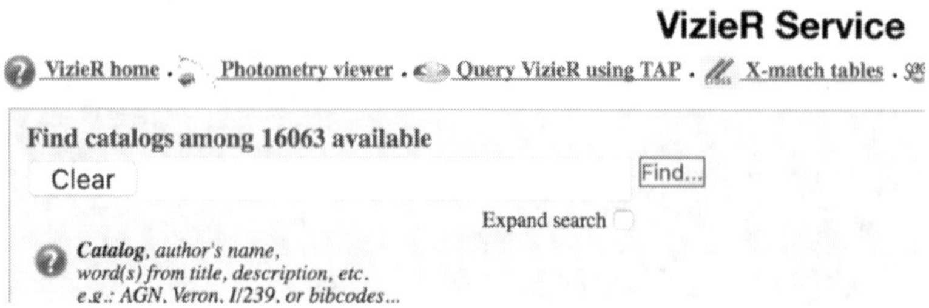

Figure 3. VizieR nowadays, a simple field to query 16,000 catalogues !

4.3 Communication and exchange revolution: Internet as a catalyst of the chain reaction... and the Web

The advent of the Internet and especially of the Web was the starting point of a new paradigm concerning the distribution of data. In August 1988, SIMBAD demos were the first to use the Internet in France. Figure 2 shows how Simbad was accessed before and after this date. In 2017, the VizieR Web page (Figure 3) offers to the user the possibility of downloading a part of a catalogue, to cross-match columns from several catalogues, etc. A work around interoperability is ongoing to make possible access to data coming from several surveys [3] through simple Web pages.

4.4 The content

We had mainly technical evolutions but technical developments are not sufficient on their own. For example 3D (Movies and TV) does not have enough content and is not seen as really better by the customers. Another example, Immersive 3D has a lot of possibilities, but it is unknown if the content will really follow. The last decades are a cemetery of others, for example object databases, etc.). The content is wanted and its added value.

5 The wave

Data is the new Eldorado [4] and this vision is strongly supported by politicians. The existing tools are often not adapted to manage such volume of data. It becomes mandatory to do more, to do it better and to do it quickly.

6 Flagships of the next decade

The next decade will be marked by an unprecedented deluge of data and by the return of artificial intelligence.

The keywords of the decade:

- Big Data, Open Data, Linked Data: everywhere
- A.I., Neuronal networks, Deep Learning, Machine Learning, robots, etc.: Chess (1997), Go (2016), Pocker (2017) battles
- "Professional" social networks

7 What's going on now ?

7.1 What we have

- Large storage and computing facilities
- Fast networks
- A large range of devices to interact, visualize
- A lot of data

We will have more and more intelligent tools able to do an increasing part of our work.

Therefore, is it too late? Will Catherine, Emmanuelle, Evelyne and Mihaela be replaced by Watson (IBM) [5] ?

=> No! [6]

7.2 Intelligent tools

Intelligent tools have success rates. Is 96 percent (even 99 percent) acceptable? Human added value lies in the part that aims to reach 100 percent and their work will therefore be increasingly shifted towards expertise. The volume of data and publicationsis constantly increasing and a few percent can represent a lot of work for humans, and probably not the less interesting.

8 So, what

In this part we give a few ideas which should be pillers of their work environment in a near future.

8.1 Helping to extract astronomical object names from papers

DJIN2, the new version of DJIN (Figure 4) is becoming an (internal) online tool implementing new technologies like GROBID [7] and is able to capture documentalist's decisions. GROBID uses Machine Learning to improve its recognition of the object names. DJIN2 is an example of a tool which helps documentalists every day and which learns every day to provide a better service.

8.2 Virtual Research Environment

A VRE is not only a nice user interface to access tools and services, it must be seen as an environment providing a well designed and customisable user interface with a large range of tools, data services (both to store and access), computing facilities, and community networking. Many experiments are carried out all over the World; VRE4EIC, EVER-EST (H2020) in Europe, Science gateways in the USA, and Virtual Laboratories in Australia, etc.). A VRE is not only a technical solution for a better working environment but also a way to collaborate, share, preserve, etc. The Research Data Alliance is working in this framework through an Interest Group which aim to bring people from several communities together to exchange knowledge (technical problems to solve, interoperability between VREs, common efforts to enable, etc.). A VRE for librarians / documentalists should be experimented with.

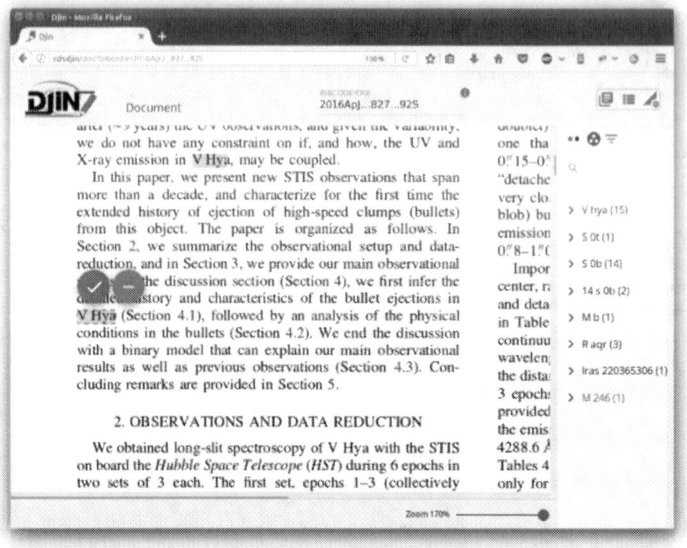

Figure 4. DJIN2, the next generation of DJIN

8.3 Professional social network

Facebook is developing Workplace [8], a social network for professionals. A social network for astronomy librarians and documentalists should develop for the exchange of information and good practices. Astronomy is a relatively small community and this will help tofacilitate. It is a way for "working together" at a larger scale.

8.4 Collaborative tools

A good way to develop expertise is to exchange information with other people in the same domain. A number of collaborative tools are available like Wikis,and source code repositories (GitHub, Gitlab, etc.). Sharing and preserving both the data and the process are key actions.

9 Conclusion

Astronomy librarians and documentalists will themselves as flagships of the coming decade. They will work with tools capturing a part of their expertise, which is also a way to preserve their expertise. It will give them more time to focus on the complex parts and the opportunity to strengthen their expertise. A key point is certainly to develop collaborations with other people of the same domain through professional social networks. Astronomy is a relatively small area and this should not be a problem. Their added value will remain precious. This will remain true on the condition that they accompany the development of the new tools.

They have to coach "Watson" to help them!

References

[1] E. Perret, T. Boch, F. Bonnarel, C. Bot, M. Buga, M. Brouty, C. Bruneau, C. Brunet, L. Cambrésy, S. Derrière et al., *Working Together at CDS: The Symbiosis Between Astronomers, Documentalists, and IT Specialists*, in *Open Science at the Frontiers of Librarianship*, edited by A. Holl, S. Lesteven, D. Dietrich, A. Gasperini (2015), Vol. 492 of *Astronomical Society of the Pacific Conference Series*, p. 13

[2] IBM, *The IBM Punched card*, http://www-03.ibm.com/ibm/history/ibm100/us/en/icons/punchcard/ (2017), [Online; accessed 31-July-2017]

[3] A. Schaaff, *Data in Astronomy: From the Pipeline to the Virtual Observatory* (Springer Berlin Heidelberg, Berlin, Heidelberg, 2007), pp. 52–62, ISBN 978-3-540-77010-7, https://doi.org/10.1007/978-3-540-77010-7_6

[4] E. U., *How Europe can gain from the rising tide of scientific data* (European Commission, 2010), http://ec.europa.eu/information_society/newsroom/cf/document.cfm?action=display&doc_id=707

[5] IBM, *IBM Watson*, https://www.ibm.com/watson/ (2017), [Online; accessed 31-July-2017]

[6] H. Rachlin, The Behavior Analyst **35**, 1 (2012)

[7] P. Lopez, *GROBID: Combining Automatic Bibliographic Data Recognition and Term Extraction for Scholarship Publications* (Springer Berlin Heidelberg, Berlin, Heidelberg, 2009), pp. 473–474, ISBN 978-3-642-04346-8, https://doi.org/10.1007/978-3-642-04346-8_62

[8] Facebook, *Workplace*, https://www.facebook.com/workplace (2017), [Online; accessed 31-July-2017]

VOResource and DataCite: Converging Metadata Schemes

Markus Demleitner[1,*]

[1] *Universität Heidelberg, Zentrum für Astronomie, Astronomisches Rechen-Institut*

Abstract. For the past 10 years, VOResource, together with its extensions, has been the metadata schema used for service description and discovery in the Virtual Observatory (VO). We are currently evolving the core VOResource schema into a version 1.1, in particular to ensure smooth interoperability with the DataCite metadata kernel. In this contribution, we will give a brief overview of how and why VOResource has supplemented Dublin Core with "actionable" metadata, as well as how the current evolution will update it to provide easy paths for VO data providers into the DataCite world.

1 Introduction

The Virtual Observatory (VO) is a large network of astronomy-related data services and data collections, held together by commonly implemented protocols and a directory of services available. This directory is called the VO Registry and is maintained by an informal association of individuals from various institutions from around the world known as the Registry Working Group (RWG) of the IVOA (International Virtual Observatory Association).

The VO Registry conceptually is the collection of metadata records (currently roughly 17000). In the next section, we briefly illustrate the most common uses for this metadata collection.

The technical realisation of the VO Registry borrows heavily from digital library technology. In the third section, we discuss how the use cases motivate the additions the VO community has made to the basic technologies.

The RWG has a strong interest in maintaining interoperability with library systems and data directories outside of astronomy, in particular regarding establishing good practices of data citation. We are therefore currently adapting our main metadata schema to facilitate workflows involving both DataCite and VO components. The fourth section discusses these recent initiatives.

We conclude with considerations on what further measures can be taken to improve interoperability between VOResource and DataCite.

2 Common VO Registry Use Cases

Currently, most VO users encounter the Registry through in-application interfaces. For instance, the popular application TOPCAT [1] lets users select services to query by matching keywords against a user-selectable set of natural language fields; the actual Registry query is hidden from the user (Fig. 1 a). Another application, SPLAT [2], uses the Registry to discover all spectral services and

*e-mail: msdemlei@ari.uni-heidelberg.de

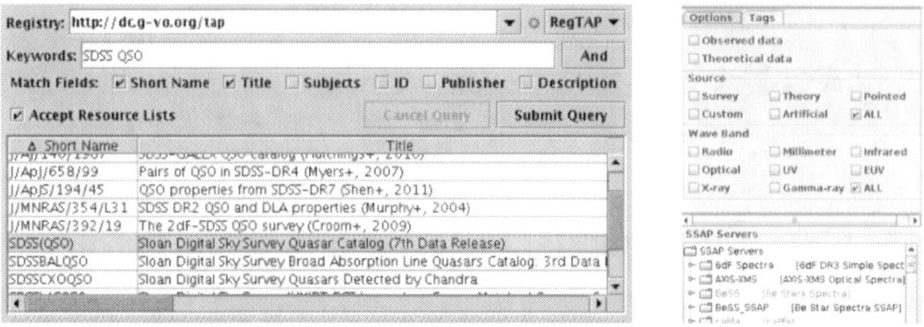

Figure 1. Examples for in-application registry interfaces. (a) TOPCAT, left; (b) SPLAT, right.

defaults to querying a subset of these services with one user interaction. This subset is selectable by defining categories of interest (Fig. 1 b). Again, end-users are shielded from the actual operation of the Registry.

The upcoming version 10 of the Aladin image processing application (cf. [3]) will feature a very interactive integrated resource discovery mode where, while scrolling over the sky, a list of services of interest to the user is updated by coverage and criteria like spectral region or the presence of certain kinds of data (proper motions, say) in the services' data schemas.

There are more conventional OPAC-like interfaces, for instance the Web Interface to the Relational Registry WIRR[1] or analogous services operated by ESA[2] or the STScI[3]. They are, however, used much less in comparison to in-application facilities.

The VO Registry has further uses in infrastructure maintenance. For instance, validators use Registry information to discover service endpoints to operate on, and to determine what queries will yield validatable results[4].

For a relatively recent review of Registry client interfaces, see [4].

3 Adaptation of Library Technology

The VO Registry is designed as a fairly conventional distributed system, where publishing registries are harvested by metadata aggregators using OAI-PMH. The relatively few extra rules are laid down in a standard called *Registry Interfaces* [5].

The central components typically are searchable registries. The normal browser-based OPACs one would typically see for libraries of texts are not really sufficient for the VO. For one, searchable registries must interoperate with the clients mentioned in sect. 2, which requires a well-defined interface in terms of both query parameters and the response schema. Also, the VO's complex metadata schema does not lend itself to simple web displays.

Therefore, in the VO the standard searchable registries support a non-library protocol called RegTAP [6], which is essentially a combination of a database schema (13 tables as of RegTAP 1.0) and the VO Table Access Protocol TAP [7], which defines how to transport database queries and their results. This obviously does not preclude the provision of searchable registries supporting non-standard interfaces, as illustrated by the web interfaces mentioned above.

[1] http://http://dc.zah.uni-heidelberg.de/wirr/q/ui/fixed
[2] http://registry.euro-vo.org/eurovo/
[3] http://nvo.stsci.edu/vor10/index.aspx
[4] e.g., https://heasarc.gsfc.nasa.gov/vo/validation/vresults.pl

As required by OAI-PMH, publishing registries in the VO can deliver their records in the OAI Dublin Core metadata schema. Service discovery, however, very typically involves queries like "Give me the access URLs of all services speaking version 1 of an image access protocol with infrared data" or "Give my database services with tables about quasars having redshifts and radio fluxes". To support such use cases, the VO Registry has the *ivo_vor* metadata schema, specified in the VOResource standard [8] and a set of extensions.

Features beyond Dublin Core include, in particular:

- capabilities – these define modes of data access, i.e., support for certain protocols; the protocols are defined using identifiers into the Registry itself. For instance, a service could say that its underlying data collection is available through a standard image search as well as through the Table Access Protocol mentioned above. In the latter capability the service could furthermore declare that the published table conforms to a pre-defined table schema ("data model"). This is actually how searchable registries are currently discovered, and is used in making collections of observational data products uniformly accessible with advanced query modes, including matches against bulk data uploaded by the client.
- interfaces – these link the capabilities to access URLs ("endpoints"). Additional metadata, such as names, types, and units for parameters supported on such an interface or the type of data returned, can be given. Interface metadata can also contain information on supported authentication methods, although only very few services actually give such metadata in current practice.
- tablesets – these describe the table structure of the data underlying a service, including name, type, unit, and physical characterisation of the columns that make up the tables. Tablesets enable the "discovery by physics" use case hinted at above.
- coverage in space, time, and spectrum – while one might suspect that this is very basic metadata for an astronomical data service, VO practice only now starts to properly support discovery by coverage. The reason for the sluggish definition of physical coverage in VO resource records is probably technological. Only the emergence of MOCs [9] – a compact, healpix-based representation of arbitrary shapes on a sphere – has made the definition and processing of spatial coverages straightforward enough for widespread adoption.

4 Convergence with DataCite

In particular with a view to facilitate consistent citation of data sources used in scholarly publications and the provenance of derived data, an overhaul of VOResource 1.0 was started. The focus was on enabling smooth interoperability between VOResource and the DataCite metadata kernel [10], which forms the basis of DOI registration. We expect to finish this update (VOResource 1.1) in 2017. It is currently under public review, the results of which are preserved in the IVOA's discussion system[5].

Major additions and changes in VOResource 1.1 with respect to DataCite interoperability include:

- records, creators, and contacts now support zero or more *altIdentifiers*. These will in general contain DOIs for records, and ORCIDs for persons, but the identifier type is intended to be inferred by the identifiers' URI schemes. The VO's own identifier scheme, *ivo:*, remains unchanged.
- VOResource had terms for *relationship* types that are incompatible with DataCite both in style (dash-joined instead of CamelCase) and content (e.g., DataCite has no served-by, VOResource's mirror-of roughly matches DataCite's IsIdenticalTo). We have taken over DataCite's terms but cannot entirely drop the old terms for backward compatibility. An RDF document now describes

[5] http://wiki.ivoa.net/twiki/bin/view/IVOA/VOResource11RFC

the relations between the terms and adds several terms not present in DataCite's vocabulary but necessary for VO operation (in particular isServiceFor and IsServedBy).

- Similarly, the terms for date roles were harmonised. This was mainly a matter of mapping word types, for instance VOResource's "creation" to DataCite's "Created".
- DataCite's model of *rights* and *rightsURI* was simply copied; VOResource 1.0's *rights* element was so restricted that it has not been used at all and thus could be replaced without adverse consequences.

As a result of these changes, an XSLT style sheet[6] can translate VOResource metadata to DataCite records almost ready for DOI minting. Additional logic is merely required to actually form the DOIs themselves.

5 Further Work

For more complete preservation of the contents of VOResource records in DataCite metadata, it would be desirable to

1. Add the extra terms *IsServiceFor* and *IsServedBy* to DataCite's *relationType* vocabulary.
2. Add the term *IVOID* to DataCite's *relatedIdentifierType* vocabulary. This is not strictly necessary, as in principle *URL* could be used with the nature of the identifier being inferred by the URI scheme. However, since the *ivo:* scheme is uncommon outside of the Virtual Observatory, an additional hint as to how the identifiers are to be resolved appears desirable.

At least for the forseeable future, an inclusion of extra VOResource metadata (capabilities, tablesets) into the DataCite metadata kernel is probably neither feasible – as seen by the size of the Registry XML schema documents compared to DataCite alone – nor desirable – as it seems unlikely generic DataCite metadata consumers will interpret the metadata any time soon.

On the VO side, the changes in VOResource 1.1 have to be translated into RegTAP's database schema. Work on this, in the form of an early working draft for RegTAP 1.1, is underway.

References

[1] M.B. Taylor, *TOPCAT & STIL: Starlink Table/VOTable Processing Software*, in *Astronomical Data Analysis Software and Systems XIV*, edited by P. Shopbell, M. Britton, R. Ebert (2005), Vol. 347 of *Astronomical Society of the Pacific Conference Series*, p. 29

[2] M. Castro-Neves, P.W. Draper, *SPLAT-VO: Spectral Analysis Tool for the Virtual Observatory* (2014), astrophysics Source Code Library, 1402.008

[3] F. Bonnarel, P. Fernique, O. Bienaymé, D. Egret, F. Genova, M. Louys, F. Ochsenbein, M. Wenger, J.G. Bartlett, Astronomy and Astrophysics Supplement **143**, 33 (2000)

[4] M. Demleitner, P. Harrison, M. Taylor, J. Normand, Astronomy and Computing **11**, 91 (2015)

[5] K. Benson, R. Plante, E. Auden, M. Graham, G. Greene, M. Hill, T. Linde, D. Morris, W. O'Mullane, G. Rixon et al., *IVOA Registry Interfaces Version 1.0*, IVOA Recommendation 04 November 2009 (2009), 1110.0513

[6] M. Demleitner, P. Harrison, M. Molinaro, G. Greene, T. Dower, M. Perdikeas, *IVOA Registry Relational Schema Version 1.0*, IVOA Recommendation 08 December 2014 (2014), 1510.02275

[6]https://volute.g-vo.org/svn/trunk/projects/registry/dois/vor-to-doi.xslt

[7] P. Dowler, G. Rixon, D. Tody, *Table Access Protocol Version 1.0*, IVOA Recommendation 27 March 2010 (2010), 1110.0497

[8] R. Plante, K. Benson, M. Graham, G. Greene, P. Harrison, G. Lemson, T. Linde, G. Rixon, A. Stébé, IVOA Registry Working Group, *VOResource: an XML Encoding Schema for Resource Metadata Version 1.03*, IVOA Recommendation 22 February 2008 (2008), 1110.0515

[9] P. Fernique, T. Boch, T. Donaldson, D. Durand, W. O'Mullane, M. Reinecke, M. Taylor, *MOC - HEALPix Multi-Order Coverage map Version 1.0*, IVOA Recommendation 02 June 2014 (2014), 1505.02937

[10] DataCite Metadata Working Group, *DataCite metadata schema – documentation for the publication and citation of research data version 4.0*, DataCite publication (2016), https://schema.datacite.org/meta/kernel-4.0/doc/DataCite-MetadataKernel_v4.0.pdf

Walking tour of the historical "Petite France" quarter (Photo: M. Wenger)

EPJ Web of Conferences **186**, 10002 (2018) https://doi.org/10.1051/epjconf/201818610002
LISA VIII

Innovative or sustainable? Future of the IBVS

András Holl[1,*], *László* Molnár[2], *László* Szabados[2], *Emese* Plachy[2], *Evelin* Bányai[3], and *Mihály* Váradi[2]

[1] *Library and Information Centre, Hungarian Academy of Sciences, Budapest, Hungary*
[2] *Konkoly Observatory, Research Centre for Astronomy and Earth Sciences, Hungarian Academy of Sciences, Budapest, Hungary*
[3] *Dept. of Physics of Complex Systems, Eötvös Loránd University, Budapest, Hungary*

Abstract. Long-term archiving of scientific information is a well known challenge. We discuss the issue of preservation for journals and their enhanced versions. We use the Information Bulletin on Variable Stars as an example. This express journal, created to satisfy the need of researchers for rapid dissemination of discoveries and results has been developed to become an enhanced journal, with several data journal features. However, the original software developed for the enhanced web version of the journal is not sustainable for the long term. We discuss the possibilities for moving the journal to a standard open source platform, and the trade-offs of the process. Long-term preservation of content, including data, is addressed as well.

1 Introduction

The Information Bulletin on Variable Stars (IBVS) is a small express journal focused on the narrow field of astrophysics: variable star research. Established as a Bulletin, it became a pioneer on the web, and became an enhanced journal.

The enhanced features of IBVS were innovative, some of them even unique, but as we had to realize, unsustainable on the long term. As IBVS is a micro journal, without extensive financial resources, and a technical editor doing technological development in his free time, it become impossible to migrate all the existing features to a new platform, when it became necessary. Editors of the IBVS decided to use an OpenAIRE grant for the renewal of the journal, and keeping some key features, abandoning certain enhanced functions, and gaining some new functionalities – like OAI-PMH compliance – the journal will move to Open Journal Systems.

2 Preservation

Preservation of digital or digitized publications, books, journal articles or volumes, needs consideration, attention, resources, but certainly is possible. Institutional or subject repositories, JSTOR and/or ADS, could provide solutions. In the case of IBVS, as it is a data journal as well, preservation of data files should be given consideration too. IBVS contributed to the preservation of the Archives

*e-mail: holl.andras@konyvtar.mta.hu ORCID: 0000-0002-6873-3425

of Unpublished Photometric Observations by digitizing some of the material which was available on paper, such as hand-written, typed or printed by early computers (note, that the data was passed on in printout form, not on magnetic tape). The data IBVS deals with is mostly photometry, which could be well represented in textual form, such as columns of numbers. These are data files we believe we could preserve for the future. Other data types are less frequently attached to IBVS papers, for example spectra are sometimes stored in textual tables and images are in PostScript or JPEG. These do not represent a very difficult problem. There are some FITS files – these are standard enough.

What about the complex services (like SIMBAD, Aladin)? IBVS is an enhanced journal, with many fancy features. Is it possible to preserve the Aladin display service IBVS uses for photometric sequences or finding charts? The advanced search features with author or object name expansion using ADS, GCVS or Sesame? These features depend on technologies, and on other parties. They might be re-written using a different programming language, but it requires a lot of effort. SIMBAD, for instance, had many successive generations – CDS was able to migrate it, preserving some important aspects, to new platforms [2]. For IBVS, we had to recognize that we have no means for re-building the enhanced features on a new platform, and might have to lose some of them.

3 History of IBVS

IBVS was conceived during the business meeting of the Commission 27 (Variable Stars) of the International Astronomical Union (IAU) held in Berkeley during the General Assembly of this international organization in 1961. In this meeting, participants accepted László Detre's (then Director of the Konkoly Observatory) suggestion concerning setting-up a new publication for rapid dissemination of information important for the variable star community. At that time, researchers of variable stars needed a quicker way of communicating new discoveries, results and announcements than it was possible through the existing journals. The first issue was published under Detre's editorship already in 1961. The IBVS has been edited and published at the Konkoly Observatory ever since. In this 'classical' era, individual issues of the Bulletin were published within a month upon receiving the manuscript.

The IBVS was most popular in the 1970-1980s: about 150-180 issues were published annually. During the almost sixty years of publishing the IBVS, the average production rate was about 110 issues/year.

During the General Assembly of the IAU held in Buenos Aires in 1991, Commission 42 (Close Binary Stars) joined Commission 27 as the sponsoring body of the IBVS. Forming an Editorial Board for the IBVS was also initiated there. Refereeing of the submitted manuscripts by international experts was introduced in 1995.

The electronic edition appeared in 1994, first using the web as a distribution platform of the PostScript issues, then in 2000 it become an enhanced journal. The print edition ceased in 2011 with issue No. 6000 being the last printed one. Since 2011 the journal has been published only electronically.

Following the restructuring of the IAU, the IBVS is now published on behalf of the Commissions G1 and G4 of the International Astronomical Union.

4 Features of the old electronic edition

The enhanced features of the journal were discussed by Andras Holl in 2012 [1]. Notable is the use of third-party tools and services for visualization of maps and the author and object name searches. The basic idea behind the enhanced version was that the LaTeX source should contain every information needed, the HTML is to be generated from the LaTeX automatically.

The enhanced HTML version of IBVS was developed with the help of CDS, using technologies developed by them.

The Bulletin is also a data journal. Authors are encouraged to submit supporting data. Data papers, for example articles that communicate observational facts like moments of minimum brightness of eclipsing variables and discovery of a new variable, are published regularly. Data files do have their own metadata, and it is possible to search for datafiles by keyword or object name.

5 The OpenAIRE grant

OpenAIRE: a FP7-funded initiative to make EU-funded research (e.g., FP7, ERC) open-access, implementing the OA initiative of the European Commission. OA publishing will be a requirement for all H2020-funded projects, therefore OpenAIRE was also aimed at funding the development the underlying infrastructure of open access publishing. One of the projects of OpenAIRE was a call for OA journals that does not collect article processing charges from the authors either. This program provided funding for APC-free journals to implement various technical improvements in their publishing services.

We submitted a successful bid to renew the journal in 2017. Our main driver was the need to replace the outdated website, and to introduce an integrated, on-line editorial workflow instead of the current one. Managing submissions between emails, ftp transfers, and Dropbox directories inevitably led to delays and occasional confusion.

6 Basic features of the new IBVS

As a first step to renew the journal, we registered IBVS at CrossRef and started assigning DOIs to the articles in late 2016, with the help of the Library of the Hungarian Academy of Sciences (MTA). We also renewed our registration with the DOAJ, and will upload article-level metadata to their database.

We selected the new 3.0 version of the Open Journal Systems software, developed by the Public Knowledge Project, for the new engine of the journal. OJS provides an online submission, review, and editorial workflow system and a publishing platform for us. The public version of the journal will be hosted by the Library of the MTA.

OJS provides many desired features out of the box. These include: OAI-PMH harvestable metadata, management of author ORCIDs, metadata submission to Crossref to register DOIs for the publication, etc.

A requirement of the OpenAIRE bid was the introduction of XML-based article versions that are well suited for machine reading. OJS uses the Lens XML viewer produced by eLife Sciences. This presented us with the greatest challenge: to migrate the archive to OJS and convert the old papers into XML. More than 6000 articles and notes have been published in the journal so far. The first 4100 issues were published in print only and have been later digitized. The rest have been submitted and published electronically too.

7 Technologies used

OJS 3.0 has a plugin for displaying articles in a new, innovative way which makes reading articles online easier. This tool is called eLife Lens which was originally developed for the eLife Sciences webpage.

IBVS uses LaTeX files as source files for its manuscripts. To display the articles with Lens they need to be converted to JATS[1] XML format. There are several LaTeX to XML converters available

[1] Journal Article Tag Suite - special set of XML elements and attributes for tagging journal articles

like LaTeXML, Tralics, LXir, Pandoc, etc. However these were not good for our purposes since we need a converter which converts to JATS XML. Furthermore IBVS has several self defined macros. There is a Lua custom writer for Pandoc on github generating JATS XML – but unfortunately there are several issues with it (e.g not recognizing the IBVS macros, not handling some special characters like &; < to its XML/HTML entity value which totally messes up the final XML document).

We decided to write our own converter to produce the proper source files for the Lens viewer. The software is being developed in C#. Briefly it parses the document into a Document Object Model (DOM) object first, then writes out the document as a JATS XML file. The converter has 3 parts: the DOM which is independent of the other two parts, the parser and the XML writer both using the DOM part, but being independent of each other.

The DOM both reflects the structure of a LaTeX document including the special IBVS macros and partly the expected structure of the XML.

The converter uses the XMLWriter C# library to write out the document in XML format which takes care, by default, of special characters like & an < and raises an error if there is a problem with the number of the end tags, that is if it cannot resolve the problem automatically.

The most complex part – and hardest to develop – is the parser. It uses a series of regular expressions to clean the document from the unnecessary LaTeX macros and commands, convert the special LaTeX letters (greek, accented etc.) to a HTML friendly format, then parse the document in a descending manner. It gets more challenging at the TeX level where it has to take care of the LaTeX markups, inline math mode etc. The program is still under development but it already can partly convert our files (e.g. markup conversion missing, some of the IBVS macros not implemented.) It already has the advantage of converting multiple files at once, parse special IBVS macros, produce a viewable document in Lens. It will be easy to extend the functionality or write new methods in the future.

Not only the "modern" LaTeX sources are to be transformed to XML, but the plain text source (produced by Optical Character Recognition) of the first 4000 issues too. These texts have neither semantic structure, nor tags. We are developing Python code to collect the metadata of these issues and to mark up the text. The most we could hope for is a semi-automatic process to be developed for the conversion.

The archive issues — either converted from plain text or from LaTeX – need to be imported to the OJS system.

8 Difficulties encountered, results achieved, plans for the future

An unanticipated difficulty in delivering the improvements agreed on in the bid was the bugs we encountered with OJS. Some features, such as issue importing or various plugins did not work as we expected them to, and we had to wait for newer releases to appear with appropriate patches.

The most daunting task in renewing the journal is the conversion and migration of the archive. Given the large number of articles, manual conversion of the files is not feasible. However, automated conversion of the files is not straightforward either. The print and electronic era of the journal present different problems.

The print issues have been digitized and are currently available as scanned images and plain text files. The format of these text files can be wildly different as they closely follow the layouts of the original printed issues. Therefore, without strong cues in the files, even the recognition of the presence of tables or equations can be problematic, let alone automatically formatting them. LaTeX-based files present different problems. Development of the respective conversion algorithms are under way.

9 Conclusion – virtues of simple, widely used technologies

Likewise, the enhanced HTML version of IBVS is a complex machinery heavily dependent on technology. Maintenance and preservation of such services is costly, and as in the present case, might become unsustainable. We might be able to continue the enhanced HTML version of the journal for a while — surely till the new platform becomes fully operational, and maybe even further. But it cannot be rebuilt with new technology, keeping the full functionality presently available.

We should plan preservation of the information services we create, such as databases and journals, beforehand. We should designate a core set, and use a simple format, which depends on technology as little as possible. Preservation should not be an afterthought.

As for IBVS, the journal will continue to serve the researcher community, in a somewhat different form. We hope that the new features and design will please the readers, and that eventually many extra services can be rebuilt using different technology, such as OJS plugins.

Acknowledgement. A.H. acknowledges the help received from CDS for the creation of the HTML version of the journal. Francois Ochsenbein's ideas and code were essential.

References

[1] A. Holl, D-Lib Magazine **18**, 4/5, 4 (2012)

[2] M. Wenger, F. Ochsenbein, F. Bonnarel et al., *The SIMBAD Database: Lessons Learned from 30 years of Experience* in *Astronomical Data Analysis Software and Systems XV ASP Conference Serie*, edited by C. Gabriel, C. Arviset, D. Ponz, and E. Solano (2006), Vol. 351 of Astronomical Society of the Pacific Conference Series, p. 662–665

Château du Pourtalès

A. Simoneau, S. Prosser, and D. Bouquin

K. Soldanova, M.E. Jiménez and R. Svašková

Mr. de Young, S. Bennett, T. de Young,
G. Coetzer and R. Botha

A.K. Robertson, A. Holl and F. Genova

(Photos: M. Wenger and M.E. Jiménez)

EPJ Web of Conferences **186**, 10003 (2018) https://doi.org/10.1051/epjconf/201818610003
LISA VIII

Elevating MAST-Data Publications with Digital Object Identifiers (DOIs)

Jenny Novacescu[1,*], *Josh E.G.* Peek[1,**], *Sarah* Weissman[1,***], *Karen* Levay[1,****], *Scott* Fleming[1,†], *Elizabeth* Fraser[1,‡], and *Jonathan R.* Hargis[1,§]

[1] *Space Telescope Science Institute (STScI), 3700 San Martin Drive, Baltimore, MD 21218, USA*

Abstract. The use of digital object identifiers (DOIs) to identify data sets used in original research allows peer reviewers and journal editors to more easily validate research methods and verify results. Fellow astronomers can duplicate results or expand on initial findings when the precise data used in a research project are identified. Precise identification of data may allow archives and observatories to better understand how the community is accessing and combining its data to reach new scientific conclusions. Earlier studies have suggested that papers with linked data are more highly cited in the literature, providing motivation for authors to adopt more stringent and thorough data citation practices.

1 Background

Since late 2015, the Space Telescope Science Institute (STScI) and the American Astronomical Society (AAS) have been working together on a pilot project which provides submitting authors based at STScI a means to assign a digital object identifier (DOI) to the data set(s) referenced in original research articles.

The Space Telescope Science Institute, which is operated by AURA for NASA, was founded in 1982 to oversee the scientific and data archiving operations of the Hubble Space Telescope (HST) and will manage the flight and scientific operations of NASA's latest flagship mission, the James Webb Space Telescope (JWST), scheduled for launch in 2018. The Barbara A. Mikulski Archive for Space Telescopes (MAST) is located at STScI. MAST collects and provides access to astronomical data from over 20 missions, with a historical focus on scientifically related data sets in the optical, ultraviolet, and near-infrared parts of the spectrum. Archived data comes from missions including Hubble, Kepler/K2, and GALEX. TESS and JWST data will also be housed in MAST.

* e-mail: library@stsci.edu ORCID: 0000-0002-8523-015X
** e-mail: jegpeek@stsci.edu
*** e-mail: sweissman@stsci.edu
**** e-mail: klevay@stsci.edu
† e-mail: fleming@stsci.edu
‡ e-mail: fraser@stsci.edu
§ e-mail: jhargis@stsci.edu

AAS was chosen as the initial partner for this pilot project because a high percentage of papers which cite MAST and specifically HST data are published in AAS journals, including Astronomical Journal (AJ), Astrophysical Journal (ApJ), Astrophysical Journal Supplements (ApJS), and Astrophysical Journal Letters (ApJL). Of the over 15,000 papers identified since 1991 that use HST observational or archival data from MAST, over 55% were published in AJ or ApJ (including ApJS and ApJL). Archival data accounts for nearly two-thirds of all publications related to the Hubble Space Telescope.

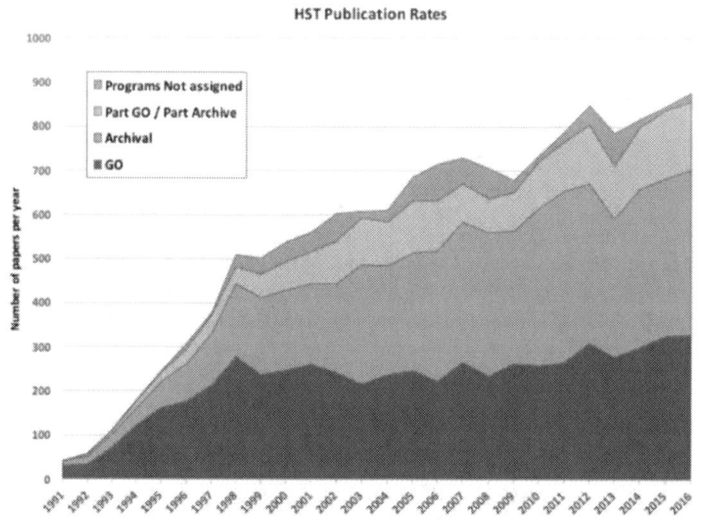

Figure 1. Hubble Space Telescope Publication Rates from 1991-2016, divided by guest observer (GO), archival (AR), combined (GO/AR), and Program not assigned [unidentifiable] data use.

Much of the text of this conference proceeding was derived or quoted in part from a 2017 unpublished, collaborative white paper available in Authorea by the named authors and AAS pilot project contributors. This white paper was drafted in preparation for the expansion of the pilot program and the LISA VIII presentation. Interested readers may contact the authors for the latest version of the working paper, *Digital Object Identifiers for Astronomical Data Archives and Journals: Some Initial Principles Developed at MAST*, which is continually updated.

2 Definitions

Digital object identifier (DOI): the American Psychological Association defines a DOI as "a unique alphanumeric string assigned by a registration agency (the International DOI Foundation) to identify content and provide a persistent link to its location on the Internet".[1] The International DOI Foundation further expands on this definition, noting that "A DOI is a Digital Object Identifier. That is a Digital Identifier of an Object, not an Identifier of a Digital Object... an actionable, interopera-

[1]What is a digital object identifier, or DOI? http://www.apastyle.org/learn/faqs/what-is-doi.aspx

ble, persistent link."[2] For further information on DOI conventions and applications in publishing and science, refer to the *Driven by DOI* brochure.

3 Goals and Objectives of the DOI Project

MAST and the STScI Library have worked collaboratively since the inception of both departments at STScI. In 2015 a DOI Working Group was formed consisting of staff from both departments, along with an associate astronomer with an interest in data citation. The Working Group has since expanded to the newly formed Data Sciences Mission Office (DSMO) at STScI. When the DOI Working Group was first created, MAST and the STScI Library recognized shared concerns with data citation in their work flows and bibliographic indexing efforts.

The pilot group was motivated to implement a DOI-based model due to earlier research by Pepe et al. (2014) [2] which showed static url data links decay over time. In this earlier study, it was found that 44% of data links from a decade earlier were broken.

A partnership with AAS was formed, and the following two goals were identified.

Primary goal: to provide authors who use MAST data a simple and elegant means to identify data analyzed in their original research publications. Doing so:

- Simplifies refereeship and data editing from the journal publisher's perspective
- Allows readers to verify, cite, and reproduce results reported, and build upon findings to produce new results
- Provides an accurate accounting of the types of data retrieved from an archive; this is important for metrics, internal reporting, and funding purposes
- Saves staff time at the bibliographer/library level and increases precision; the author as opposed to the bibliographer is in the best position to identify the data used

Secondary goal: To create a model for best practices in data identification and encourage other astronomical archives and our publishing partners to adopt the use of data DOIs in their own workflows.

Related to this second goal, STScI and AAS are operating on the premise that data identification will become increasingly important with the rise of survey telescopes and proliferation of astrophysical data, as outline by Zhang and Zhao (2015)[3].

4 Principles

The DOI-data set pilot was designed with the following mutual principles in mind for STScI and AAS:

4.1 Journal publishers and archives must actively solicit DOIs for data identification.

The data-hosting archive must provide a simple and elegant tool within its interface if authors are expected to use the service.

[2] Driven by DOI https://www.doi.org/driven_by_doi/DOI_Marketing_Brochure.pdf

4.2 Minimal integration is best. This model allows for interoperability among other publishers/archives in the future.

DOIs are passed between MAST and AAS by simple cut-and-paste into a webform. There is no handoff of data happening behind the scenes. This has the major benefit of easy federation, and no additional standards. The expectation is that in the future AAS Journals will work with other archives to integrate DOIs into articles; similarly MAST will partner with other journals. We could not see a motivation for a more complex system.

4.3 Fixed DOIs vs. custom DOIs should be made available.

Fixed DOIs refer to High-Level Science Products (HLSPs) such as catalogs, surveys, and entire Kepler quarters, which are housed in MAST. In many cases, specific large data sets are used as a whole in a manuscript. Authors have the ability to identify HLSPs using pre-assigned, fixed DOIs. Having a set of fixed DOIs for popular products eliminates the problem of minting multiple DOIs to refer back to the same observations. By assigning a consistent identifier in advance to an HLSP, MAST has given authors a means to provide a persistent link to the data and encourages accurate citation and proper acknowledgment.

In contrast, an author may also need to refer to sets of previously unrelated observations. The MAST DOI Portal tool[3], built by S. Weissman and T. Donaldson, allows users to concatenate observations and make a new DOI at will. Custom DOIs allow the author to aggregate observations and data sets within the entire MAST Portal which might otherwise appear unrelated. Authors are given flexibility to create one DOI for all observations analyzed in a research publication or multiple DOIs for different sets of observations.

4.4 Data DOIs are not first-class citable objects on their own.

While DOIs are often used to link to and identify first-class citable objects, e.g. journal articles, MAST and AAS are in agreement that data DOIs are not first-class citable objects on their own, and thus do not show up in the bibliography.

The logic behind this decision is that an author wishing to return to the data set for future study should cite the paper in which the data set was first identified and analyzed, along with the full context of the data use. Data set DOIs are thought of as "permalinks" for data, and are used to point to existing data in much the same way that simple URLs or DOIs are used to point to journal articles today.

4.5 DOIs refer to the described data set.

Because the intention of the DOI is to describe the data set analyzed in the paper, we do not force the DOI to be forever fixed in the case where the author made some kind of mistake in DOI generation. Our protocol is for MAST to check with the AAS before changing the content of the data set to which the DOI is linked, but we generally believe users should be allowed to edit their DOIs via a mediated process to match the content of their papers.

[3]MAST DOI Portal https://mast.stsci.edu/portal/Mashup/Clients/DOI/DOIPortal.html

5 DOI Minting Process During Paper Submission

In an effort to encourage more observatories and journal publishers to adopt DOI implementation, we are outlining the simple workflow we have established in consultation with the AAS.

5.1 Initiating the form

The author begins paper submission on the eJournal (EJ) Press website to submit to AAS titles. The EJPress submission form asks whether data from the MAST archive was used in the publication.

See further discussion below in *Pilot Outcomes and Lessons Learned* regarding how the question is phrased on the submission form. At this time, only submitting authors whose email domain ends in @stsci.edu are prompted. STScI is currently working with the AAS to expand the domains and invited institutions.

5.2 Determining eligibility

The author specifies whether MAST data was used.

"No" reroutes the author back to finish the paper submission process on EJPress. "Yes" routes the author to an MAST DOI Home page.[4]

5.3 Selecting the DOI type

From the MAST DOI Home page, authors are asked if they used:

1. a collection of specific, curated observations (custom DOI);
2. data from a High-Level Science Product (fixed DOI type);
3. a catalog, e.g., Kepler/GALEX (fixed DOI type); or
4. a large, clear sub-section of a catalog, e.g. a quarter of Kepler long cadence data (fixed DOI type).

Authors who select options 2, 3, or 4 are prompted to select a pre-assigned, fixed DOI(s) from a list. Authors who select option 1 are directed to the custom version of the MAST Portal which allows them to select observations used in their research. A future release of the MAST Portal may allow users access to their download history, but at this time users must re-create their search in order to locate the observations used in their research. As noted, authors have the liberty to mint a single DOI for all observations or a subset of DOIs for different sets of data.

5.4 Metadata assignment

When creating a custom DOI, the author submits basic metadata such as their name, manuscript title, and an optional description of their data. Other metadata, such as date created and data set ID, are auto-assigned.

STScI is investigating future uses and expansion of DOI metadata fields, though most relevant metadata about the individual observations such as instrument and wavelength are already stored in MAST and do not need to be replicated.

[4]MAST DOI Home https://archive.stsci.edu/doi/search/index.html

5.5 Form submission

Once the custom DOI(s) is created, the author is taken back to the EJPress submission form where they cut and paste the DOI(s) and complete the manscript submission. The author also receives an automated email with their information and summary of the DOI metadata.

At this point, the author has completed their end of the process.

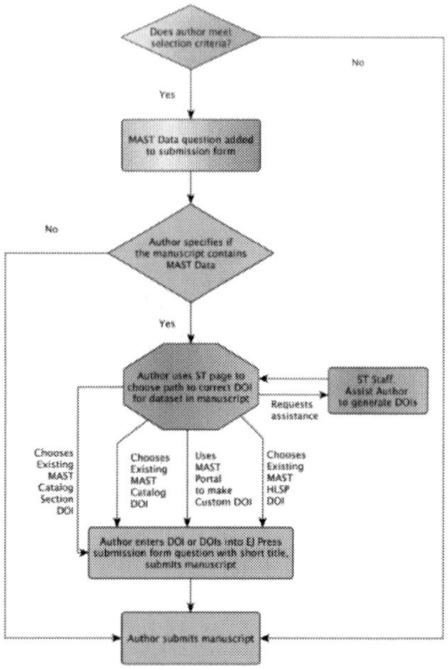

Figure 2. A workflow diagram for an author interacting with the EJPress submission form when submitting to a AAS journal. Green indicates EJPress site functions; yellow indicates author actions on the EJPress site; blue indicates actions on the MAST DOI site.

Another consideration in developing workflows between archives and publishers is the role of the journal publisher and/or data archive in reporting non-compliance. STScI and the AAS came to an agreement that AAS staff would initially track instances of non-compliance so they could report back to STScI. STScI is then responsible to follow up with authors to find out why they elected not to mint a custom DOI or select a fixed DOI, or were unable to do so. Tracking non-compliance at the start of the pilot is essential as it allows your institution to contact authors who were eligible to mint a DOI, but chose not to. Non-compliance also helps the institution find out what part of the process prevented or deterred the author from using the service so the user interface can be enhanced or simplified as needed. Was the initial question about data use unclear, for example? Were instructions confusing? What technical challenges did the submitting author experience when trying to create a DOI(s)? Starting with your own institute allows you to use local staff as a test bed for the DOI service, but you should keep in mind that internal staff may be more savvy at navigating your data archive than the general community.

6 Pilot Outcomes and Lessons Learned

Since the launch of the pilot in early 2016, fifteen STScI first-authored papers used the MAST DOI Portal to identify data sets and observations used. Two of the fifteen authors made use of HLSP pre-assigned DOIs and the remaining thirteen created custom DOIs. One of the thirteen custom DOIs was used to collate over 14,000 rows of data, demonstrating that this process allows for succinct citation of large and complex data sets. With marketing and user-assistance on the part of the local DOI Working Group, the overall compliance rate for eligible papers was 75%. Non-compliance occurred with less than five submissions. In these cases, STScI authors submitted AAS manuscripts that contained MAST data but elected not to use the DOI minting service.

Authors of the earlier papers were confused by the way the initial question was phrased on the EJPress submission form. Our original question was: *Does your manuscript* ***directly refer to data*** *from Hubble, Kepler, GALEX, IUE, or other data in MAST?*

Our revised question is now: *Does your manuscript* ***use or analyze data*** *from Hubble, Kepler, GALEX, IUE, or other data in MAST?*

Acknowledgments

The authors thank Jill Lagerstrom, former Chief Librarian at the Space Telescope Science Institute, for her ground work on the initial DOI planning, protoyping, and implementation. We wish to also thank our contacts at the American Astronomical Society: Julie Steffen, Director of Publishing; August (Gus) Muench, Journals Data Scientist; and Greg Schwarz, Journals Data Editor. We extend our appreciation to Tom Donaldson and Amanda Marrione in the Archive Sciences branch at STScI for their work developing and testing the DOI minting infrastructure in MAST and Randy Thompson in the Archive Sciences branch for his efforts on web integration.

This research has made use of NASA's Astrophysics Data System (ADS) Bibliographic Services.

References

[1] E.A. Henneken, A. Accomazzi, *Linking to Data: Effect on Citation Rates in Astronomy* in *Astronomical Data Analysis Software and Systems XXI*, edited by P. Ballester, D. Egret, and N.P.F. Lorente (2012), Vol. 461 of Astronomical Society of the Pacific Conference Series, p. 763–766
[2] A. Pepe, A. Goodman, A. Muench et al. , PLOSl One **9**, 104798 (2014)
[3] Y. Zhang, Y. Zhao, Data Science Journal **14**, 11 (2015)

EPJ Web of Conferences **186**, 11001 (2018) https://doi.org/10.1051/epjconf/201818611001
LISA VIII

AASTeX and the challenges of publishing in an increasingly interactive environment

Greg Schwarz[1,*]

[1] *Data Editor*
American Astronomical Society

Abstract. In 2016 the American Astronomical Society (AAS) released two new versions of its LaTeX classfile, AASTeX. These were the first changes in over 11 years and included many new features to enhance an author's ability to present their science in a format conducive to publishing in the AAS environment. While LaTeX is an excellent way to convey the written word, it lacks robust support for many desirable features including collaborative editing, large table support and figure interactivity. Bridging the gap between the old methods of writing and reviewing a printed manuscript and the new features the AAS has available and is working on implementing in a published HTML article is an ongoing challenge. This talk will highlight the new features of AASTeX and discuss how AAS publishing will move forward.

1 Introduction

LaTeX is a programming language for manuscript typesetting that is well suited to displaying mathematics, making it a natural fit for astronomy and astrophysics articles. In the early Internet, the American Astronomical Society (AAS) envisioned a future where authors could electronically submit a LaTeX manuscripts to save time and effort. To take advantage of this, the first LaTeX macros were written in 1989 for authors to format a manuscript for the AAS Journal's exact specifications. During the 1990s many additional versions were created both by the AAS and third parties. These LaTeX upgrades were coincident with the AAS's first electronic HTML journals in the mid 1990s. In 1999 the AAS publication board decided to push the envelope of HTML publishing by expanding beyond the then standard of text, tables and figures. The AAS Journals now offer many different data and narrative options. These distinctive HTML features give authors much more flexibility to control the narrative when describing their results. They also are very powerful for readers by allowing unique interactions with the data. Authors also have more ways to provide access to the data used in their articles. Since these attributes only appear in the final HTML article it is challenging to present these items with only LaTeX manuscript to collaborators when composing the paper and also to the science editor and reviewer during peer review.

This article discusses the latest upgrades to AASTeX, the custom designed classfile for LaTeX used by all the AAS Journals. While the upgrade provides a richer template for authors it lacks

*e-mail: greg.schwarz@aas.org ORCID:0000-0002-0786-7307

much of the functionality that is only available in a web environment. The latter portion of this article describes the available HTML only attributes and ends with a discussion on where the AAS publishing will go in the future.

2 New AASTeX features

In early 2016 the AAS released AASTeX v6.0. It was the first new classfile update in 11 years. Version v6.1 was released in early 2017. This list is a summary of some of the new features in the latest versions.

1. Based on the popular emulateapj classfile.
2. Front end changes including:
   (a) watermarking,
   (b) author/affiliation/collaboration upgrades, and
   (c) front matter truncation.
3. Two new author revision mark up styles.
4. New Figure commands,
   (a) figure set mark up (see Section 3.2) and
   (b) support for figures consisting of multiple files.
5. Improved table features including:
   (a) automatic column numbering,
   (b) math mode in designated columns,
   (c) the ability to hide columns,
   (d) decimal alignment, and
   (e) splitting wide tables.
6. Improved software citation,
   (a) an upgraded bst file for creating 1st class references with bibtex and
   (b) a new software command to highlight code used in the work.

Many of these improvements were based on author suggestions. More information about the most recent version of AASTeX can be found at http://journals.aas.org/authors/aastex.html.

3 HTML only article features

The AAS Journals offer a wide range of options for authors to provide information that simply could not be published years ago or at best only existed temporarily on personal websites. This section briefly discusses these HTML only features, all of which are highlighted in a new tab called "Article Data." Figure 1 shows an example which illustrates how easy it is for a reader to identify and find this HTML only information.

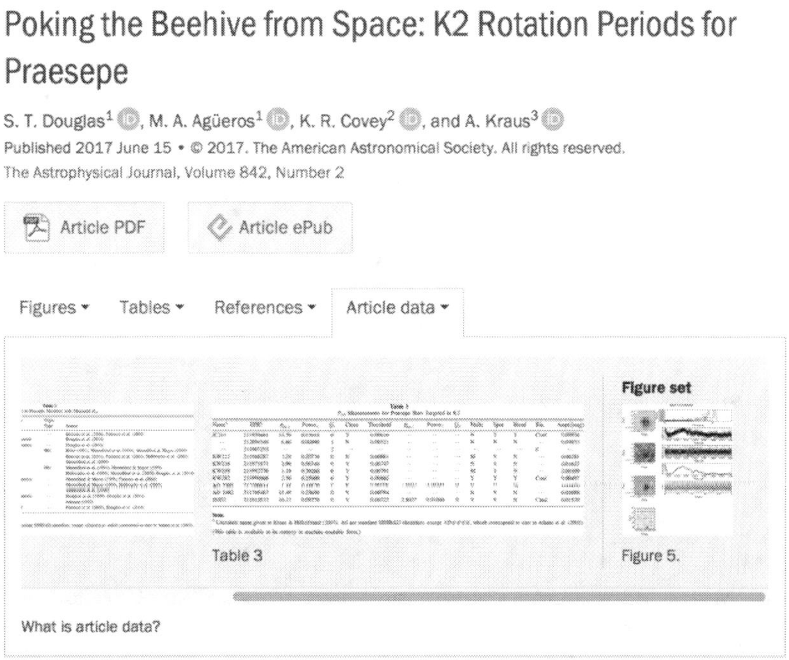

Figure 1. Some data products available under the "Article Data" tab of Douglas et al. 2017, ApJ, 842, 2. This article has multiple MRTs and a Figure Set.

3.1 Machine readable tables

Astronomers are very good at distilling large, raw data into useful catalogs. These catalogs often have characteristics that make them impossible to show as regular LaTeX tables such as extreme length or to wide for a print page. To address this issue the first non-LaTeX option, machine readable tables (MRT), began in 2000. A MRT consists of a meta-data header attached to a formatted data file. The meta-data uses the same standards and styles developed by CDS for their VizieR catalogs. In 17 years more than 10,000 MRTs have been created in over 5,000 articles making it the most used data feature. Many of these tables have been incorporated into CDS as VizieR catalogs.

3.2 Figure sets

When authors create a series of similar figures it can be desirable to group them together especially when there are a significant number of figures, e.g. many tens to thousands of figure files. Prior to electronic publishing it was not feasible to print these figures. AASTeX now contains specific figure set commands that let the publisher create an electronic flipbook of all the set figures. Figure 2 shows an example of a Figure Set and illustrates how a reader can scroll through a large group of similar figures. Since 2005 the AAS has published over 750 individual figure sets with on order 10,000 component figures.

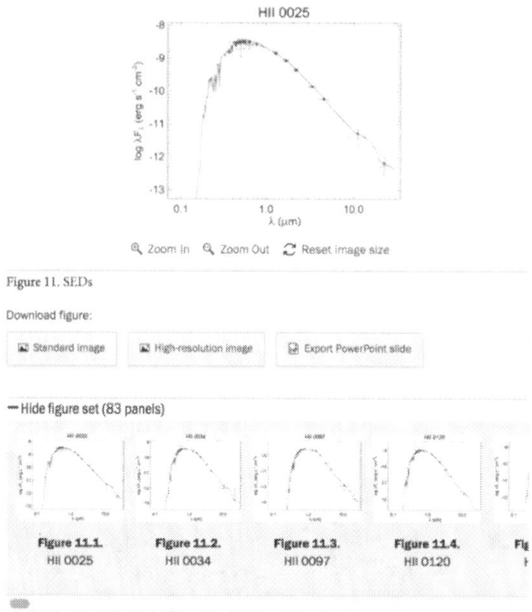

Figure 2. Example of a Figure Set from Somers and Stassun 2017, AJ, 153, 3. The Figure Set contains 83 component figures of various stellar SED fits. Clicking on the thumbnail in the lower scroll bar section displays the full version above.

3.3 Data behind the Figure

Governments agencies have increasingly mandated that authors provide the data used in the articles produced using their funding. In 2010 the AAS created Data behind the Figure (DbF) to give authors another way to incorporate the underlying data into their articles to help satisfy these mandates. Since figures can be astronomical images or vector drawn graphics, a DbF can be either a MRT or a FITS format. Over 250 articles have had DbFs.

3.4 Streaming video and Interactive figures

Authors have been allowed to include animations with their article since the AAS started publishing its journals electronically. The movie was available for download and a reader had to play it locally thus a specific video standard was enforced for all submission to ensured that the experience would be the same for all readers. This added an extra burden for the author nor was it a great experience for the reader. Now that streaming services are ubiquitous the HTML article has been updated so that all videos are streamed inline much like an embedded YouTube video. In addition, the video standards have been relaxed which makes it easier to use author produced movies. We currently have over 2,500 animations in the AAS Journals primarily generated by the solar physics community.

Interactive figures allow the reader to manipulate the information contained in an image. This also adds clarity for complex figures and helps further the author's narrative. Interactive control is provided via a dynamic JavaScript and HTML framework. The most common example of an interactive figure is a 3D model. The majority of the 17 interactive figures currently published are 3D figures. Since this is a new technology the AAS offers limited support on an ad hoc basis. Figure 3 shows an example of an interactive figure.

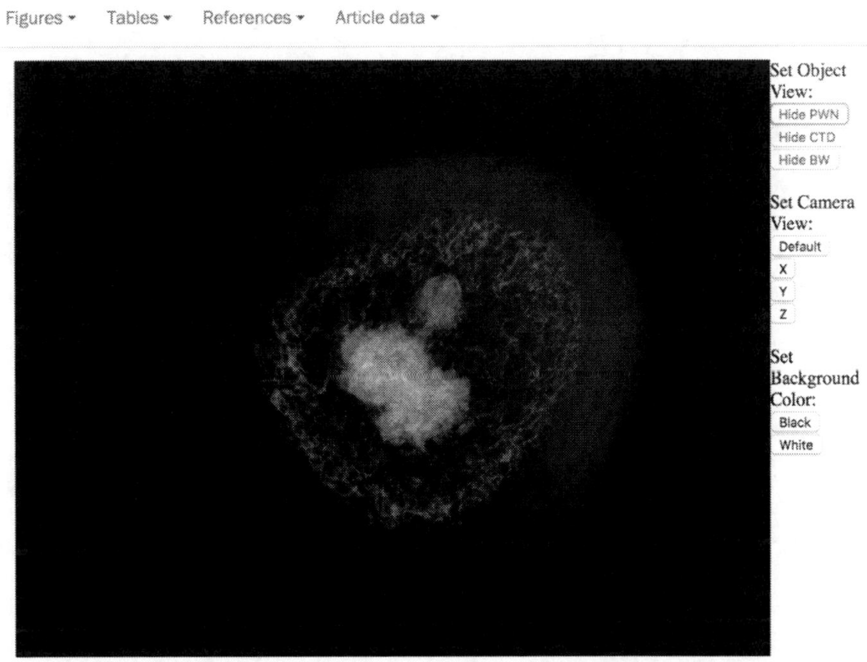

Figure 3. Example of an interactive figure from Kolb et al. 2017, ApJ, 844, 1. In this custom created figure, the JavaScript controls on the right set the initial orientation and turn on/off some aspects of the image. 3D movement is obtained by moving the mouse within the image.

3.5 External repository support

While the AAS can capture and publish a large range of data products not everything authors produced can be hosted. Examples include extremely large files (> GBs), databases and active source code. For these cases the AAS has developed recommendations and tutorials for using 3rd party repositories. Authors are requested to use repositories that issue DOIs for their holdings so that they can be properly cited and will be protected against link rot. Currently we recommend Zenodo, DataVerse, and Figshare but will soon be including the Canadian CANFAR repository, gSTAR center at Swinburne University, Australia, and the joint AAS/University of Arizona Astrolabe project.

Another repository project involves the MAST archive at STScI. When select authors submit to the AAS peer review system they are asked if the paper used any MAST data, e.g. HST. If yes, then the author is sent to MAST where the a DOI is minted for the MAST data collection associated with the paper, see Figure 4. Authors are instructed to included this MAST DOI in the accepted manuscript so it can be highlighted under the "Article Data" tab. This MAST project has been active for a subset of articles but will go live for all AAS Journals in 2018. Using the experience working with MAST, the AAS plans to partner with other data centers to provide a similar functionality.

Figure 4. Query to authors about MAST data products presented at submission (left) and the external MAST site where authors collect their data for DOI minting (right). While participation is voluntary, the AAS believe that the advantages of interconnectivity between the literature and archives will accelerate acceptance of these data collection links.

4 Data review

The adoption of these data products outlined in Section 3 is related to how long they have been available. In an effort to speed up adoption and increase general awareness the AAS data editors now conducted a data review of all submitted manuscripts. The data review does the following.

1. Search for vector drawn graphics figures that contain highly useful data that would be easy to capture for DbFs (Section 3.3). Simple examples include light curves, spectra, etc.

2. Encourage authors to follow the new software policy (http://journals.aas.org/policy/software.html) by properly tagging and citing the code used.

3. Check for supplementary data not optimally stored, such as a personal website, and recommend better alternatives (Section 3.4).

4. Check that animations meet new streaming requirements and remind authors that this functionality exists in cases where videos could be made from the simulations (Section 3.4).

5. Suggest better ways to set up an article and potentially save publication charges. A common example is merging similar figures into a Figure Set (Section 3.2).

5 Future

The AAS believes that there is great benefit to both the author and the reader to have these data items integrated into the article thus the future work includes better collection of meta-data, more author interactive figure tutorials, and HTML proofing.

The current trick is how to integrate these HTML only items into a system based on writing and reviewing manuscripts in LaTeX which is inherently paper, or PDF, based. Some improvements to the AASTeX classfile will help but we are also investigating different ways to write manuscripts. As a first step in this direction, the AAS has partnered with the collaborative writing tools of Authorea and Overleaf. With these tools authors can easily collaborate with co-authors and track all changes. In addition, Authorea and Overleaf authors can do a direct submission of the manuscript to the AAS peer review system when they are ready for publication. All of these collaborative tools will save time and effort in constructing their papers. Another interesting, albeit long term investigation would be Jupyter notebooks based articles.

Regardless of where the future takes us, the AAS remains committed to providing the best publishing experience for both authors and readers.

The runners team: A. Rots, M. Veldre, L. Cambrésy, A. Tisseau des Escotais, D. Bouquin, T. Chen, M. Buga (Photo: M. Brouty)

EPJ Web of Conferences **186**, 11002 (2018) https://doi.org/10.1051/epjconf/201818611002
LISA VIII

Scientific Writing Studio: How publishers can help authors to write smarter and faster

Agnes Henri[1,*] and *Charlotte* Van Rooyen[1,**]

[1] *EDP Sciences, 17, av. du Hoggar - BP 112 - 91944 Les Ulis Cedex A, France*

Abstract. Researchers are under pressure: publish or perish is one of their main challenges. Writing is hard and scientific writing is harder, and any help Publishers can provide to assist with the process is greatly appreciated, especially by younger authors. EDP Sciences and Astronomy & Astrophysics (A&A) have worked together to support their authors and developed a solution called **Scientific Writing Studio (SWS)** designed to make the article writing process easier. **SWS** is a tool that allows authors to collaboratively write their paper, then submit and then follow all stages of the publication process within a single unique interface. This paper explains the background to the development of the **SWS**, and describes the current benefits of the SWS as well as outlining future developments.

1 Background

A&A has been publishing the best astronomical research for over 45 years and is the first choice for publication of important special issues including those from the missions of Planck, Herschel, Rosetta and Gaia (Fig. 1). This means there are often many contributors to the same paper (Fig. 2).

The Editors-in-Chief of A&A conceived the idea of the **SWS** in 2013.

The Editors Dreamland (Fig. 3) was to create a collaborative writing tool where authors could work with their co-authors using journal templates and eventually be able to submit their articles directly to the Journal.

2 Development

The writing studio work group tested a number of online collaborative LaTeX writing tools, and considered ways of adapting these tools to the specific needs of A&A. The key requirements at that time were: the usability of the interface, the ability to promote the A&A brand, and the ease of integrating A&A specific tools and services into the interface.

*e-mail: agnes.henri@edpsciences.org ORCID: 0000-0001-7928-2060
**e-mail: charlotte.vanrooyen@edpsciences.org

Figure 1. Some of the special issues published by Astronomy & Astrophysics

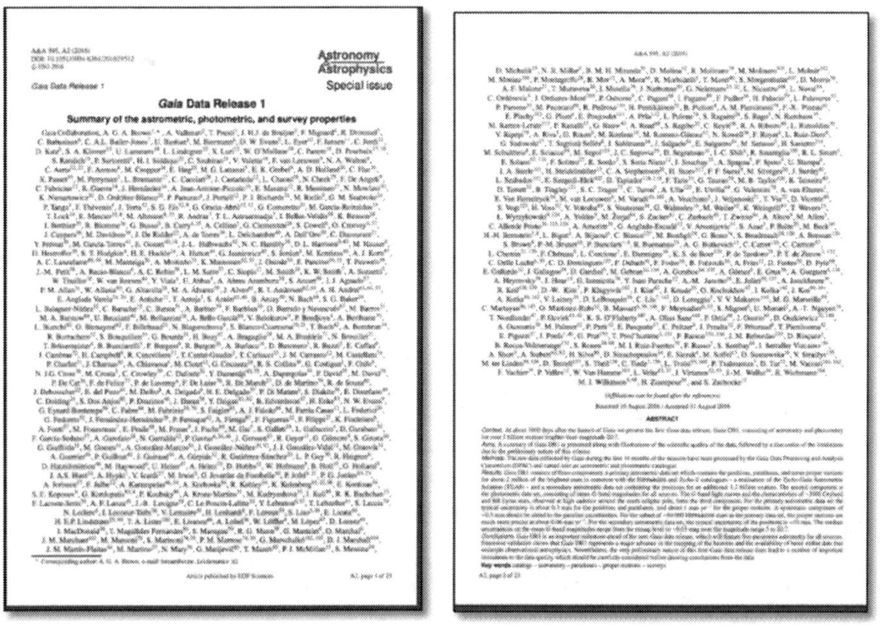

Figure 2. The many contributors to the Gaia Data Release 1.

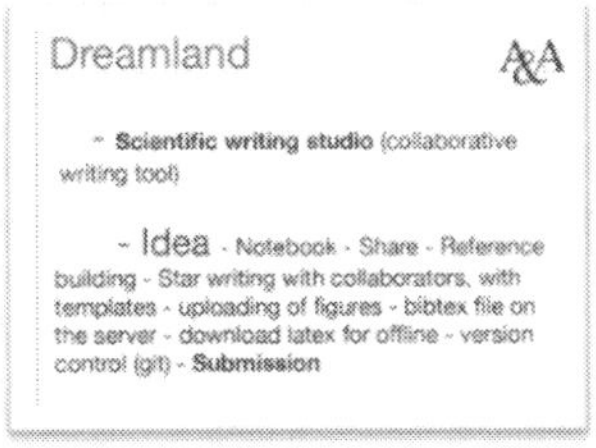

Figure 3. Writing Studio Dreamland.

The working group chose ShareLaTeX and a prototype was built using the ShareLaTeX open-source code supported under the open-source license that it uses. All changes and revisions were published to a Github public repository. In addition Dropbox sync functionality was provided under a license agreement to make this available to all users of the writing studio.
The tool underwent rigorous user-testing, gathering feedback from editors and beta testers to help design the best solutions and features to ensure the best user experience. The ambition was for the **SWS** to eventually provide a simple tool not just for authoring an article, but also to be able to submit to the journal. Additionally, it was required that the A&A language editors should be able to use the tool to submit their corrections and edits.

3 Making Scientific Writing Studio different

This tool has been designed and developed specifically for authors of A&A and those writing in the disciplines of astronomy and astrophysics. Specific features were developed for the **SWS**, for example EDP Sciences worked closely with CDS to include a tool in SWS that automatically detects and highlights stellar objects (see Fig. 4). Login using an ORCID iD was also integrated in order to authenticate login and ensure no ambiguity over author names, which is increasingly a problem in the scholarly research community.

4 Launch

The SWS was launched at the end of 2015 and now has well over 600 accounts with nearly 800 articles having been worked on. The Editors' ambition for a collaborative tool has been achieved as

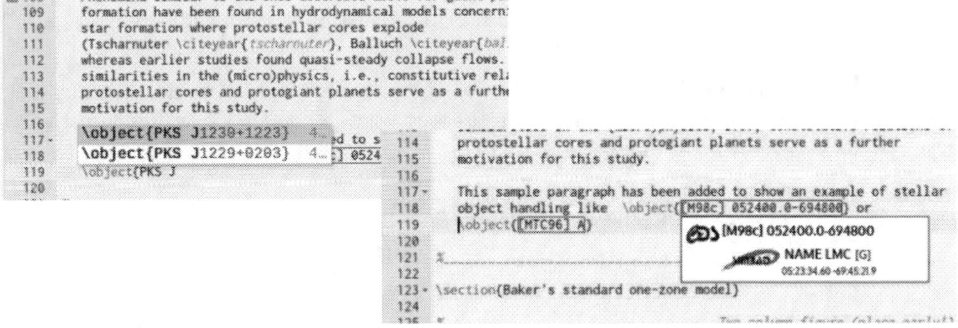

Figure 4. Links to stellar objects.

Table 1. Some numbers

	May 17	August 16
Number of active accounts (connected at least once):	511	323
Number of projects (total):	738	355
Projects with at least one declared collaborator:	170	70
Maximum number of collaborators on a project:	16 (1 project)	8
Number of files with more than 100 modifications:	332	92
Chats around the projects:	208	103

up to 16 authors at a time have been working on a single paper, and nearly 200 with more than one collaborator! (Table 1).

5 Overview of capabilities

The **SWS** has been developed to allow authors to spend less time with LaTex and process, and more time on presenting and defending their science in collaboration with their co-authors. Today, the **SWS** offers authors the following benefits:

- A collaborative LaTeX solution: Author and co-authors can work on a unique version of the paper with A&A macros already included.
- Easier links to stellar objects: Authors get auto-completion for stellar objects and additional information using tools developed with the CDS.
- Chat with co-authors: Authors are able to ask questions and chat with co-authors for direct communication.
- Dropbox integration: Articles can be edited offline and are automatically synchronized with the Dropbox online copy.
- ORCID integration: allows ease of login recognition to Writing Studio.
- Ease of Submission: automatic upload of article to submission system, with key fields pre-filled ready for checking.

All of these means **SWS**, provides a stronger bond between the journal and the author by making the writing of their articles easier, and that means they want to come back to the journal with the results of their research in the future (Fig. 5).

Figure 5. Login page for the Scientific Writing Studio.

6 Future Developments

Next steps are to:

- Integrate a citation helper to make building reference lists easy for the end user.
- The ability for an author to import from their ADS library papers they have bookmarked.
- Develop a comprehensive user guide.

7 Conclusion

By understanding the needs of researchers in a particular discipline, and working closely with the associated organisations in that field, as well as understanding the publishing process and services and tools to support publication, Publishers can assist authors to present their science in collaboration with their co-authors in an efficient way leaving more time for their research.

The **SWS** is bringing A&A closer to their goal of a fully integrated and user-friendly workflow from writing through to submission and dissemination of an article (Fig.6).

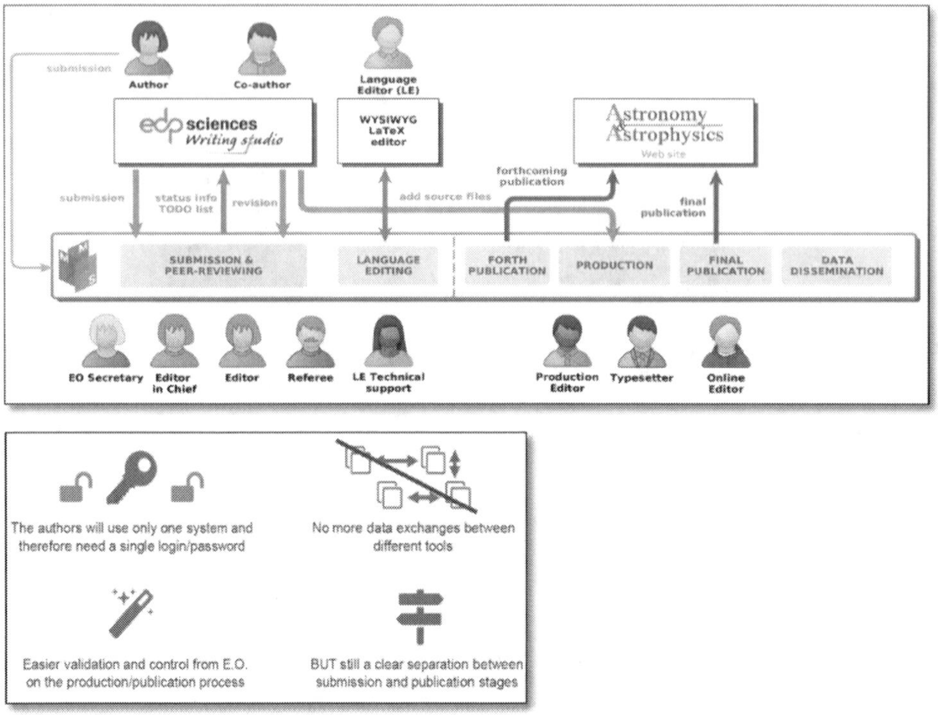

Figure 6. A&A workflow.

CDS - Interfaces for published and reference data

Mark Allen[1,*] and the CDS Team

[1] *Université de Strasbourg, CNRS, Observatoire astronomique de Strasbourg, UMR 7550, F-67000 Strasbourg, France*

Abstract. The Centre de Données astronomiques de Strasbourg (CDS) provides added value to published and reference data, enabling scientific research with the data through innovative services and tools. As a long standing data centre it is important to maintain and update the processes that have been developed over many years, in particular to manage the increasing volume of publications and data. CDS presentations at the LISA conference have highlighted these developments, and a poster supporting these presentations showed the current CDS user interfaces. Here we highlight how developments in the processes and underlying databases result in new features in the CDS user interfaces.

1 Introduction

The Centre de Données astronomiques de Strasbourg (CDS) is an astronomy reference data centre with the core role of providing added value to published and reference data, and enabling scientific research with the data through innovative services and tools. The processes for the ingestion of data into the CDS databases are constantly revised and adapted to new requirements and new technologies [1]. At this LISA meeting we have reported on evolutions, improvements and updates in the areas of:

- Cross-identification of astronomical objects for the SIMBAD database (Brunet)
- The new CDS documentalist task of curating pixel data in HiPS format (Buga)
- The VizieR associated data service for data (spectra, images etc.) connected to a journal article (Ocvirk)
- The certification of CDS services with the Data Seal of Approval (Landais)
- Shared nomenclature and identifiers for telescopes and instruments (Perret)
- The rapid development and future visions of technologies that facilitate the documentalits' work (Schaaff)

2 Organisation of CDS work

The CDS work is carried out by an integrated team of scientists, software engineers and documentalists with a deep level of interaction and coordination between the different types of activities. Figure 1 provides an overview of all of the activities of the CDS. The reference services are: SIMBAD, the

*e-mail: mark.allen@astro.unistra.fr ORCID: 0000-0003-2168-0087

reference database for the nomenclature and bibliography of astronomical objects (outside the solar system) [2]. VizieR, the reference service for astronomical catalogues including tables and data published in academic journals [3], the Aladin interactive sky atlas, the CDS X-Match large astronomical catalogue cross-correlation service [4], and a new integrated CDS Portal[1].

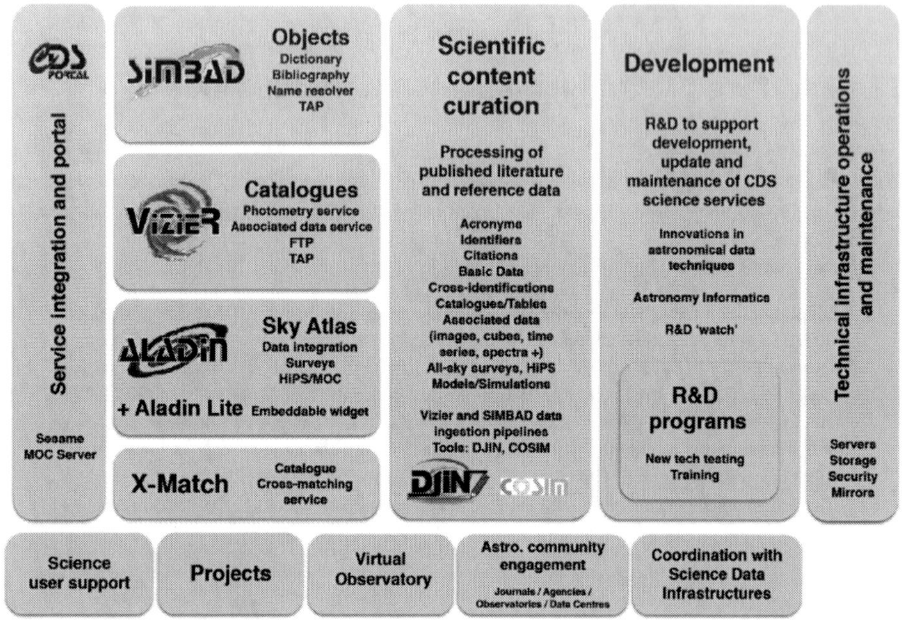

Figure 1. An overview of CDS activities

3 CDS Service Interfaces

Astronomers use the CDS services via the various interfaces. The CDS web pages[2] provides the starting point for access to all of these interfaces, in particular the individual web interfaces of the different services. In addition to these web interfaces there are also many programmatic interfaces to CDS services that allow calls to these services by computer programs or scripts, which are often used by external services. For example, calls to the SIMBAD name resolver which is implemented in the search field of many observatory/mission archive services. Other examples include the access to CDS services via standard IVOA[3] protocols such as the TAP (Table Access Protocol) interfaces to SIMBAD and VizieR that enables complex queries of these databases using the ADQL language, and also the SSA (Simple Spectral Access) and SIA (Simple Image Access) interfaces to the VizierR associated data service. The Aladin sky atlas has two main interfaces, one a desktop application, and the other called Aladin Lite [5] which is a javascript widget that can be embedded in web pages.

Recently CDS has released a new CDS Portal, which is an integrated interface to CDS databases, allowing uses to search across multiples services from a single query box. This has allowed us to

[1]http://cdsportal.u-strasbg.fr
[2]http://cds.unistra.fr
[3]International Virtual Observatory Alliance, http://ivoa.net

explore and test new ideas for the interfaces, and to benefit from many improvements in the underlying database infrastructure and metadata.

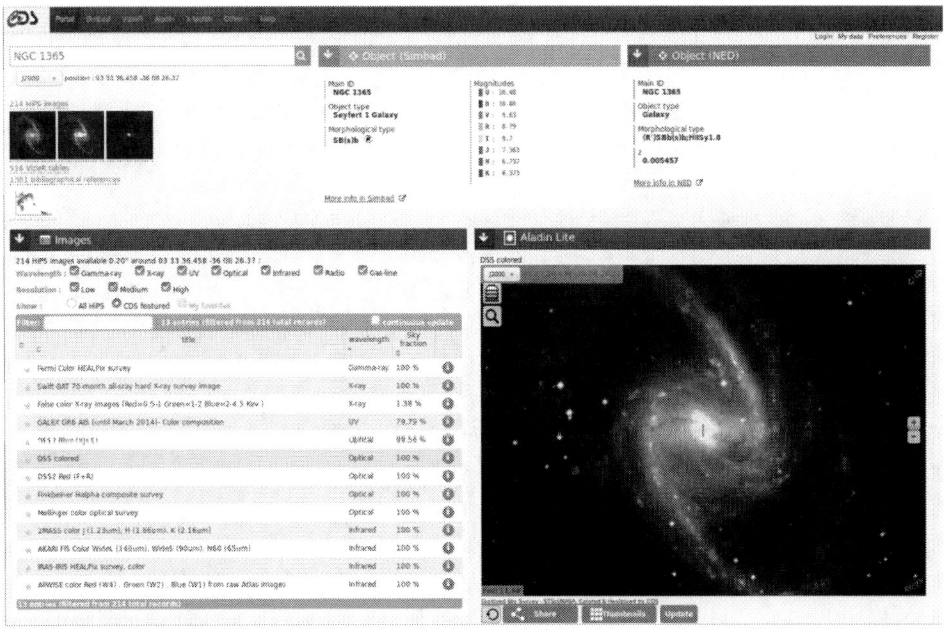

Figure 2. CDS Portal query results for Objects and Images

Figure 2 shows the result of query for information on the galaxy NGC 1365. Reference data from SIMBAD (and also NED) are provided with links to more details. The image section makes use of the new HiPS network of all-sky images [6], and its implementation in Aladin Lite. The metadata about the surveys allows for filtering of the hundreds of such surveys.

Figure 3 shows another part of the new CDS Portal for catalogues, with the result of the coordinate based query. The detailed metadata that is curated during the ingestion of catalogues into VizieR allows complex searching capabilities, with the new CDS Portal enabling interactive filtering via the use of facets. Searches can be constrained by wavelength, astronomy keywords, sky fraction and the presence of associated data files. Data may be selected and used for quick-look analysis in the portal including overlay on the Aladin Lite panel.

4 Summary

The reference data that is carefully ingested into the CDS services by the team of documentalists, engineers and astronomers, is made searchable and usable via the various interactive and programmatic interfaces. Continual adaptation and improvement in the ingestion processes is necessary for managing the increasing volume of data associated with journal publications, and to respond to the scientific needs of the astronomical community. Recent advances in object cross-identification, handling of very large pixel data sets and associated data, all contribute to the quality and interoperability of the reference data, allowing it to be used in innovative new ways.

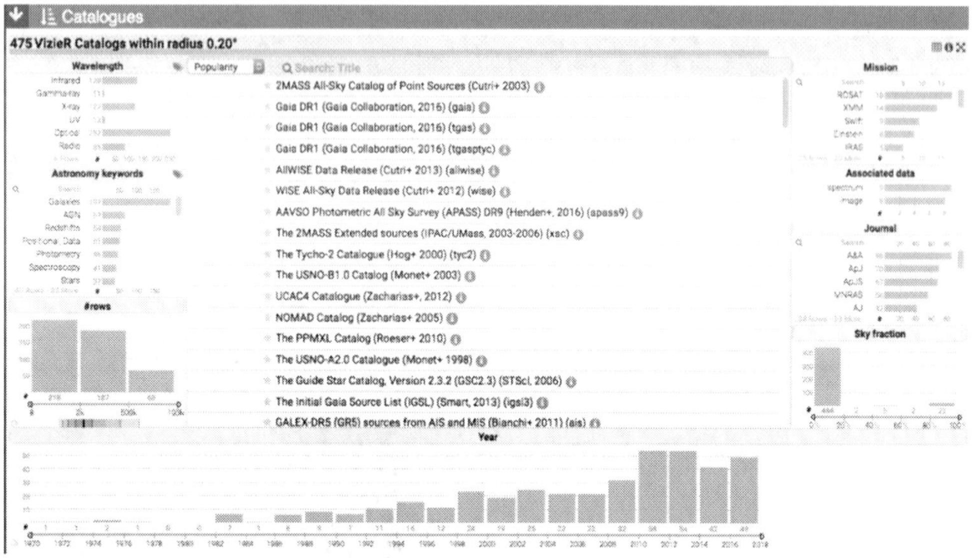

Figure 3. CDS Portal query results for Catalogues

References

[1] E. Perret, T. Boch, F. Bonnarel et al., *Working Together at CDS: The Symbiosis Between Astronomers, Documentalists, and IT Specialists* in *Open Science at the Frontiers of Librarianship*, edited by András Holl, Soizick Lesteven, Dianne Dietrich, and Antonella Gasperini (2015), Vol. 492 of Astronomical Society of the Pacific Conference Series, p. 13–21

[2] M. Wenger, F. Ochsenbein, D. Egret et al., A&AS **143**, 9 (2000)

[3] G. Landais, T. Boch, M. Brouty, et al., *Management of Catalogs at CDS* in *Open Science at the Frontiers of Librarianship*, edited by András Holl, Soizick Lesteven, Dianne Dietrich, and Antonella Gasperini (2015), Vol. 492 of Astronomical Society of the Pacific Conference Series, p. 57–60

[4] F. Pineau, T. Boch, S. Derriere, & Arches Consortium, *Towards a Next-Generation Catalogue Cross-Match Service* in Astronomical Data Analysis Software an Systems XXIV (ADASS XXIV), edited by A.R. Taylor and E. Rosolowsky (2015), Vol. 495 of Astronomical Society of the Pacific Conference Series, p. 61–64

[5] T. Boch, P. Fernique, *Aladin Lite: Embed your Sky in the Browser* in *Astronomical Data Analysis Software and Systems XXIII (ADASS XXIII)*, edited by N. Manset and P. Forshay (2014), Vol. 485 of Astronomical Society of the Pacific Conference Series, p. 277–280

[6] P. Fernique, M. G. Allen, T. Boch, et al., A&A **578**, A114 (2015)

Registration time (Photo: M.Wenger)

EPJ Web of Conferences **186**, 12002 (2018) https://doi.org/10.1051/epjconf/201818612002
LISA VIII

Progress with implementing the new research data management system at HartRAO

Glenda Coetzer[1,2,★], *Roelf* Botha[1], and *Lorette* Jacobs[2]

[1] *Hartebeesthoek Radio Astronomy Observatory, South Africa*
[2] *University of South Africa, South Africa*

Abstract. The Hartebeesthoek Radio Astronomy Observatory (HartRAO) participates in global radio astronomy and fundamental astronomy (space geodesy) research activities. Data and data products produced by HartRAO's expanding range of on-site and off-site instrumentation must be archived and stored at HartRAO and made accessible to the scientific community. The data management and storage systems currently being used for managing fundamental astronomy data are not capable of handling the large volumes of data and have become obsolete. This necessitated the design and implementation of a next-generation Geodetic Research Data Management System (GRDMS), which complies with internationally accepted data service standards. We present the top-level conceptual model of the GRDMS and progress to date with developments of various sub-systems, data structuring and organisation within the sub-systems.

1 Introduction

The Hartebeesthoek Radio Astronomy Observatory (HartRAO) participates in radio astronomy and fundamental astronomy (space geodesy) research activities. Radio astronomy activities focus predominantly on the use of HartRAO's 26 m radio telescope to conduct radio astronomy observations in single-dish mode and Very Long Baseline Interferometry (VLBI) observations in conjunction with other radio telescopes across the globe. HartRAO's fundamental astronomy activities include the use of HartRAO's 15-m and 26-m radio telescopes, in conjunction with other radio telescopes globally, to conduct astrometric and geodetic VLBI observations. Collocated on the HartRAO site are various Global Navigation Satellite Systems (GNSS) reference stations, two Satellite Laser Ranging (SLR) systems, a Doppler Orbitography and Radiopositioning Integrated by Satellite (DORIS) system, various meteorological instrumentations as well as a gravimetric and seismic instrument. In the near future, a Lunar Laser Ranging (LLR) system and VLBI Global Observing System (VGOS) radio telescope will become operational at HartRAO. We are also expanding our network of instruments with the installation of GNSS, seismic and meteorological instruments across Southern Africa. Together with HartRAO's participation in the African VLBI Network (AVN) these will expand HartRAO's fundamental astronomy data collection and research activities.

HartRAO is the only operational facility of its kind in Africa and provides a vital link to the global geodetic network of data providers [1]. HartRAO is responsible for the generation, correlation, storage

★e-mail: glenda@hartrao.ac.za ORCID: 0000-0002-0788-5660

and dissemination of technique-specific data and data products to international service providers, such as the Crustal Dynamics Data Information System (CDDIS), University NAVSTAR Consortium (UNAVCO), International Laser Ranging Service (ILRS) and International VLBI Service for Geodesy and Astrometry (IVS), to name but a few. Some data and data products, such as Global Positioning System (GPS) and single-dish observation data, are stored on-site. Other data, for example VLBI data, are shipped to international correlators and data centres [2].

HartRAO's current fundamental astronomy data management and storage system is not capable of managing the predicted additional large volumes and complexity of data resulting from the expansion. The current system has many drawbacks - it is outdated, segmented and distributed; it has limited functionality and capacity to handle different data types and different user requirements; different approaches are followed for handling datasets; data are stored on different servers; access to the data holdings are via ftp or http, which have many broken links and are not frequently updated. This necessitated the design and implementation of a next-generation Geodetic Research Data Management System (GRDMS) that will be able to cater for all of HartRAO's fundamental astronomy data, whilst complying with internationally accepted data service standards.

2 Overview of the GRDMS

The main objectives of the new system is to organise, structure, archive and store geodesy and geodynamics related data and data products in a central data bank; disseminate data and data products to global data service providers and the growing research community; maintain metadata and provide metrics for reporting purposes. Professionals from different disciplines - academic/scientific, information technology and communications (ITC) and library and information science (LIS) - are working together to design and implement a new GRDMS [1]. To meet the set objectives and not to 'reinvent the wheel', it was decided that the new data management system will feature components (see Figure 1) similar to that of the CDDIS and UNAVCO [3, 4]. The same data structures and file-naming convention will be used for all datasets. Each dataset will receive persistent interoperable identifiers. User access will be facilitated by an interactive web-based graphic user interface (GUI). The complete system will be divided into functional units, which can operate relatively independently of each other - this helps to avoid critical overall failures and also enables faster repairs of broken / damaged components. Security is also a major concern and the systems have been designed so that, from the public domain, only limited, read-only or no direct access is possible to mission-critical systems and the main data archives.

Redundancies on all functional units and processes are required to prevent data loss and system downtime, therefore a clone of the complete GRDMS will be housed at a different location. It will be transparent to the stations as well as to the end-user which data centre is being used.

2.1 Functional overview of the GRDMS

The overall functional concept of data collection and distribution is depicted by Figure 1. The main components are the functional units as follows:

- Virtual Private Networking is used to connect all stations and instruments to the Data Collection Virtual Machine (VM), Vector, which handles the data collection station monitoring.
- The Data Tools VM then pulls the raw data from the Data Collection system, Quality Check (QC) it and stores it in the Archive, in the correct folder and file format and naming convention.

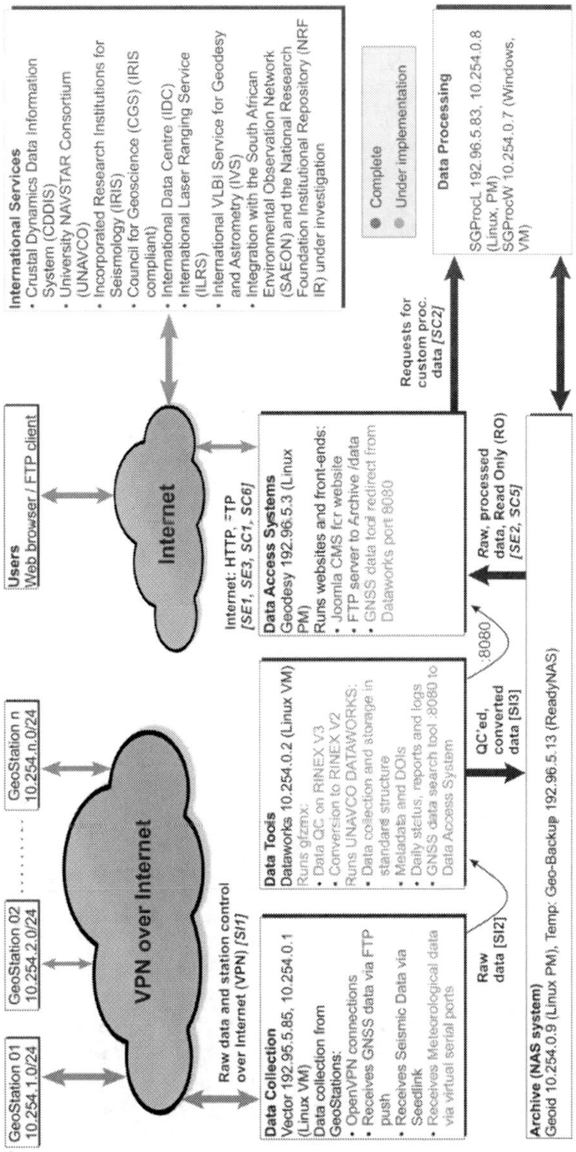

Figure 1. The GRDMS functional diagram (square brackets denote steps in the data cycle, see Section 2.2).

- The Data Access System provides various tools (websites, FTP access etc.) to let users gain access to the raw data, as well as request custom processed data products. This system also streams the data to the international partners and service providers.
- The Data Processing Systems consists of a Linux Physical Machine (PM) and a Windows VM. These have read-only access to the raw data in the Archive and can also store data products to the

 https://doi.org/10.1051/epjconf/201818612002

archive. The processing can occur by requests via the Data Access System or from a user logged into one of the systems.

- The Archive is a storage cluster that can be expanded dynamically, it also has an off-site fail-over duplication.

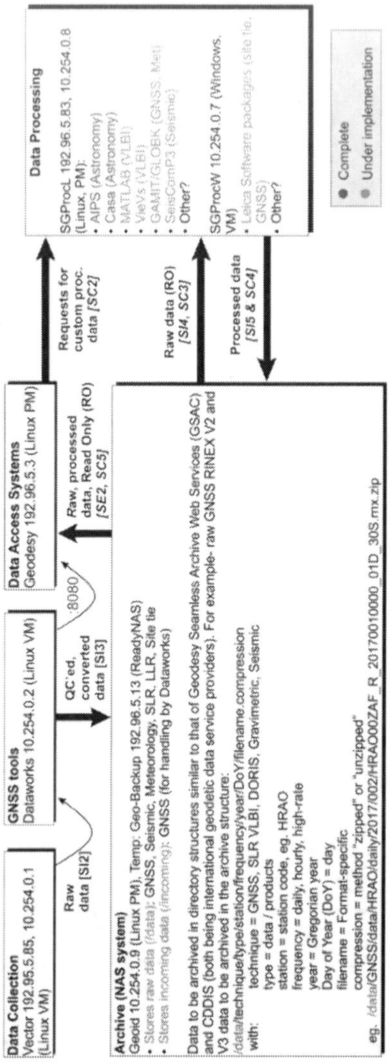

Figure 2. The GRDMS Archive internal data structure and the Data Processing software packages (square brackets denote steps in the data cycle, see Section 2.2).

The internal structure of the Archive and software on the Data Processing systems is depicted in Figure 2.

2.2 Logical processes of the GRDMS

The proposed logical processes are divided in three types: automated Internal Steps [SI], Steps for External requests [SE] and Steps for Custom data requests [SC] - these are depicted in Figures 1 and 2 as [SI], [SE] and [SC] and will be used in the filenames and procedures executing these steps.

Internal Steps [SI] data cycle:

- SI1: The Data Collection Unit controls different instruments at different stations and receives raw data from the stations.
- SI2: Raw data is collected by the Data Tools Unit to quality check it and then pre-process it into required file types, formats and filing structures (eg. the Rinex V3 data are quality checked and converted to Rinex V2 data).
- SI3: Checked and pre-processed data are sent and stored in specified formats and structures in the Archive Unit.
- SI4: Checked and pre-processed data are collected by the Data Processing Unit to process it to predetermined basic data products.
- SI5: The basic data products data are sent and stored in specified formats and structures in the Archive Unit.

External Steps [SE] data cycle:

- SE1: The scientific community can interact with the Data Access System and request data via a website (HTTP) and/or FTP.
- SE2: The Data Access System obtains the requested data from the Data Archive Unit.
- SE3: Requested data is packaged into a single compressed file and made available.

Custom Steps [SC] data cycles (for requests which are simple enough to be handled automatically by the system):

- SC1: Requests are submitted on the website via a special interface.
- SC2: Requests are translated to a script and pulled by the relevant Data Processing Unit.
- SC3: The Data Processing Unit obtains the required raw data and processes it.
- SC4: Processed data is sent to the Archive Unit for storage.
- SC5: The Data Access System retrieves processed data.
- SC6: Results are sent to the requesting user.

2.3 Data structure within the GRDMS

The GRDMS data structures will be similar to those of the CDDIS. An example of the structure of raw technique-specific Rinex data is explained in Table 1.

At least the "8.3.Z" file-naming convention will be used for all GNSS data for example, see Table 2.

Table 1. An example of the structure of raw technique-specific Rinex data in the format: /data/technique/type/station/frequency/year/DoY/filename.compression

Descriptor	Description	Example value(s)
technique	Technique abbreviation	gnss, slr, vlbi, doris, gravity, seismic
type	Data Type	Rinex
station	Technique specific station Code	HRAO, MATJ
frequency	File Frequency	daily, hourly, high-rate
year	Gregorian year	2017
DoY	Day of Year	028
filename	Technique-specific	see Table 2 for example
compression	Compressed? and type	.Z, .gz, .zip

Table 2. An example of the structure of a GNSS Rinex2 filename "8.3.Z" eg.SSSSDDD0#.YY.Z

Descriptor	Description	Example value(s)
SSSS	Site code	HRAO, MATJ
DDD	Day of Year (DOY)	028
0	Sequence number	0
#	session code	a
YY	two-digit year	17
Z	compressed?	.Z

3 Progress with the design and implementation of the GRDMS

The sections marked in green within Figures 1 and 2 are complete, the sections in Orange within these figures are under implementation.

The following processes and software implemented in different sub-systems have been completed:

- within the Data Collection unit, open Virtual Private Network (VPN) connections were established
- the system receives GNSS data via ftp push and receives seismic data via Seedlink
- Joomla Content Management System (CMS) for websites was installed on the Data Access System unit and ftp server to archive data
- Astronomical Image Processing System (AIPS) and Casa software were installed on the Data Processing unit

Functionality still to be implemented are:

- the receipt of meteorological data via virtual serial ports
- quality checks on Rinex Version 3 data and the conversion of Rinex Version 3 data to Version 2 data
- the installation and running of UNAVCO Dataworks software to manage the collection and storage of data in standard predefined structures
- the collection of metadata and assigning DOIs to datasets
- daily status reports and logs and the installation of GNSS data search tools to redirect data from the Dataworks port
- the installation of MATLAB, VieVs, GAMIT/GLOBK, SeisComP3 and Leica software packages

The new seismic network has a specially designed data processing and storage systems, which arrived at HartRAO during 2016. These systems are currently being integrated into the GRDMS. HartRAO has also obtained a Dell Westmere processing cluster, containing of 1152 CPU cores (14.8 Teraflops) and 3456 GB RAM. This cluster will form an additional functional section of the Data Processing system, focused towards processing of VLBI data. Implementation is expected to occur during 2018.

4 Progress with the design and implementation of the GRDMS

Data and data products resulting from HartRAO's expanded range of on-site and off-site instruments must be archived and stored at HartRAO and made accessible to the scientific community. We have highlighted the drawbacks of the current data management and storage system, which served as a motivation for the design and implementation of a new research data management system, currently under implementation. We presented progress to date on various sub-systems (units) as well as progress on establishing data structure and organisation within the GRDMS sub-systems.

Acknowledgement

The authors would like to acknowledge funding awarded by the National Equipment Programme (NEP) of the National Research Foundation (NRF) for funding the development of the co-located academic network.

References

[1] G.C. Coetzer, R.C. Botha, L. Combrinck et al., *A New Geodetic Research Data Management System at the Hartebeesthoek Radio Astronomy Observatory* in *Open Science at the Frontiers of Librarianship*, edited by András Holl, Soizick Lesteven, Dianne Dietrich, and Antonella Gasperini (2015), Vol. 492 of Astronomical Society of the Pacific Conference Series, p. 22–30

[2] Behrend, D., Data Science Journal **12**, WDS81-WDS84 (2013)

[3] Noll, C., CDDIS. IVS2015-2016 Biennial Report (2017)

[4] UNAVCO, UNAVCO GSAC WS: Web Services for Geodesy Data Repositories (2013year)

EPJ Web of Conferences **186**, 12003 (2018) https://doi.org/10.1051/epjconf/201818612003
LISA VIII

Managing Institutional Bibliographies using the ADS API:

A new workflow using Google Sheets

James Damon[1,2,⋆], *Edwin* Henneken[1,3], and *Alberto* Accomazzi[1,4]

[1]*NASA ADS*
[2]*orcid.org/0000-0002-1069-2376*
[3]*orcid.org/0000-0003-4264-2450*
[4]*orcid.org/0000-0002-4110-3511*

Abstract. Curating institutional bibliographies with the ADS web interface is currently a manual process that scales with the number of search terms. Long author lists and institutions with multiple sub-organizations or name variations increase the workload. Review work is monotonous and can take significant time depending on the size of the institution and the frequency of reviews. Consequently, bibliographies generated in this way are costly and may suffer from human error. We propose a semi-automated workflow that uses an iterative approach to discovery with ADS's new search engine and a recently developed Google Sheets add on. First, affiliation strings from a user created spreadsheet are searched with the ADS API and for each result the matched affiliation and the paired author are retrieved. Next, each author name string is searched and items where that author is paired with an empty affiliation field are retrieved. The results from both queries are then compiled into output sheets with pertinent information for manual review. Finally, the selected items can be added to an ADS library from the Google Sheets interface. The tool can also use previously rejected affiliation strings to flag false positives in subsequent queries. Curators do not need to have extensive technical skills in order to use the workflow and they can help improve the ADS by opting to share ORCIDs, author synonyms, and affiliation synonyms.

1 Introduction

Managing the bibliography for the Harvard-Smithsonian Center for Astrophysics has always been based on an initial author query with high recall but low precision. Each record in the results set is manually reviewed in order to maximize precision in the final selection. Finally, the selected records are added to the bibliography in a batch.

Maintaining the possibility of a methodical hand-curated process was a key design consideration during the creation of the tool. The designers opted to provide larger results sets with many fields so curators could make more informed decisions while maintaining control of the process.

The reduction in time cost and the added utility stem from the use of Google Sheets and the ADS API.[1] Spreadsheet software is commonly used for data manipulation. By building this tool, we are allowing curators to leverage existing skills and reduce the time needed to effectively manage a

⋆e-mail: jdamon@cfa.harvard.edu ORCID: 0000-0002-1069-2376

bibliography. Additionally, the tool can be used at varying scales. It was designed with institutional level bibliographies in mind but it can also be used for labs or individual authors. Furthermore, it allows the work to be split up among multiple collaborators.

2 Using the Tool

While it is up to individual users to take advantage of the tool's features during the completion of tasks, the designers had certain workflows in mind during the creation of the tool. The underlying assumptions and an example workflow are outlined in the following section.

2.1 Results Sets

When performing an ADS query for a bibliography, the two types of text strings that are used in the search are author name and affiliation. While ORCID is the recommended format for author lists, its adoption was recent enough that most bibliographies should still include full author name queries in the workflow. ADS search includes results for author name queries from known author name synonyms so it may not be necessary to search for variations of an author's name. The affiliation text string used by authors from some institutions may not be consistent so unlike the author name search, it is best to identify the possible variations. Each individual variation may then be searched or alternatively, if the different versions of the string have some overlap that is not likely to cause a large increase in false positives, that portion of the string may be a good candidate for use during affiliation queries.

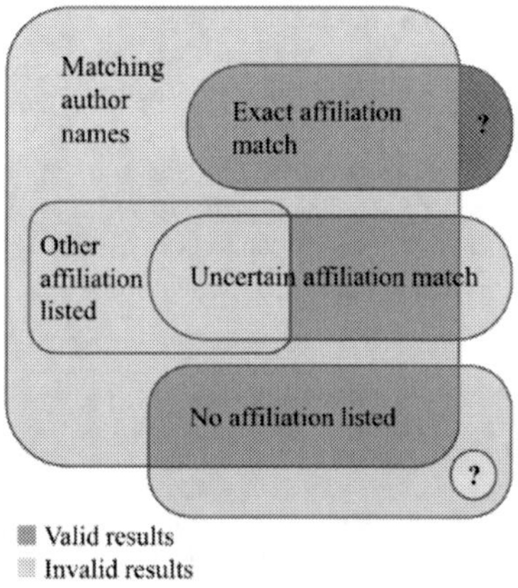

Figure 1: Results Sets

The ADS API is able to search author and affiliation strings in the same query however, it cannot currently search author and affiliation pairings.[2] In order to address this, for both affiliation and author name queries, the paired author name or affiliation is retrieved by the Google Sheets add on as well and included in the search results for the curator to review.

The results sets from affiliation and author name queries overlap as illustrated in figure 1. Exact affiliation matches and exact author name matches can be added automatically however other cases are up to the curator's discretion. Exact affiliation matches with unexpected authors may be the result of an out of date author list or authors falsely claiming an affiliation. Uncertain affiliation matches with exact author name matches may be the result of variations on common affiliation strings whereas uncertain affiliations paired with unexpected authors are more likely to be negative results that should not be included in the bibliography. Author name matches with no affiliation data require the curator to find and inspect the particular item in order to determine to validity of the record. Finally, there is a theoretical set of records that will not match on author names or affiliation due to typos or uncommon name/affiliation variations. For most bibliographies, these records would be ignored because of the extreme effort required to find them.

Table 1: Example affiliation variations

Original Affiliation String	Ambiguity	Example False Positive
Harvard-Smithsonian Center for Astrophysics	Exact	-
Harvard Smithsonian Center for Astrophysics	Exact	-
Harvard-Smithsonian Ctr. for Astrophysics	Exact	-
Harvard-Smithsonian CfA	Exact	-
Harvard College Observatory	Exact	-
HCO	Uncertain	H.C.Ã~. Universitetsparken
Harvard Astronomy Department	Exact	-
Smithsonian Astrophysical Observatory	Exact	-
Smithsonian Astr. Obs.	Exact	-
Smithsonian	Uncertain	Smithsonian Institution
SAO	Uncertain	SÃ£o Paulo

2.2 Example Workflow

Individual workflows will vary. The ideal workflow for the creation of a bibliography will be different from the workflow used to maintain a bibliography. Furthermore, the curator must determine an optimal level of diligence based on the time they are able to commit and the underlying purpose of the bibliography. Regardless of the specifics of the chosen workflow, the tool will be of most use for affiliation and author name queries. An example workflow follows and is represented by figure 2.

For new institutional bibliographies it may be useful to do an initial author query to search for variations on the name of the institution. Once the curator has confidence in the list of affiliation names that should be used to positively identify items for the bibliography, an affiliation query is run on each string and the results are viewable in the spreadsheet.

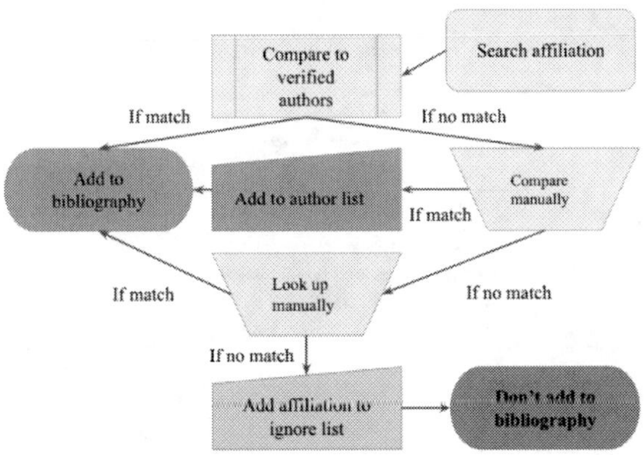

Figure 2: Example Workflow

When using an author list, the curator may compare the author names that are paired with the affiliation string of each result and add items that match both. Results from exact affiliation queries may be reviewed for unexpected items and then added as appropriate. Uncertain affiliation results should be reviewed more carefully and may include unusual affiliation name variations that the curator should be aware of.

If the curator has an author name list, they can perform an author query to make the bibliography more comprehensive. While some author names are unusual or unique, author queries tend to have higher recall and less precision overall due to frequent negative results from queries of common names. During the review of affiliation query results, many records would have been reviewed by the curator. In order to reduce the duplication of effort, the curator can use the spreadsheet to identify results from the author query that were already reviewed during the affiliation query and ignore them. The curator may opt to review the affiliation strings that are paired with the author name results in order to find previously unknown affiliation strings or typos. For author name results that do not have an affiliation, stringent curators may decide to use the links to items on the ADS's web interface to look for affiliation information on the original paper.

Once all the desired papers have been identified by the curator, they may be used to generate an ADS library from within the Google Sheets interface.

3 Evaluation

In order to compare the time commitment needed to maintain a bibliography using the classic interface vs using the Google Sheets add on, student colleagues timed themselves while performing example tasks. Each task involved including or excluding records for a bibliography.

Task 1 used the new bibliography tool. Task 1a was selecting records from the set of results that had affiliation strings. Task 1b was rejecting a subset of records with certain bibstems from a results set . Task 1c was reviewing results that had empty or missing affiliation strings in the bibliography tool.

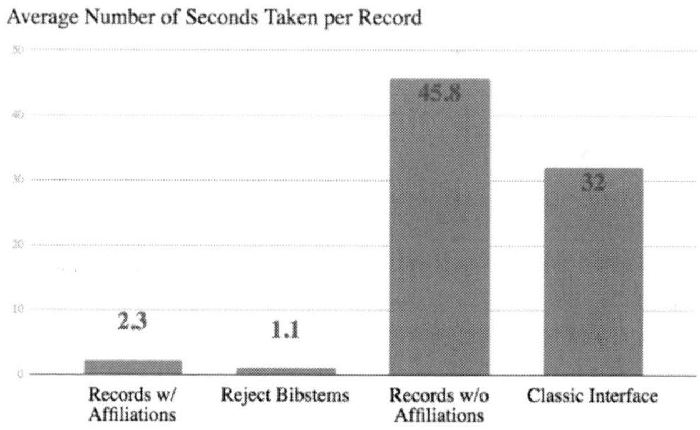

Figure 3

Task 2 was to use the ADS Classic interface to select records. The average time taken for each type of task can be seen in figure 3.

4 Results

By splitting review work into three discrete tasks and reducing the time needed for two out of three of the tasks, the tool offers an improvement over the previous method of bibliography generation as seen in the simulation results in figure 4.

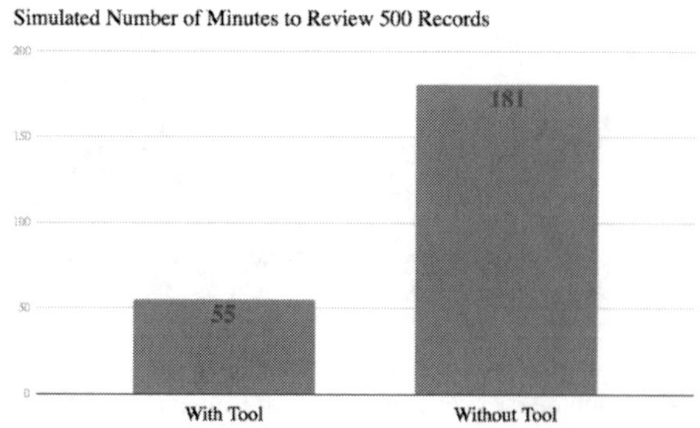

Figure 4

5 Future Work

Features currently under consideration or in process include bibgroup

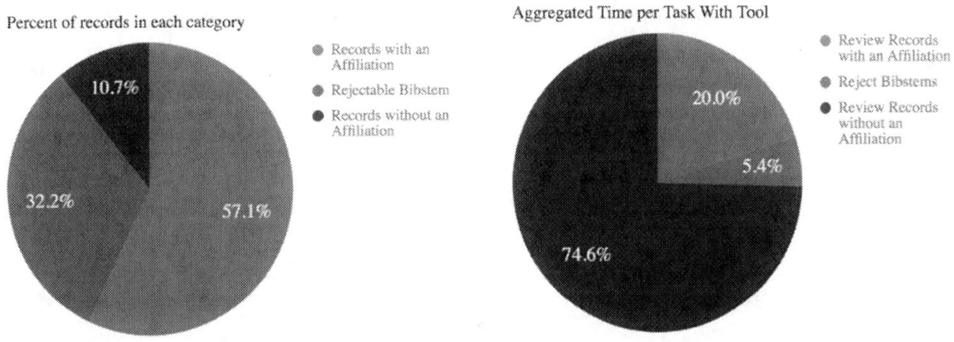

(a)   (b)

Figure 5

based queries, user defined searches, and formal synonym feedback mechanisms. Future development of fully featured spreadsheet based information retrieval interfaces may also be worth exploring.

Acknowledgements

NASA ADS is operated by the Smithsonian Astrophysical Observatory under NASA Cooperative Agreement NNX16AC86A

References

[1] A. Accomazzi, M. J. Kurtz, E. A. Henneken, et al., *ADS: The Next Generation Search Platform* in *Open Science at the Frontiers of Librarianship*, edited by András Holl, Soizick Lesteven, Dianne Dietrich, and Antonella Gasperini (2015), Vol. 492 of Astronomical Society of the Pacific Conference Series, p. 189–197

[2] C. S. Grant, D. M. Thompson, R. Chyla, et al., *Enabling Meaningful Affiliation Searches in the ADS* in *Open Science at the Frontiers of Librarianship*, edited by András Holl, Soizick Lesteven, Dianne Dietrich, and Antonella Gasperini (2015), Vol. 492 of Astronomical Society of the Pacific Conference Series, p. 208–211

Bibliographical references: From publishers to SIMBAD

Thomas Delacour[1,⋆], *Soizick* Lesteven[1], *Gilles* Landais[1], *Aline* Eisele[1], *Magali* Neuville[1], *Evelyne* Son[1], and *Philippe* Vonflie[1]

[1] *Centre de données astronomique de Strasbourg (CDS)*

Abstract. The SIMBAD astronomical database hosted by the CDS provides basic data, cross-identifications, bibliography and measurements for astronomical objects outside the solar system.
The CDS receives the bibliographic meta-data of the articles published in the main astronomical journals directly from the publishers. How we receive the data and their format vary from one publisher to the next. These data are first extracted and stored in files with a standardised format. Then, to avoid errors or misprints, we perform different tests on these data:

- Author names are compared to a reference list maintained at CDS, and the keywords are compared with the AAS list
- Astronomical objects are verified by checking their name in the SIMBAD database
- A completion test checks that all of articles of a journal volume are present

The next step identifies whether an astronomical object appears inside a title, a keyword or an abstract, and if so, we add a link to the object in SIMBAD. Once all of the verifications and corrections have been made we add the meta-data into SIMBAD. We also add other information such as the number of different astronomical objects studied in the paper, the presence tables and their links to VizieR, any new acronyms, as well as some comments. New developments are in progress to automatically extract the data from the tables in the articles (that have not been processed by, or provided to VizieR) .
In addition, each night automatic checks are executed to list the new data and to test the coherence of these data in SIMBAD.

Introduction: The context

The objective of the Centre de Données astronomiques de Strasbourg [1] is to collect, homogenize, distribute and preserve astronomical information published in astronomical journals for the usage of the whole community. The articles published in the astronomical literature represent the thread of our activity. Before looking for data inside the paper, we have to create the bibliographical entries for the SIMBAD and VizieR databases.

By special agreements with the main journals, the CDS receives meta-data of articles in electronic form. How we receive the data and their format vary from one publisher to the next, and also vary over time. The first operation is the extraction of the required information, and the second step is to store

⋆e-mail: thomas.delacour@astro.unistra.fr ORCID: 0000-0002-4941-7478

this data in a unique and specific format. Receiving all of these data allows us to be more exhaustive, more efficient and assure a better quality service. For a small number of journals documentalists still add bibliographic data manually.

1 The bibliographical data

In astronomy, each article is designed by the bibcode [2]. We associate this bibcode to the article DOI, the copyright, the numbers of pages, the complete list of authors with their affiliations, the title, the abstract and the keywords. All this information allows us to advertise and locate the papers.

1.1 Extraction/storage

We began our collaborations in 1994 with the A&A journal, which was quickly followed up with the AAS journals [3]. In 25 years, the data provided and their formats have evolved greatly. To take into account all these changes, we continually upgrade our software so that they are always adapted to the received data. Currently, we receive data mainly in XML format, sometimes in ASCII. The useful information is extracted from the files and is stored in parfiles.
The parfile format is used to store the data because it can be easily edited by a human and parsed by a program. A parfile consists on a series of blocks, each of them representing an article, separated by a new line. All the lines inside a block start with the symbol % followed by a letter which indicate what the line contains (Abstract, Title, Author...).

```
%R 2017A&A...600A..57C
%F A+A.ori/m2017-041/v600/aa29522-16.xml
%J-57
%M 15
%DOI 10.1051/0004-6361/201629522
%c {copyright} ESO, 2017
%A Chiaberge, M.
%A Ely, J.C.
%A Meyer, E.T.

...
%K galaxies: active
%K quasars: individual: 3C 186
%K galaxies: jets
%K gravitational waves
```

Figure 1. Parfile format

For each journal, volume per volume or issue per issue, all of the information is stored in parfiles.

1.2 Checks and SIMBAD update

To ensure the quality of the data and to avoid errors that can occur during the data extraction, we carry out a multitude of checks. For each issue or volume we process, we check the completeness of it to be sure that no articles are missing, and when we have an erratum we add this information to the related article. We also maintain a correlation list between the bibcodes and the DOIs. All author names are compared to the list of all authors already existing in our databases. If an author is not found, it must be verified to check if it is a new one or a misprint. In the same way we check the keywords with the list a keywords from the AAS, and when a keyword does not match any in the list the program will display a message. When an individual object is given in a keyword we first validate the identifier and then add a link to SIMBAD. Once all the data have been checked and corrected we add them into SIMBAD. It is important to point out that no corrections are made to the published data. We only correct errors made during our transcriptions. When there is an error on an object, we add a macro latex that will allow us to link the misnamed object directly to the right object in SIMBAD. In the other cases, we add a note to the data in SIMBAD. The journal editors are contacted, if needed, to signal an error.

At that stage, the documentalists in charge of updating SIMBAD begin to link the articles to the SIMBAD astronomical objects [4].

2 Table extractions for VizieR

Published articles may sometimes contain small astronomical data tables which should be integrated to Vizier [5]. Until recently, documentalists used the pdf file of the article provided by journals and copy-pasted the table information directly into a text editor. But this is not a convenient way to do it because it can modify the formatting of the table, some characters are not recognized and it demands a lot of work to get the content as it is in the pdf. The tables can now be processed from xml files provided by the publishers. A program has been written to retrieve files, extract, process and convert the tables into ASCII files. Documentalists can then use those files to create the standard Readme description for Vizier.

To download xml files the program is given a bibcode, get the corresponding doi and use it to connect to a publisher site. Once files are stored, the program will search for the xml then start to parse it. Tables are written in mathml or latex format so it has to be convert to ASCII before documentalists can use them. The program parses two times. The first one is to determine the shape of each table. This avoids problems in cases where the number of rows and columns are not consistent. The second time it will extract data, and mathematical symbols and as such are transformed so they can be easily integrated. The program will try to keep the shape of a table and write rows and columns separated by a space.

Besides data in tables, information about the associated article (authors, title, abstract, references, table description...) are also extracted. Each table is then written separately in ASCII and csv files. Sometimes a table cannot be processed into ASCII correctly due to its format or data, in which case a csv file may be used instead to manipulate data easily. A Readme file containing information about the article is generated along with a summary of different tables and reference articles. After creation of all files documentalists can verify the information and data by comparing them to the original tables and article. Once verifications are done the data are put in Vizier and can be consulted on the website along with tables.

In further updates of SIMBAD, we add other information to the bibliographical data such as the number of different astronomical objects studied in the paper, the presence of one or more tables and the links to VizieR, the creation of a new acronym, and also some comments.

3 Daily and weekly tests

Information processing at CDS is a chain [6]. To monitor the progress of the work among all the team members and assure the quality of the data, we have developed automatic test processes and we send the results to the concerned documentalists. Several tests are launched every day and/or week automatically.

Concerning the data we receive from the publishers, every night a program compares the present bibliographical content with the one from the day before. It checks for new references entered in SIMBAD, the deleted references, the number of references, and new authors names. It checks also the objects names in the keywords and the object names tagged by the authors in some journals (A&A). An email is sent to the CDS documentalists who correct, if necessary, the bibliographical data every morning.

Furthermore, each reference has an associated status that indicates that work has been done on it. The status generally goes with a detailed comment that indicates what has already been done. At the end of the week the program counts the total of every status and makes a comparison with the file containing the status of the last week. It reports to documentalists the number of status added, removed and the modifications of comments.

When data are modified in SIMBAD, tests allow verification of links validity and their correction if needed.

4 Conclusion

Receiving all of these data from the publishers allows us to be more exhaustive and more efficient. Bibliographical data go through a lot of steps from the publishers to SIMBAD. These processes and verifications are essential to maintain the quality and coherence of the database. Data are checked automatically or manually until their integration in SIMBAD which makes mistakes rare, moreover permanent checks ensure the correctness over time. Improvements are in progress to homogenize the database with the new abstract, keywords and DOI added to SIMBAD. Developments for table extraction continues with the objective to process more journals and return better results.

References

[1] F. Genova, D. Egret, O. Bienaymé, et al, A&AS **143**, 1 (2000)

[2] M. Schmitz, G. Helou, C. Lague, et al, Vistas in Astron., **39**, 272 (1995)

[3] F. Ochsenbein, J. Lequeux, Vistas in Astron., **39**, 227 (1995)

[4] M. Buga, C. Bot, M. Brouty et al., *How Documentalists Update SIMBAD* in *Open Science at the Frontiers of Librarianship*, edited by András Holl, Soizick Lesteven, Dianne Dietrich, and Antonella Gasperini (2015), Vol. 492 of Astronomical Society of the Pacific Conference Series, p. 47–50

[5] G. Landais, T. Boch, M. Brouty, et al., *Management of Catalogs at CDS* in *Open Science at the Frontiers of Librarianship*, edited by András Holl, Soizick Lesteven, Dianne Dietrich, and Antonella Gasperini (2015), Vol. 492 of Astronomical Society of the Pacific Conference Series, p. 57–60

[6] E. Perret, T. Boch, F. Bonnarel et al., *Working Together at CDS: The Symbiosis Between Astronomers, Documentalists, and IT Specialists* in *Open Science at the Frontiers of Librarianship*, edited by András Holl, Soizick Lesteven, Dianne Dietrich, and Antonella Gasperini (2015), Vol. 492 of Astronomical Society of the Pacific Conference Series, p. 13–21

The Cost of Astronomy

Publishing fees in astronomy: Is something rotten in the case of Denmark?

Bertil F. Dorch*

[1] *University Library of Southern Denmark, SDU, Campusvej 55, DK-5230 Odense, Denmark*

Abstract. Using Scopus and national sources, I have investigated the evolution of the cost of publishing in Danish astronomy on a fine scale over a number of years. I find that the number of publications per year from Danish astronomers increased by a factor of four during 15 years: naturally, the corresponding potential cost of publishing must have increased similarly. The actual realized cost of publishing in core journals are investigated for a high profile Danish astronomy research institutions. I argue that the situation is highly unstable if the current cost scenario continues, and I speculate that Danish astronomy is risking a scholarly communication collapse due to the combination of increasing subscription cost, increased research output, and increased direct publishing costs related to Open Access and other page charges.

1 Introduction

In March 2016, The New York Times [5] wrote that "Journal publishers collectively earned 10 billion USD last year, much of it from research libraries, which pay annual subscription fees ranging from 2,000 to 35,000 USD per title if they don't buy subscriptions of bundled titles, which cost millions. The largest companies, like Elsevier, Taylor & Francis, Springer and Wiley, typically have profit margins of over 30 percent"

In the same article, Peter Suber, the Director of Harvard's Office of Scholarly Communication was quoted saying that "the prices have been rising twice as fast as the price of health care over the past 20 years, so there's a real scandal there to be exposed". One solution, Suber was quoted saying, "was to persuade researchers to publish in Open Access journals like those under the umbrella of the Public Library of Science,or PLOS, co-founded by Dr. Eisen at Berkeley. But that financial model requires authors to pay a processing charge that can run anywhere from 1,500 to 3,000 USD per article so the publisher can recoup its costs".

In 2014, The Welcome Trust [2] published details of how much the research foundation spent on Open Access publishing during 2012–2013 in an attempt to make the debate around the costs of open access publishing more evidence-based: The Welcome Trust spent approximately 3.9 million GBP on APCs for 2126 articles in the year 2012–2013. Overall, the average APC for publication in hybrid journals is 2,727 USD.

A study, that is in a way a supplementary to that of [2] was presented by Gray [6] to show that the APCs paid to publish Open Access content, and the indirect costs of handling Open Access mandates

*e-mail: bfd@bib.sdu.dk ORCID: 0000-0003-2594-6778

Table 1. Core journals of Danish astronomy defined as journals with contributions from authors coming from Denmark during 2010–2014 according to Scopus (left column), and the average annual number of publications in the corresponding journals (right column).

Journal	Average annual number of publications
Astronomy and Astrophysics	55.5
Astronomical Journal	3.1
Annales Geophysicae	2.7
Astrophysical Journal*	57.3
Astronomische Nachrichten	3.8
Experimental Astronomy	1.1
Proceedings of the International Astronomical Union	6.6
ICARUS journal	1.1
International Journal of Modern Physics A-D	7.5
Journal of Cosmology and Astroparticle Physics	8.6
Monthly Notices of the Royal Astronomical Society*	29.7
Publication of the Astronomical Society of Japan	0.5
Publications of the Astronomical Society of the Pacific	1.1
Physics of the Earth and Planetary Interiors	1.3
Planetary and Space Science	1.8
Solar Physics	2.2
Space Science Reviews	1.1

costs may be unexpectedly high for individual institutions: the study added an estimate for the other publication charges currently paid by research institutions, a significant element which has been neglected by recent studies. When page and color charges are included, the total cost to institutions as of 2013–2014 is around 18.5 percent over and above the cost of journal subscriptions—11 percent from APCs, 5.5 percent from indirect costs, and 2 percent from other publication charges. For the Great Britain as a whole, this represents a total cost of publication around 213 million GBP against a conservatively estimated journal spend of 180 million GBP, with non-Open Access APCs representing around 3.6 million GBP [6].

Recently, the Max Planck Digital Library White Paper on Open Access by [3] makes the case for a large-scale transformation of the current scholarly publishing system to an Open Access model. The White Paper advocates that there need to be "a shared understanding that the money currently locked in the journal subscription system must be withdrawn and re-purposed for open access publishing services".

Gauffriau *et al.* [4] conducted an analysis showing that the funds spent on journal licenses by universities in Denmark is sufficient to pay for the number of publications from the same universities should they be subjected to an APC. In accordance with the Max Planck Digital Library White Paper, the authors found that the average cost per publication of 3,600 Euro in Denmark is only a little lower than a similar calculation by [3] yielding a global level cost per publication of 3,800 Euro [4].

2 Danish astronomy as a case

To investigate the cost of scholarly publishing on an even finer scale – and to include all direct publication costs, as argued by e.g. [6], I used astronomy and astrophysics in Denmark as a case. It can be argued that astronomy in several ways is both a traditional field of natural science with roots back

Table 2. Publishing fees associated with core journals of Danish astronomy (article charges in Euros, 2017).

Journal	Publisher	APC (Euro)	Gold OA
Astronomy and Astrophysics	EDP	-	400
Astronomical Journal	IOP	763	-
Astrophysical Journal	IOP	763	-
Journal of Cosmology and Astroparticle Physics	IOP	-	1600
Monthly Notices of the Royal Astronomical Society	Oxford	-	2175
Publications of the Astronomical Society of the Pacific	Chicago	1000	-
Solar Physics	Springer	-	2200

to the origin of the scientific method and beyond, but also a first-mover e.g. with respect to adopting Open Science practices including Open Access all the way back to the birth of the Internet. Thanks to arXiv.org, within the field of astronomy and astrophysics, Open Access has generally been achieved through Green Open Access self-archiving.

The core journals of Danish astronomy are defined here by considering all astronomy "documents" with contributions from authors coming from Denmark during 2010--2014 according to Scopus (46 journals in total), and selecting the set of journals with recurrent authorships (i.e. no one-hit wonders) resulting in the journals listed in Table 1.

During the period 2000–2014 the total number of publications per year increased from around a hundred articles to several hundred, cf. Fig. 1 (left). Mostly, the increase comes from publishing in the identified core journals, which also happen to be journals with related publishing fees, i.e. either Golden Open Access APCs or non-Open Access article charges, cf. Fig. 1 (right). Only a minor part of the articles are published in Platinum Open Access journals (journals without associated APCs or fees).

On the one hand, even though most core astronomy journals allow or even encourage self-archiving, some of them do in fact offer a Golden Open Access option corresponding to an APC. On the other hand, some core astronomy journals require APCs to be paid that are not related to Open Access. An example of the latter is ApJ from The American Astronomical Society, which is published by IOP Publishing. Article charges are calculated by The American Astronomical Society by counting units of information in the digital form that the author supplies called "digital quanta", i.e. units of information in digital form that can include words, figures, tables, digital-only components etc.

This way of calculating article charges makes the cost of an article a unique value, as opposed to the typical Open Access APC which is fixed for a particular journal or publisher, cf. Table 2.

In order to get a handle on the typical non-Open Access article charge for a paper published in ApJ, I calculated the average cost of publishing in this journal by the DARK Cosmology Center at University of Copenhagen during 2012–2015. These four years, scholars at DARK together with international colleagues published a substantial number of articles in ApJ and were billed by IOP for a total of 32 of these articles costing DARK all in a sum of 181,980 DKK. The average cost per article per year ranged between 5,033 and 9,100 DKK corresponding to an overall average of 763 Euro.

As an example for comparison to DARK's 32 ApJ articles during four years, a total 1030 articles with authors from Denmark was published in The Astrophysical Journal during a quarter of a century, 1990–2016. Of these, 775 articles also had co-authors from the USA, 411 articles have co-authors from the UK, 327 have co-authors from Germany, 264 have co-authors from Italy, 206 have co-authors from France etc. (Scopus, March 2016). The average is thus almost 40 ApJ articles per year (nearly 160 papers per four years) corresponding to potentially tens of thousands of Euros per year.

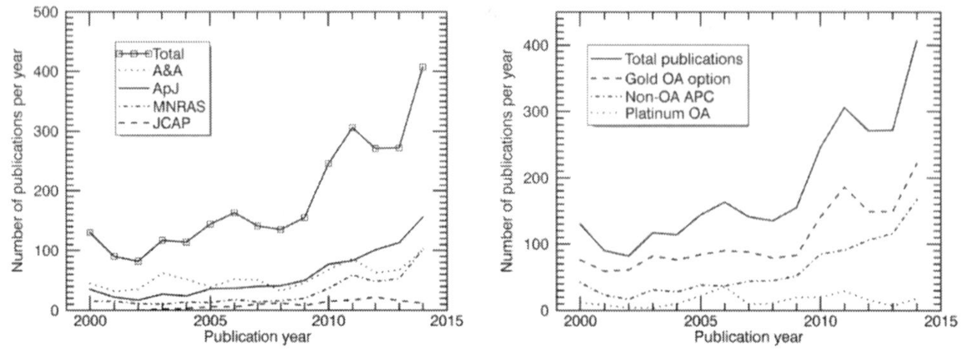

Figure 1. Left: The total number of publications per year from 2000 to 2014 (solid squared curve). Also shown are the number of publications per year in core journals A&A; ApJ, MNRAS and JCAP. Right: The total number of publications per year broken down into publications with a Gold Open Access option, a non-Open Access article charge (APC) and Platinum Open Access (no APC's or fees).

3 Discussion and outlook

Non-Open Access or Toll Access publishing in astronomy and astrophysics is traditionally supplemented by Green Open Access through authors' self-archiving into the open repository arXiv.org at Cornell University Library. This method of achieving Open Access has been the modus operandi for astronomy and astrophysics even prior to the invention of the term Open Access around the millennium: arXiv.org has now been in operation for more than a quarter of a century.

The estimated (maximum) potential annual cost of APCs related either Golden Open Access publishing or traditional Toll Access publishing in astronomy and astrophysics — based on the publication record in Denmark during 15 years — are the same order of magnitude, cf. Fig. 2.

Consequently, the potential cost of publishing in astronomy and astrophysics in Denmark is on the order of a several 100,000 Euros per year, including both Open Access and non-Open Access publishing.

Publishing in astronomy is of course funded in several different ways, e.g. through international collaboration, and potential costs — future or present — are not only realized in Denmark. However, it remains to be seen, if the current publishing system is sustainable in the Open Access era, and if the realization of increasing costs will quench the apparent growth of Danish astronomy as reflected by the publication output from 2000 and thereafter. Additionally, the important factor of subscription costs needs to be included when assessing the evolution of the said economy.

Acknowledgements
I wish to acknowledge the DARK Cosmology Center at University of Copenhagen for providing me with data regarding the center's publishing in ApJ. Furthermore, I would like to acknowledge SDU, The University of Southern Denmark, the organizing committee of the LISA 8 conference, and the International Astronomical Union for supporting LISA 8.

 https://doi.org/10.1051/epjconf/201818612005

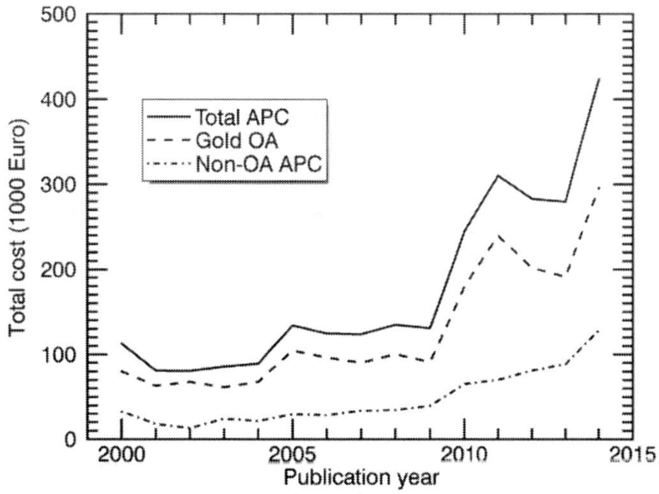

Figure 2. The potential (maximum) total cost of Danish astronomy publishing per year from 2000 to 2014 (solid curve). The potential cost per year is broken down into Gold Open Access fees (dashed curve) and non-Open Access article charges (dashed-dotted curve).

References

[1] Van Noorden, R., Nature **495**, 426–429. doi:10.1038/495426a (2013)

[2] B.-C. Björk, D. Solomon, **Developing an Effective Market for Open Access Article Processing Charges** (Report from Wellcome Trust, United Kingdom, 2014)

[3] R. Schimmer, K. K. Geschuhn, A. Vogler, **Disrupting the subscription journals' business model for the necessary large-scale transformation to open access** (Max Planck Digital Library Open Access Policy White Paper, Germany, 2015)

[4] M. Gauffriau, A. V. Larsen, K. H. Lianez, **Transition from the License to read model to Open Access publishing. A Danish study** (DEFF Report, Denmark, 2016)

[5] K. Murphy, "Should all research papers be free?", The New York Times, **March 12** (2016)

[6] A. Gray, Publications **3**, 248-262 (2015)

EPJ Web of Conferences **186**, 12006 (2018) https://doi.org/10.1051/epjconf/201818612006
LISA VIII

Negotiating with the scientific publishers

A new experience for the Italian Astronomical Libraries

Antonella Gasperini[1,*] and *Laura* Abrami[2,**]

[1] *INAF-Osservatorio astrofisico di Arcetri, largo Enrico Fermi 5, I-50125 Firenze, Italy*
[2] *INAF-Osservatorio astronomico di Trieste, via G.B. Tiepolo 11, I-34143, Trieste, Italy*

Abstract. In 2016, Italian Astronomical Libraries decided not to renew their subscriptions to electronic journals directly with the most important scientific publishers but to adhere, wherever possible, to the negotiating activities conducted by CARE, a coordinating group within the Libraries Commission, of the Conference of Italian University Rectors (CRUI), the Italian association of state and private universities. Participating in a consortium would have meant better economic conditions for INAF and more favourable contract terms for the final user. Advantages and disadvantages of direct negotiations with the publishers or of participation in a large consortium are analyzed here.

1 Introduction

In 2002, with the merging of the twelve Italian astronomical observatories into a single institution and the birth of INAF-National Institute for Astrophysics, a centralized national astronomical Library Service was created with the aim to promote collaboration among observatory libraries. The first and the most relevant activity was the cooperative acquisition of journals, that was carried out in 2005 [1].

In the subsequent years, most of the subscriptions were made through agents by call for tender and only a small part by direct contracts with the publishers. With the disappearance of the printed version of the journals and the gradual affirmation of the online-only subscriptions, the number of journal titles included in the call for tender was reduced and the direct contracts with the publishers increased.

The opportunity to take part in an existing consortium for the purchase of electronic resources had been considered since 2006, but the applied conditions were not advantageous for research institutes focused on one discipline, as the National Institute for Astrophysics. Furthermore, the agreements proposed by the publishers forced the Institute to subscribe to multidisciplinary packages including a very high number of off-topic titles.

2 Participation in the CARE consortium

CARE[1] is a coordinating group within the Libraries Commission of the Conference of Italian University Rectors (CRUI), the Italian association of state and private universities. CARE manages the

*e-mail: antonella.gasperini@inaf.it ORCID: 0000 0002 1666 6558
**e-mail: laura.abrami@inaf.it
[1] http://www.crui-risorselettroniche.it/

negotiations with international publishers for most of the Italian academic and research libraries with the aim to have a stronger bargaining power. It works to:

- facilitate the acquisition and the management of electronic resources economically
- obtain better contractual conditions and the security in the access and preservation
- strengthen the position of the universities and research institutes against the monopoly of the international publishers
- promote the collaboration with other library consortia internationally

Initially CARE managed contracts with the publishers only for Italian universities. Only later some research institutes expressed their interest in taking part in this consortium. INAF was one of them and, even if there were no well-established previous experiences, our Institute decided to participate in CARE and to accept possible uncertainties and risks.

At the end of 2016, Italian Astronomical Libraries, after a long period of information gathering and analysis, decided not to renew their subscriptions to electronic journals directly with the most important scientific publishers but to adhere, wherever possible, to the negotiating activities conducted by CARE.

3 INAF subscriptions

In 2017, INAF subscribed to more than 800 journals[2], as well as some databases, and the subscriptions were managed in three different ways:

1. through direct contracts with the following publishers for the e-journals and databases: ASP, EDP, IOP, OSA, SPIE. It was a solution adopted for the publishers not in negotiation with CARE and for those negotiated by CARE but with unfavourable terms and conditions
2. through CARE consortium: AAAAScience, Annual Reviews, Elsevier, IEEE, Nature, Springer, Wiley. These online-only subscriptions were negotiated by CARE with terms and conditions favourable for INAF
3. through an agent selected by a tendering procedure: this solution was chosen especially for subscriptions including print editions.

In comparison with 2016, the renewal of some subscriptions through CARE brought immediate advantages such as the increase of subscribed titles and cost containment. For some publishers, for example Annual Reviews and Springer, both aspects were particularly relevant (see Fig. 1).

4 Pros and cons

An evaluation of the favourable and negative aspects of INAF's participation in CARE contracts can be made after one year experience.

The positive features that emerged from the work done can be summarize as follows:

- greater bargaining power with the publishers
- same economic conditions but acquisition of more resources (Springer, Annual Reviews)

[2]http://www.beniculturali.inaf.it/eng/libraries/online-journals-and-resources/

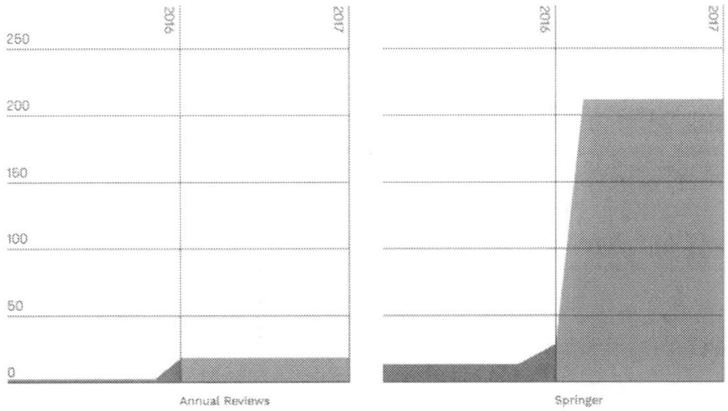

Figure 1. Increase of subscribed titles

- same economic conditions with lower price increase for the years ahead
- administrative (invoicing and VAT fulfillment) simplification: INAF has to pay only a debit note to CRUI.

On the other hand, the following critical issues need to be considered:

- "big-deal" negotiations with multidisciplinary scientific and technological publishers. Astronomy is too specific a matter (e.g. CRUI signed a big contract with IOP, but the package does not contain The Astrophysical Journal and The Astronomical Journal)
- acquisitions of single titles and not thematic packages. Some publishers offer this possibility but in most cases the economic conditions (subscription and price increase) are the same as with the direct contracts (to be considered also the fee for CRUI)
- payment of a small fee to CRUI.

5 Conclusion

The cooperation between academic and research libraries is fundamental not only for the cost containment and a better organization of the e-resources, but also to confront the aggressive policies of the scientific commercial publishing. However, the particular specificity of the astronomical libraries, which is mainly focused on a single discipline, does not permit adherence to the big contracts (covering a too large range of titles).

Reference

[1] A. Gasperini, L. Abrami, E. Olostro Cirella, *Astronomical Interlibrary Cooperation: The Long and Difficult Plan for Coordinated Acquisition of Journals – the Italian Case* in *Library and Information Services in Astronomy V: Common Challenges, Uncommon Solutions*, edited by S. Ricketts, C. Birdie, E. Isaksson (2007), Vol. 377 of Astronomical Society of the Pacific Conference Series, p. 397–404

EPJ Web of Conferences **186**, 12007 (2018) https://doi.org/10.1051/epjconf/201818612007
LISA VIII

The NOAO Publications Tracking Program: Developing Policies & Procedures for Usage, Acknowledgment, and Citation of Data

Sharon Hunt[1,*] and *Dara J.* Norman[1]

[1] *National Optical Astronomy Observatory, 950 N. Cherry Ave., Tucson, AZ 85719, USA*

Abstract. The National Optical Astronomy Observatory (NOAO) has a complex publications tracking program, which is coordinated by the NOAO Librarian. We track publications that use data from multiple observatories (Kitt Peak National Observatory [KPNO] in Tucson, Arizona, and Cerro Tololo Inter-American Observatory [CTIO] in La Serena, Chile, as well as non-NOAO telescopes to which we grant community access at other observatories); multiple telescopes and instruments; multiple observing programs, including survey programs; and multiple datasets in the Science Data Archive (SDA) at NOAO. The growing SDA and new user services, such as the Data Lab, ensure that the use of NOAO archival datasets will become even more widespread and important for scientific discovery and research. Accurate and complete publication tracking is essential for reporting to our funding agencies, developing our programs, and determining our scientific contributions and impact. Setting policies to guide authors in specifying and acknowledging usage of NOAO data is an essential element in our ability to determine and document relevant publications. Developing systems to track the use of our telescope data and archival datasets is another critical element. New policies to track the use of our telescope data are being developed, new systems to track the use and reuse of datasets within our particular institutional context are being explored, and methods of linking our publications data with our dataset usage data are being devised. Updated policies for the proper and ethical usage, acknowledgement, and citation of our archival data are being written and reviewed.

1 Introduction

A high-quality telescope bibliography is an essential tool for developing and reporting on observatory programs; at the same time, acknowledgments of data usage are essential for making explicit the scientific contributions of an observatory, particularly to funding agencies. The National Optical Astronomy Observatory (NOAO) has sought to develop our publications tracking policies and procedures to improve the efficiency and accuracy of our publications tracking, to provide guidance to telescope data users on appropriate acknowledgments in publications, and to develop and adhere to standards of professional ethics relating to data usage.

* e-mail: hunt@noao.edu ORCID: 0000-0002-5858-2002

 https://doi.org/10.1051/epjconf/201818612007

2 Identifying and Documenting Publications That Utilize Data from NOAO Facilities and Services

NOAO has a multifaceted publications tracking program, as we track the usage of telescope data from multiple sites, telescopes and instruments, survey programs, and data products and services. We maintain telescopes at two sites (Kitt Peak National Observatory in Tucson, Arizona, USA, and Cerro Tololo Inter-American Observatory in La Serena/Cerro Tololo, Chile). We also grant community-access telescope time on several non-NOAO telescopes. We collect astronomical data from multiple telescope/instrument combinations and survey programs and store and make accessible data in the Science Data Archive at NOAO.

Our literature search and analysis process is complex, utilizing multiple keywords and searching strategies due to the wide scope of our program and the variant designations (e.g., KPNO vs. Kitt Peak, CTIO vs. Cerro Tololo; Mayall vs. Kitt Peak 4m, etc.) used by authors for our resources. We also contend with the issue of incomplete documentation on the data used in some publications. Because of these factors, we face the need to analyze closely many publications to weed out false positives and determine the specifics of data usage. The documentation of the many parameters relating to use of our data (telescope/instrument combination; survey program; dataset; investigators) is time-consuming and labor-intensive.

Our revised policies and procedures address these challenges in identifying and documenting relevant publications in two ways: providing explicit and readily accessible acknowledgements statements and utilizing the NOAO Proposal ID as a source of details of data usage.

2.1 Acknowledgments Guidelines

Standardized and comprehensive acknowledgment of the use of NOAO data by authors is vital to properly tracking publications. In addition, ethical standards would suggest that there should be recognition of other people's work through proper acknowledgment of data usage. Our first goal was to develop explicit guidelines and statements on how authors can stipulate and acknowledge use and reuse of our telescope data, data products, and archival datasets. We then made these policies readily accessible through our NOAO Library Publications Acknowledgements web page. This resource serves as a centralized repository for NOAO publications policies and procedures, with links from all appropriate web pages (e.g., Science Data Archive at NOAO; proposals for telescope usage web pages; and publications listings) so that authors can easily find acknowledgments information and so that acknowledgments statements are standardized in all publications. This web page contains acknowledgements statements specific to particular telescopes, programs, and data products and services and thus provides explicit guidance to authors on our expectations relating to the use of NOAO data and their responsibilities as users of this data.

2.2 NOAO Proposal Identifiers

The NOAO Proposal Identifier (NOAO Prop. ID) is a unique number given to each proposal requesting time on our telescopes. See Figure 1 for an example of a NOAO Prop. ID web page and the information it provides.

This unique identifier offers a mechanism to streamline and enhance the literature searching and analysis process. We now request the usage of this identifier in the acknowledgments section of publications. It provides us with the needed details on investigators, dates, telescopes, datasets, and instruments and negates the necessity of individually capturing specifics of data usage from a particular publication. It often supplies more information on the data usage than may be explicitly mentioned in a publication, and it lessens the amount of time and effort required to track and analyze publications.

Proposal Information for 2014A-0073

PI: Mukremin Kilic, University of Oklahoma, kilic@ou.edu
Address: Physics Department, 440 W Brooks Street, Norman, OK 73019, USA

CoI: Sara Barber, University of Oklahoma
CoI: Buell Januzzi, NOAO
CoI: Arjun Dey, NOAO
CoI: Peter Stetson, National Research Council of Canada

Title: A Search for Habitable Planets Around White Dwarfs

Abstract: A large fraction of white dwarfs (WDs) may host planets in their habitable zones. Here we propose to take advantage of the unique capability of DECam on the CTIO 4m to perform a pre-LSST transit survey that is capable of detecting the first planet in the habitable zone of a WD. We propose to obtain DECam time-series photometry of the CFHT Legacy Survey (CFHTLS) Wide2 field to search for planetary, substellar, and stellar companions to more than 500 WDs. Thanks to the 3 square degree field of view of DECam, for the first time in history, a large number of WDs can be observed at the same time, which is essential for discovering transits. We will use the original data taken 8 years ago for the CFHTLS and our stacked DECam images to identify the WDs based on their colors and proper motions. Given the observing window from the ground, if every WD has an Earth-like planet in its habitable zone, we expect to find up to 3 planets in this initial survey. In addition to planets, we will also constrain the frequency of eclipsing substellar and stellar companions to WDs, which has important implications for the LSST. Our stacked dataset will also provide a significantly improved WD luminosity function for the Galactic disk and halo.

Figure 1. NOAO proposal web page, providing information on observing program, investigators, and telescope/instrument combinations.

The NOAO Prop. ID ties the data used in a publication to specific investigators and will be attached to any re-use of this data in another publication. It thus serves to make the contribution of the investigator explicit and easily trackable.

We are developing new systems to track the use and reuse of datasets within our particular institutional context. The NOAO Prop. ID is the mechanism through which we can link our proposals database, datasets in the Science Data Archive at NOAO, and our publications database, allowing us to combine and analyze in a seamless manner the details of NOAO data usage (Figures 2 and 3).

Proposal ID	Survey ID	Release date	Observing date	UT	PI	RA	Dec	Telescope	Instrument	Filter	Exposure	Observation type	Observing mode	Processing	Product
2014A-0073	?	2015-08-03	2014-02-02	2014-02-03 05:03:21.505	Kilic	135.758700	-4.583333	ct4m	decam	g DECam SDSS c0001 4720.0 1520.0	30.0	object	imaging	RAW	image

Figure 2. Science Data Archive at NOAO web page with information on observing program

3 NOAO Data Policies

Usage of NOAO archival datasets is becoming even more widespread and important for scientific discovery and follow-up, making proper acknowledgment complex. Policies on standards of professional

Figure 3. Citation from NOAO online telescope bibliography using data from observing program of NOAO Prop. ID 2014A-0073. The NOAO Prop. ID was listed in the acknowledgments section of this publication. The citation, NOAO proposal web page, and Science Data Archive at NOAO are linked through the NOAO Prop. ID.

Belardi, C., ... **Dey, A.,** et al. 2016, *MNRAS*, 462, 2506, The DECam minute cadence survey - I
2016MNRAS.462.2506B
Blanco +DECam; NOAO Prop. ID 2014A-0073, PI: M. Kilic; Science Data Archive at NOAO

Figure 4. Citation from NOAO online telescope bibliography using data from observing program of NOAO Prop. ID 2014A-0073. The NOAO Prop. ID was listed in the acknowledgments section of this publication. The citation, NOAO proposal web page, and Science Data Archive at NOAO are linked through the NOAO Prop. ID.

ethics and conduct with respect to data usage are not always known and adhered to. Furthermore, revised policies are needed to reflect our expansion into open-access digital science platforms, services, and software (e.g., Data Lab, ANTARES).

We seek to ensure that NOAO data policies are consistent with (or ahead of) other observatories and data archives/centers and that they reflect proper ethical standards for data use and citation. We regularly review policies of other observatories and data archives/centers and revise and add statements for new data products, services, and software as needed. We propose modifications and updates to suggested responsible use of archival data, possibly including acknowledgment of original PI(s), for review by relevant managers. We propose to work with publication groups to modify instructions to journal referees to look for and point out failures to properly acknowledge data use, software, and services.

4 Conclusions

Since we have developed our NOAO Library Publications Acknowledgments web page, we have seen steadily increasing usage of our suggested acknowledgments statements in publications. Providing a readily accessible reference source detailing our expectations regarding acknowledgments has aided authors in specifying the usage of our data. This in turn has streamlined our publications tracking process. Our documentation of NOAO data usage is more efficient and more complete with the increased usage of NOAO Prop. IDs in publication acknowledgments sections, as the Prop. ID provides us with a wealth of information on the observing program that gathered the data used in a specific publication. The use of our standardized acknowledgments statements more readily identifies NOAO's and specific investigators' contributions.

Acknowledgments

The National Optical Astronomy Observatory (NOAO) is operated by the Association of Universities for Research in Astronomy (AURA) under cooperative agreement with the National Science Foundation (NSF).

"Boletín de los Observatorios de Tonantzintla y Tacubaya" 1952 to 1972: Preliminary results of a bibliometric study

María Elena Jiménez-Fragozo[1,*] and *Hector* Aceves[1]

[1] *Instituto de Astronomía. UNAM. Ensenada. México*

Abstract. We analyzed the scientific production of astronomers published in the first Mexican journal of astronomy, the *Boletín de losObservatorios de Tonantzintla y Tacubaya* (BOTT), from 1952 to 1972. The bibliometric behavior of the BOTT is provided, identifying the most productive authors, article usage, as well as impact and visibility.

1 Introduction

Editing the publications of the Mexican astronomical community started in1880 with the first volume of the *AnuarioAstronómico Nacional*, which continues to be published today. Ten years later, in 1890 publication of the *BoletíndelObservatorioAstronómico de Tacubaya* started,endinged in 1896. In 1952, fifty six years later and in a different Mexican historical time, the *Boletín de losObservatorios de Tonantzintla y Tacubaya* [BOTT] was published [1].

The first issue of the BOTT was published in January 1952 with a 9 page article in Spanish, with an abstract written in English, by the Mexican astronomer Guillermo Haro. The first five issues copies of the BOTT were mimeographed, a very popular, artisanal and inexpensive way to print at that time and it was disseminated to the international astronomical community by postal mail (Fig. 1).

The BOTT was totally financed by the *Instituto de Astronomía* of the Universidad Nacional Autónoma de México (UNAM). The back of the title page of volume 5, number 35 (1970), specifies that the BOTT was partially sponsored by the *Instituto Nacional de la InvestigaciónCientífica* and for the first time included the names of the editorial board members: Guillermo Haro, Paris Pismis and Silvia Torres-Peimbert.

The BOTT was published from January 1952 to December 1972, and it was the main publication where Mexican astronomers published their research results. It was founded and edited throughout the 20 years mainly by Guillermo Haro, "to publish only the original astronomical research, which, were not being published elsewhere" [2]. The main objective of this work is to show preliminary results of the BOTT bibliometric behavior.

2 Methodology

The BOTT was not a peer-reviewed publication, the works werenot included in any commercial databases, thereforethe sources used for this research were NASA's Astrophysics Data Systems and

*e-mail: jimenez@astro.unam.mx ORCID: 0000-0002-6654-5148

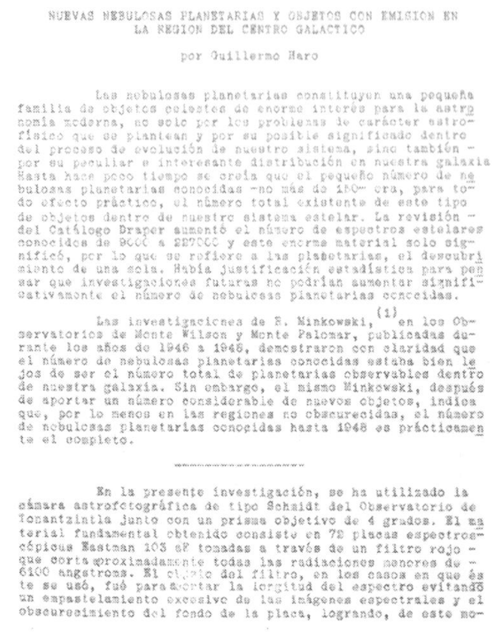

NUEVAS NEBULOSAS PLANETARIAS Y OBJETOS CON EMISION EN LA REGION DEL CENTRO GALACTICO

por Guillermo Haro

Las nebulosas planetarias constituyen una pequeña familia de objetos celestes de enorme interés para la astronomía moderna, no solo por los problemas de carácter astrofísico que se plantean y por su posible significado dentro del proceso de evolución de nuestro sistema, sino también por su peculiar e interesante distribución en nuestra galaxia Hasta hace poco tiempo se creía que el pequeño número de nebulosas planetarias conocidas -no más de 150- era, para todo efecto práctico, el número total existente de este tipo de objetos dentro de nuestro sistema estelar. La revisión del Catálogo Draper aumentó el número de espectros estelares conocidos de 9000 a 227000 y este enorme material solo significó, por lo que se refiere a las planetarias, el descubrimiento de una sola. Había justificación estadística para pensar que investigaciones futuras no podrían aumentar significativamente el número de nebulosas planetarias conocidas.

Las investigaciones de R. Minkowski, (1) en los Observatorios de Monte Wilson y Monte Palomar, publicadas durante los años de 1946 a 1948, demostraron con claridad que el número de nebulosas planetarias conocidas estaba bien lejos de ser el número total de planetarias observables dentro de nuestra galaxia. Sin embargo, el mismo Minkowski, después de aportar un número considerable de nuevos objetos, indica que, por lo menos en las regiones no obscurecidas, el número de nebulosas planetarias conocidas hasta 1948 es prácticamente el completo.

En la presente investigación, se ha utilizado la cámara astrofotográfica de tipo Schmidt del Observatorio de Tonantzintla junto con un prisma objetivo de 4 grados. El material fundamental obtenido consiste en 72 placas espectroscópicas Eastman 103 aF tomadas a través de un filtro rojo que corta aproximadamente todas las radiaciones menores de 6100 angstroms. El objeto del filtro, en los casos en que éste se usó, fué para acortar la longitud del espectro evitando un empastelamiento excesivo de las imágenes espectrales y el obscurecimiento del fondo de la placa, logrando, de este mo-

Figure 1. First issue cover and first article title page in the BOTT.

the BOTT itself, both in the printed version available at our Library as well as digital versions available online.[1] The BOTT published 176 papers in 6 volumes with 1,698 references and received 2,407 citations until April 2017. In order to facilitate our study three databases were made, one for scientific production, a second one for the references and a third one for citations (results which are not included in this work). Standardizing authors, titles and institutions in order to appear under a single and unique denomination was done. The software used to analyze the data was Microsoft Excel®

3 Results

Publication date and the number of articles published by volume number were found to be irregularly distributed in time. Some of the volumesincluded just one article. However, in 1970 the publications reached 21 articles, being the highest number of articles published in a year in the BOTT (Fig. 2).

Principal author analysis shows that 102 of the 176 articles published in the BOTT were written by 6 authors; a 57.95% of the total scientific production reported (Fig. 3, *left*)

A strong collaboration between Mexican and American astronomers, from prestigious observatories and university departments, was observed. Collaborations with other nations included astronomers from the former Soviet Union, N. Ireland, Argentina and Chile (Fig. 3, *right*).

[1] www.astroscu.unam.mx/bott

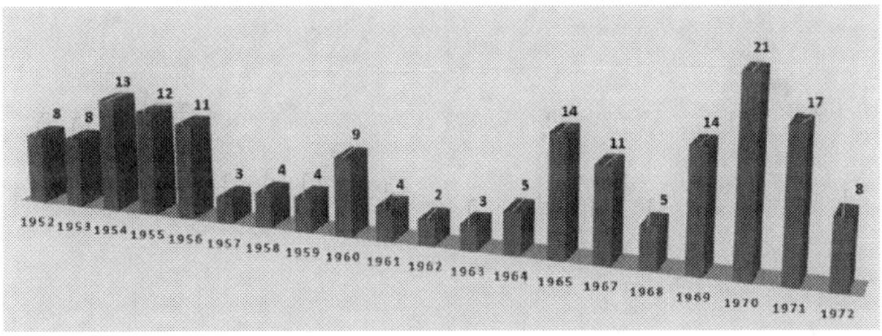

Figure 2. Number of articles published per year.

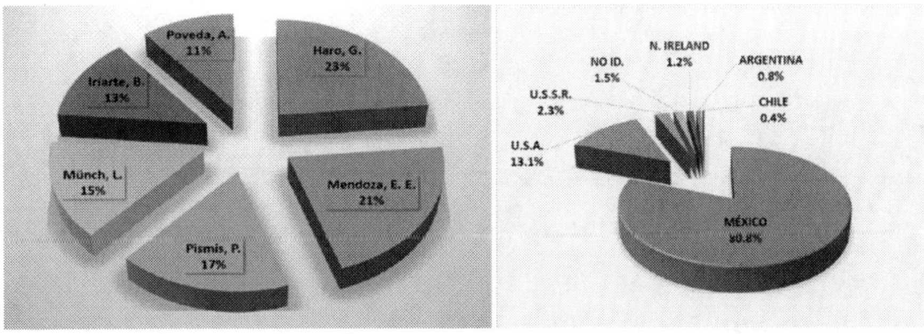

Figure 3. (*Left*) First author distribution. (*Right*) Distribution of BOTT publications by authors' affiliation.

The BOTT published original works on mainstream topics in astronomy and astrophysics. Six articles received 1,123 cites, which means 46.7% of the 2,407 citations received. The distribution of Authors, Article Title, Number of Citations received and percentage is shown in Fig. 4 .

AUTHOR	ARTICLE TITLE	CITATIONS	%
Peimbert, M.	Chemical Abundances in Galactic HII Regions	393	16.33
Poveda, A.	Run-away Stars as the Result of the Gravitational Collapse of Proto-stellar Clusters	175	7.27
Johnson, H. L.	The colors, Bolometric corrections and effective temperatures of the bright stars	174	7.23
Mendoza, E. E.	Multicolor Photometry of Stellar Aggregates	155	6.44
Mitchell, R. I.	Photoelectric Photometry of Cepheid Variable Stars	119	4.94
Haro, G.	Preliminary Note on Blue Galaxies with emission lines	107	4.45
		1123	46.66

Figure 4. Distribution of the six most cited articles published in the BOTT.

The Bradford's law was applied to the total data sample of references cited on the BOTT. More than half of the references (59.48%) are in a core list of six journals titles, that included the BOTT, as shown in Fig. 5. The latter indicates that astronomers who published in the BOTT used mainstream science journals of astronomy and astrophysics.

The first scientific document published in the BOTTwritten by Mexican women astronomers, was written by Guillermina and Graciela González (who were sisters) in 1952 entitled: *"Estrellas de altaluminosidaden las vecindades de las constelaciones del Cisne y Cefeo"* (NASA ADS title: `A finding list of high luminosity stars around Cygnus and Cepheus`); see Figure 6.

Journals	Docs.	No. docs.	Cumulative Journals	Cumulative docs.	% Cumulative docs.	%Cumulative Journals	JOURNAL TITLES
1	585	585	1	585	34.45	0.40	The Astrophysical Journal
1	199	199	2	784	46.17	0.80	Boletín de los Observatorios Tonantzintla y Tacubaya
1	67	67	3	851	50.12	1.20	The Astronomical Journal
1	64	64	4	915	53.89	1.59	Monthly Notices of the Royal Astronomical Society
1	60	60	5	975	57.42	1.99	Publications of the Astronomical Society of the Pacific
1	35	35	6	1010	59.48	2.39	The Astrophysical Journal Supplement Series

Figure 5. Journals used by astronomers publishing in the BOTT.

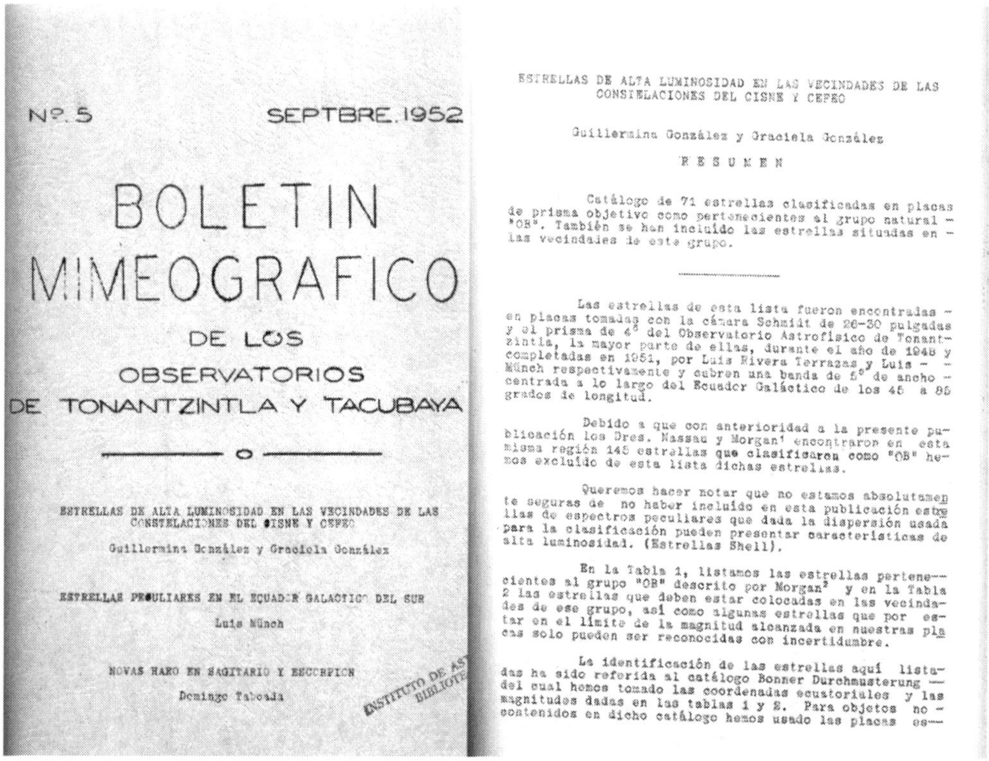
Nº. 5 SEPTBRE. 1952

BOLETIN MIMEOGRAFICO DE LOS OBSERVATORIOS DE TONANTZINTLA Y TACUBAYA

ESTRELLAS DE ALTA LUMINOSIDAD EN LAS VECINDADES DE LAS CONSTELACIONES DEL CISNE Y CEFEO

Guillermina González y Graciela González

ESTRELLAS PECULIARES EN EL ECUADOR GALACTICO DEL SUR

Luis Münch

NOVAS HARO EN SAGITARIO Y ESCORPION

Domingo Taboada

ESTRELLAS DE ALTA LUMINOSIDAD EN LAS VECINDADES DE LAS CONSTELACIONES DEL CISNE Y CEFEO

Guillermina González y Graciela González

R E S U M E N

Catálogo de 71 estrellas clasificadas en placas de prisma objetivo como pertenecientes al grupo natural "OB". También se han incluido las estrellas situadas en las vecindades de este grupo.

Las estrellas de esta lista fueron encontradas en placas tomadas con la cámara Schmidt de 26-30 pulgadas y el prisma de 4° del Observatorio Astrofísico de Tonantzintla, la mayor parte de ellas, durante el año de 1948 y completadas en 1951, por Luis Rivera Terrazas y Luis Münch respectivamente y cubren una banda de 5° de ancho centrada a lo largo del Ecuador Galáctico de los 45 a 85 grados de longitud.

Debido a que con anterioridad a la presente publicación los Dres. Nassau y Morgan[1] encontraron en esta misma región 145 estrellas que clasificaron como "OB" hemos excluído de esta lista dichas estrellas.

Queremos hacer notar que no estamos absolutamente seguras de no haber incluído en esta publicación estrellas de espectros peculiares que dada la dispersión usada para la clasificación pueden presentar características de alta luminosidad. (Estrellas Shell).

En la Tabla 1, listamos las estrellas pertenecientes al grupo "OB" descrito por Morgan[2] y en la Tabla 2 las estrellas que deben estar colocadas en las vecindades de ese grupo, así como algunas estrellas que por estar en el límite de la magnitud alcanzada en nuestras placas solo pueden ser reconocidas con incertidumbre.

La identificación de las estrellas aquí listadas ha sido referida al catálogo Bonner Durchmusterung del cual hemos tomado las coordenadas ecuatoriales y las magnitudes dadas en las tablas 1 y 2. Para objetos no contenidos en dicho catálogo hemos usado las placas es-

Figure 6. First document published in the BOTT by Mexican women astronomers.

The BOTT had a strong presence of women with 5 as principal authors. The gender distribution of principal authors in the BOTT indicates that 31 articles of the 176 articles were written by a woman; that is a 17.6%. Women authors' distribution is shown in Fig. 7.

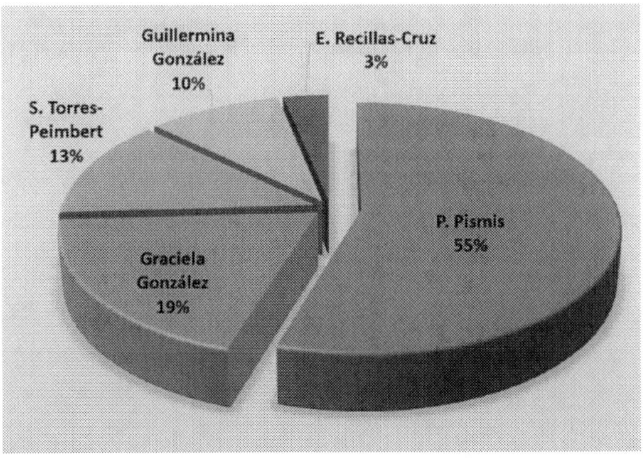

Figure 7. Distribution of the BOTT publications by women as first author.

4 Final Remarks

- An emerging relationship with Latin American and international astronomers was initiated by the BOTT; this was confirmed by the results of the study of the Revista Mexicana de Astronomía y Astrofísica 1976-2000 by Mata Acosta [1]
- The BOTT also published the first research articles on optics in México [3].
- A complete citation analysis is ongoing and a paper is being prepared for publication.
- The BOTT and Mexican astronomy was very open to female research, as indicated by the publication of works by Pismis, the González Sisters, Torres-Peimbert (current president of the IAU) and others. A study on the participation of women in astronomy in México at the middle of the previous century is underway.
- A study of the recent Mexican scientific production in astronomy and astrophysics is planned in order to search patterns of coauthorship and international collaborations.

Acknowledgments

This research has made use of NASA's Astrophysics Data System Bibliographic Services. We thank the *Instituto de Astronomía* at Ensenada, of the Universidad Nacional Autónoma de México for support to carry out this work.

References

[1] G.V. Mata Acosta, *Comportamientobibliométrico de la Revista Mexicana de Astronomía y Astrofísica 1976-2000.* (UNAM, BA Thesis, Mexico, 2001)

[2] S. Torres-Peimbert, O. López-Cruz, M. Peimbert *Historictical Note* in *Highlights of Boletín de los Observatorios de Tonantzintla y Tacubaya*, edited by S. Torres-Peimbert, O. López-Cruz (2011), Vol. 39 of Revista Mexicana de Astronomia y Astrofisica Conference Series, p. xi-xii

[3] A. Cornejo Rodrígue , *The First Years of Optics in Mexico and the role of the Boletín de los Observatorios de Tonantzintla y Tacubaya on its development* in *Highlights of Boletín de los Observatorios de Tonantzintla y Tacubaya*, edited by S. Torres-Peimbert, O. López-Cruz (2011), Vol. 39 of Revista Mexicana de Astronomia y Astrofisica Conference Series, p. 109–115

EPJ Web of Conferences **186**, 12009 (2018) https://doi.org/10.1051/epjconf/201818612009
LISA VIII

Categorisations of object types in SIMBAD

Anaïs Oberto[1,*], *Cécile* Loup[1], *Mark* Allen[1], *Caroline* Bot[1], *Laurent* Cambrésy[1], *Sébastien* Derrière[1], *Françoise* Genova[1], *Ada* Nebot[1], *Pierre* Ocvirk[1], *Arnaud* Siebert[1], and *Bernd* Vollmer[1]

[1] *Université de Strasbourg, CNRS, Observatoire astronomique de Strasbourg, UMR 7550, F-67000 Strasbourg, France*

Abstract. Astronomical objects may be classified into types in many ways, and the evolution of such categorisations changes with new discoveries and progress in astrophysical understanding. The SIMBAD database contains information on astronomical objects that have been studied in the published literature, including a field that specifies astronomical object types. As a record that is derived entirely from the literature, a given astronomical object in SIMBAD may have multiple object types, and the list of object types must be maintained and updated. The SIMBAD object type list currently contains some 200 types, that are organised into a hierarchy based on astrophysical concepts. The hierarchical structure also includes relations between object types, and this facilitates searches of SIMBAD to obtain lists of all of the astronomical objects in a given category independently of the publisher or the year of publication. We will explain the organisation of astronomical object types in SIMBAD and how they may be used in queries of the SIMBAD database, and visualised on all-sky maps.

1 Introduction

The SIMBAD database [1] contains a lot of information on astronomical objects. Among these the object type is of crucial importance. It can be just a wavelength (e.g. infrared or X-ray source), or give the basic nature of the object (e.g. star, galaxy, or molecular cloud), or give a much more detailed description on the nature, geometry, or evolutionary status of the object (e.g. Seyfert 1 or 2 galaxy, or Wolf-Rayet star). In practice, SIMBAD lists a main object type and a list of secondary types for each object. It is fundamental that the CDS team can give the most detailed object type as a main type to the users, which means that we have to rely on a robust algorithm to decide when the main type should be changed and when contradictory classifications should be checked.

Historically, the list of object types in SIMBAD was mostly phenomenological, and organized as a linear hierarchy. This list is still the one given to the users on the web site when clicking on 'Object types." It is also used by the cross-identification program COSIM (section 5). Objects with the highest priority for the main object type are at the bottom, while objects with the lowest priorities are at the top. Object types at the same level in the historical hierarchy are assumed to be incompatible, which lower the cross-identification score of COSIM.

This historical list will be replaced in the coming months. It has been reorganized as much as possible on astrophysical concepts, especially the evolutionary stage for the stars, taking advantage of

*e-mail: anais.oberto@astro.unistra.fr ORCID: 0000-0001-7141-1542

the wide expertise of the astronomers attached to the CDS team. For instance, the variable stars are not categorized separately anymore as in most cases they correspond to an evolutionary status or/and initial mass (e.g. a RR Lyrae is on the Horizontal Branch, and an Orion variable is an YSO). This new list is still hierarchical, though the hierarchy has changed, but the compatibilities between object types are much more complex than before. This should be of great help to better determine the "best" type and optimize cross-identifications with COSIM.

The hierarchical structure is made of relations, and can be used in SIMBAD in different sorts of use cases:

- Simple description of the astronomical object's classification
- Search in all types to obtain a list of objects that belongs to a category
- See the whole hierarchy and links between object types
- Visualise the localisation in the sky of a specific (or generic) type, and cross those information
- Compare two object types according to the levels and branches in the hierarchy tree

2 Web usage

Common usage of object types in SIMBAD is a text shown in the web page. For instance, the "Messier 1" object, also called "Crab Nebula", has the specific type "SuperNova Remnant" that is the expanding shell left by a stellar supernova explosion.

SIMBAD also gives more object types, that could be used to complete the main one. It helps to understand that this object was part of a group of observations within a general pipeline type (generally linked to catalogues).

3 SQL usage

The relations between object types can be structured in tables of the SIMBAD database. In order to do that, SQL is generated to create all the hierarchy. This can be used by the astronomers to query directly SIMBAD and filter objects on a category. For instance, the figure 1 shows the schema inside the database that could be used by anyone who wants to query SIMBAD.

Thanks to such a schema, other databases or services will be allowed to query SIMBAD programmatically and exploit at best the hierarchy of object types for solving complex queries.

4 Graphical visualisation

Documentation can be automatically generated from the stored information of this hierarchical organization, and can be displayed as a nice graph (thanks to the graphviz [1] open source tool) as you can see on figure 2.

In this kind of graph, the links between objects are more obvious, and on the left part you can see that we show different levels of granularity.

The generation is done automatically with a program reading information of links in the database and generating a file in "dot" format, that is used to generate this graph. So that, any modification in the SIMBAD hierarchy will automatically be reported in the graphical documentation.

 https://doi.org/10.1051/epjconf/201818612009

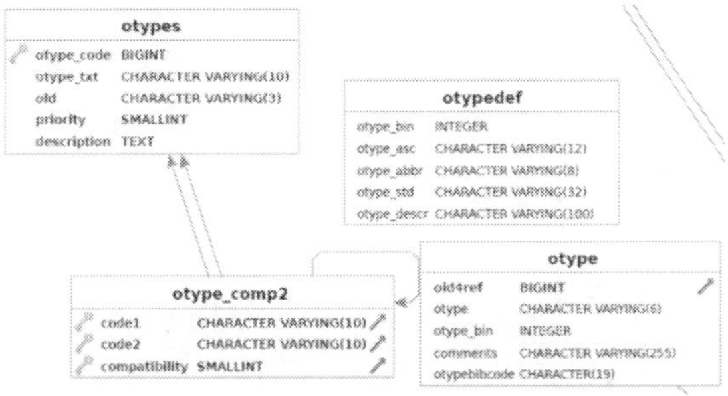

Figure 1. A table (here "otype_comp2") will manage the links of compatibility between 2 object types. Then, for instance, anyone would be able to search easily young stars using the hierarchy and thanks to a new specific function like "otype_childs" with this type of query : `SELECT * FROM basic WHERE otype IN otype_childs('YSO')`

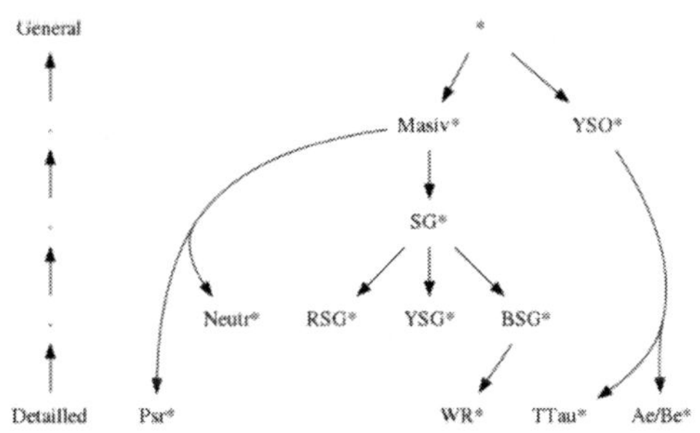

Figure 2. Graph automatically generated (partial hierarchy : a sub- tree under "Stars")

5 Cross-matching object types

Each SIMBAD object is a composition of all observations on many telescopes and satellites. That is why we could have secondary object types, depending on the points of view. In any case, before the insertion of astronomical objects in SIMBAD, we should compare these object types to check if there is no discrepancy according to the types hierarchy. The "COSIM" software (Comparisons of Objects for SIMbad [2] : internal cross-identification tool) is used to compare astronomical objects coming from a table, and compute a score of compatibility on each data, and specially on object type:

```
... Search around ICRS 20:16:31.91+37:39:09.1
... 2 object(s) found from coo
    1/1: PoC/*iC ( -10)  60.0"D ( 2.9)
+++ 2/2: **/*iC  (   1)   8.8"B (  0.0)
```

COSIM is going up and down in this hierarchy to quantify the relation between object types. We also have links between object types to know if they are compatible or not, outside the hierarchy. Both of these relations are used to compute a score for cross-identification.

In the previous example, the compatibility of a star in a cluster is much higher with a double star than the compatibility with a part of a cluster.

6 Full sky usage: SIMBAD object types in a map (MOC[3])

Once the information is well organised in the database, astronomers can use it to query directly SIMBAD and visualise the coverage of objects on one or more categories.

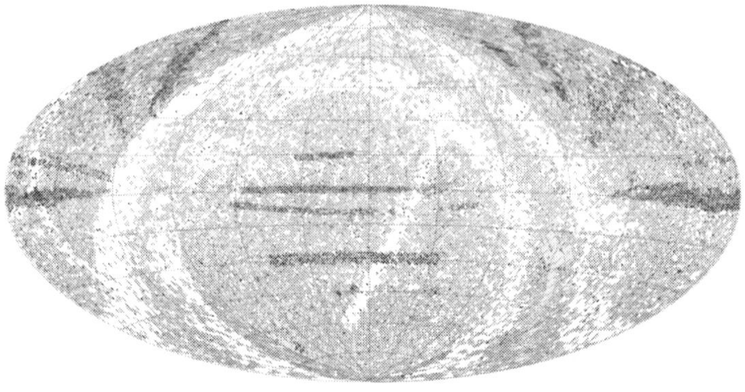

Figure 3. Visualisation of localisation of Infra-Red (orange) and Seyfert (blue) objects in SIMBAD

7 Conclusion

This work is still in progress, and should not be taken "as is." It opens a new perspective to improve in the long term the classification of objects and work has begun to share information with the astronomical ontology [4].

References

[1] M. Wenger, F. Ochsenbein, D. Egret, P. Dubois, F. Bonnarel, S. Borde, F. Genova, G. Jasniewicz, S. Laloë, S. Lesteven et al., Astronomy and Astrophysics Supplement **143**, 9 (2000), `astro-ph/0002110`

[2] C. Brunet, A. Oberto, M. Brouty, M. Buga, C. Loup, E. Perret, B. Vollmer, F. Woelfel, *COSIM: the necessary evolution of a cross-identification tool along with data evolution*, in *These Proceedings* (2018)

[3] P. Fernique, T. Boch, T. Donaldson, D. Durand, W. O'Mullane, M. Reinecke, M. Taylor, *MOC - HEALPix Multi-Order Coverage map Version 1.0*, IVOA Recommendation 02 June 2014 (2014), `1505.02937`

[4] S. Derriere, A. Preite-Martinez, A. Richard, L. Cambrésy, P. Padovani, *Ontology of Astronomical Object Types Version 1.3*, IVOA Note 03 March 2010 (2010)

EPJ Web of Conferences **186**, 12010 (2018) https://doi.org/10.1051/epjconf/201818612010
LISA VIII

New audiences for old books: Using astronomy archives in school outreach activities

Siân Prosser[1,*] and *Sheila* Kanani[1]

[1] *Royal Astronomical Society, Burlington House, Piccadilly, London W1J 0BQ, United Kingdom*

Abstract. The Royal Astronomical Society promotes the study of astronomy and geophysics and has recently committed to increasing public engagement with new audiences. The librarian, Sian Prosser has worked with the outreach, education and diversity officer, Sheila Kanani, to put the scientific archives of the Society at the heart of a primary school workshop telling the story of Caroline Herschel.

1 Introduction

The Royal Astronomical Society exists to advance, and to record the history of, our understanding of astronomy and geophysics. It does so through various activities, including maintaining a library and archive to preserve its collections and make them accessible to a widening audience, supporting research, education and outreach work. Since 1820 the Royal Astronomical Society has developed a unique collection of books and manuscripts created or donated by its members. The library aims to widen access to its distinctive collections, bringing in new audiences. In particular, the librarian has worked with the outreach officer to devise a school workshop that provides access to primary sources in the RAS archive. The workshop focuses on the life and work of Caroline Herschel, featuring science demonstrations, drama, art and a chance to read the astronomer's handwritten observations.

2 Caroline Herschel

Caroline Herschel's remarkable life and achievements made her an ideal focus for a primary school workshop. In 1785 she became the first woman in Britain to receive an annual pension from the King in recognition of her scientific work. King George III granted her the pension for assisting her brother William, which is the work for which she is most often credited, but she made significant contributions of her own to astronomy, discovering eight comets and many more nebulae. In 1828 she was awarded the Gold Medal of the Royal Astronomical Society for her 'unparalleled' cataloguing work, and she and Mary Somerville became the first women to be made honorary members of the Society in 1835. These were remarkable achievements for a woman who was born in Hanover in 1750 and brought up as her family's unpaid servant before William Herschel arranged for her to live with him in Bath, England, in 1772, where she initially trained as a musician before being drawn into full-time work as her brother's assistant. Caroline Herschel's observation notebooks are now part of the Royal Astronomical Society archive, and play an important role in the primary school workshop.

* e-mail: sp@ras.org.uk ORCID: 0000-0002-9610-8678

 https://doi.org/10.1051/epjconf/201818612010

Figure 1. Portrait of Caroline Herschel from 1847.

3 Workshop format

3.1 Science and Drama

Aimed at students aged 9-11, the visit falls into three parts, each lasting 30 minutes. The RAS building is not large, but it's possible to accommodate a whole year by running all three sessions simultaneously as a carousel of activities on each floor of the building. The outreach officer takes the students on a tour of the solar system, finishing with a dry-ice comet-making demonstration. The students have an audience with Caroline Herschel (played by an actor from Spectrum Drama) who narrates her journey from domestic drudge to musician and scientist. In the library the children look at drawings of William Herschel's telescope, and examine one of the speculum mirrors he made, before turning to Caroline Herschel's observation notebooks.

3.2 Caroline Herschel's observation notebooks

The image below shows Caroline's first sighting of what would turn out to be her third comet, now known as Comet C/1790 A1 [1]. The children find her handwriting fairly easy to read out loud, in spite of unfamiliar letters and words. In a group we interpret the text and the image, for example, why is the comet sketched in a circle, and what is the impact of weather on astronomical observations? We then read the rest of the notebook in which Caroline Herschel records the progress of the celestial object, firming up her hypothesis that it is indeed a comet. The children then sketch comets inspired by photographs of Halley's comet dating from 1910, learning more about cometary cycles in the process.

Figure 2. Tour of the solar system and comet-making demonstration.

Figure 3. An actor playing the role of Caroline Herschel.

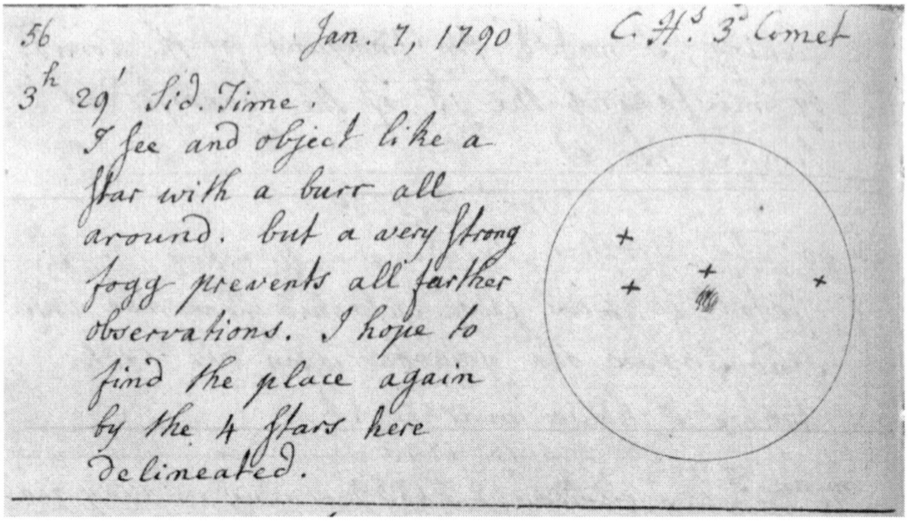
56 Jan. 7, 1790 C. H.s 3d Comet
3h 29' Sid. Time.
I see and object like a star with a burr all around. but a very strong fogg prevents all farther observations. I hope to find the place again by the 4 stars here delineated.

Figure 4. Caroline Herschel's discovery of Comet C/1790 A1 (Herschel) (ref: RAS MSS Herschel C.1/1.2).

4 Feedback and next steps

"The activities were hands-on and educational and varied in objective so the children were engaged and learning all the time," according to one teacher. Other feedback from teachers includes the suggestion to make the library activity more challenging by asking children to plot the course of Caroline Herschel's 3rd comet on paper. The librarian and the outreach officer also aim to develop similar workshops based on other astronomers whose fascinating observations, books and instruments can be found in the collections of the Royal Astronomical Society. So far over 250 children have taken part in the workshops, encouraging a passion for astronomy and geophysics using resources that students can't easily access in the classroom, and it is hoped that many more children will be able to take part in future. The RAS is grateful for support from ESERO UK.

Reference

[1] Caroline Herschel, Book of Observations. From July 7, 1788 to Augt 15, 1797. RAS MSS Herschel C.1/1.2.

EPJ Web of Conferences **186**, 12011 (2018) https://doi.org/10.1051/epjconf/201818612011
LISA VIII

A Model for Using DataCite DOIs in Observatory Bibliographies

Arnold Rots[1,*], *Raffaele* D'Abrusco[1], and *Sherry* Winkelman[1]

[1] *Harvard-Smithsonian Center for Astrophysics*

Abstract. The use of DOI persistent identifiers has become an attractive mechanism for citing datasets in articles. However, taking the users' interests into account requires careful consideration of the way in which we apply these identifiers.
Our objective is the application of DOIs in such a way that individual datasets are properly cited while presenting the citations to the reader in a user-friendly manner.
This is achieved by making judicious use of the metadata structure provided by DataCite.

1 Introduction

The Chandra Data Archive's bibliographic database has been using IVOA persistent identifiers (PID), issued under the authority of the ADS, to link its datasets to articles in the literature. This type of identifier was established in 2002 by the NASA astrophysics data centers and has served us well, as it had enabled us to provide complete literature-data linking at a fine-grained level, resulting in the most complete observatory bibliography currently in existence. However, at this point it is time to transition to international standard identifiers that have come into existence after we started developing our bibliography. The obvious choice is the DataCite DOI. However, it is not an entirely trivial conversion. Just like each publication has its own unique PID, it is essential that each dataset receive its unique PID. The problem that arises is the aggregation of PIDs for any particular article. Currently, the ADS performs the aggregation function for the IVOA-type identifiers, but that will not be feasible for the DataCite DOIs. Aggregation is necessary, since dataset PIDs need to refer to a landing page and separate landing pages for each dataset would be extremely user-unfriendly. The solution, adopted by some, to mint a single PID for each article that holds all the datasets is not acceptable, since it violates the requirement that each dataset have a unique PID. The solution is to mint two types of PIDs, one for articles, the other for datasets, that refer to each other using standard DataCite DOI metadata elements. A related initiative is the Scholix project[1], an initiative under the auspices of the Research Data Alliance (RDA) that has the potential to provide comprehensive data-literature linking services and should be closely monitored.

2 Requirements

In order to satisfy the interests of the dataset providers and the readers, we identify three simple requirements.

*e-mail: arnold@rots.net ORCID: 0000-0003-2377-2356

[1] http://www.scholix.org/home

2.1 Each Dataset Needs its own DOI

This is a logical requirement, akin to requiring that each article in a given journal volume have its own DOI, so that its citations and other statistics can be tracked. And it is (obviously) simple to implement.

2.2 A Single Landing Page per Datatset

It shall be possible to reach a single page that lists (and provides links to) all articles that cite a given dataset. This is simple to implement, too, by including the references to the articles' DOIs in the metadata of the dataset's DOI (see Requirement 1).

2.3 A Single Landing Page per Repository per Article

It shall be possible to reach a single page that lists (and provides links to) all datasets that are cited in a given article. This is where it gets tricky. If we only have the DOIs specified by Requirement 1, we need an aggregation service that has the capability to aggregate all datasets cited in an article, collect their links, and pass that list on to a landing page. Since this service cannot be provided by the DOI repository, we need an independent mechanism. The solution is found in minting a separate series of DOIs, one per article, that contains in its metadata all the links to a given repository's cited DOIs. Note that sole reliance on this type of DOI does not allow for quick retrieval of dataset-based statistics, as datasets would then lack a DOI-based identity.

3 Implementation

We propose an implementation model satisfying the requirements above through the combination of three types of DOIs, the first of which is the DOI assigned to the article by its publisher.

3.1 Article

We assume each article has been issued a CrossRef DOI (or a BibCode; they should be interchangeable) by its publisher.

3.2 Article-based List of Cited Datasets

This is a DOI that contains references to the article and to each of the datasets cited in the article, allowing a single landing page to be provided for all cited datasets from a single repository. It provides necessary information for constructing an article-based landing page of cited datasets through:

- A single metadata element `IsCitedBy` points to the publisher's DOI
- One or more metadata elements `HasPart` point to individual datasets DOIs.

3.3 Dataset

This is a DOI that contains references to the article and to the list of datasets cited in the article, allowing a direct way of collecting linkage and statistics by dataset. It provides necessary information for constructing a dataset-based landing page of articles citing the dataset through:

- One or more metadata elements `IsPartOf` that point to the article based DOIs defined above that reference this dataset; these metadata elements are the reciprocal elements of the `HasPart` metadata elements in those DOIs.
- One or more metadata elements `IsCitedBy` that point to the publisher's DOIs corresponding to the ones cited in the article-based DOIs it is part of.

4 User Interface

The user/reader can view the citation information in two ways:

4.1 From the Article

The reader is directed from the article to a landing page providing access to all datasets from a specific repository, based on the `HasPart` metadata associated with the DOI specified in Section 3.2.

4.2 From the Dataset

From the dataset in the repository the user is directed to a landing page providing access to all articles that cite that particular dataset, based on the `IsPartOf` metadata associated with the dataset's DOI as specified in Section 3.3 and the `IsCitedBy` metadata in the article-based DOI, or just by the `IsCitedBy` metadata element in the dataset's DOI.

4.3 Observatory Bibliometrics

The observatory can gather its publication statistics from the dataset-based DOIs' metadata, although one would expect that information also to be available in the observatory's databases.

This work has been supported by NASA under contract NAS 8-03060 to the Smithsonian Astrophysical Observatory for operation of the Chandra X-ray Center.

Conclusion time: A. Accomazzi, U. Grothkopf, A. Holl and J. Steffen (Photo: M. Wenger)

EPJ Web of Conferences **186**, 12012 (2018) https://doi.org/10.1051/epjconf/201818612012
LISA VIII

Measuring Research Impact of Astronomers/Astrophysicists by using Astrophysics Data System Beta: A Powerful New Interface: A case study with Special Reference to Prof. Jayant V. Narlikar

Hemant Kumar Sahu[1,*] and *Surya Nath* Singh[2]

[1] *Inter-University Centre for Astronomy and Astrophysics, Ganeshkhind, Pune – 411007, India*
[2] *Head/Group leader, Information Centre & Library, National Institute of Virology (ICMR), Pune – 411001, India*

Abstract. This paper highlights qualitatively and quantitatively research and presents a valuable overview of new citation enhanced databases in the context of research evaluation for the productivity of Prof. Jayant V. Narlikar. He has total 472 research publications in the fields of Astronomy and Astrophysics (AA) published from 1961-2015. In addition, this paper provides an overview of the citation-enhanced databases viz Astrophysics Data System Beta: A powerful new interface for performing citation analysis. Previously, scientometrics had been used to measure the publication productivity of Prof. Jayant V. Narlikar using the Astrophysics Data System (ADS). The scope of this paper is limited to Astrophysics Data System Beta. The result indicates that most of his papers are published in peer reviewed journals having the highest Impact Factor. The average number of publications per year is 8.74 and with the maximum papers published during 1981-1990 and 1996-2000. The total number of citations for his publications is 3516 covering 665 of his papers published during 1961-1970. He had many research collaborations, specifically with Prof. F. Hoyle (87 papers), Prof. G. Burbidge (38 papers), Prof. N. Wickramasinghe (22 papers) and Prof. T. Padmanabhan (21 papers).

1 Introduction

The great Indian Padmashree Prof. Jayant V. Narlikar is a well-known Astrophysicist and a proponent of Steady State Cosmology. He has developed the Hoyle-Narlikar theory of gravity with Sir F. Holye, which is a machian theory of gravity that fits into the quasi steady state model of the universe (Wikipedia, 2016). Recently he broke new grounds in space research. His research on space object is considered noteworthy. In addition to scientific publications, Prof. J.V. Narlikar is a writer of popular science articles, science fictions, short stories in English, Hindi and Marathi and also well known for communication of science through radio, and TV programs. He has been honored with many national and international awards. In this age of competition, quality and quantity have value. To value the publications, it is essential to measure the publication productivity of particular scientists. Scientometrics analysis is a useful method to collect information as well as an opportunity to access collections and services. Knowledge and survival are two sides of a coin in teaching and research.

*e-mail: hksahu@iucaa.in ORCID: 0000-0002-5588-8604

Hence, the analysis and evolution of research is essential to find out its importance. There are various methods to measure the research in the form of articles, books, conference proceedings, and personal communications among the researchers. These are popularly known as Bibliometrics, Scientometrics Webometrics (also Cybermetrics), Informetrics, Productivity patterns, etc. It aims to estimate the varying contributions of scholarly work to the advancement of knowledge. The number of citations received by a publication is seen as a quantitative measure of the resonance and impact that this publication has produced in the scientific community [1]. The listings of references in publications are a convention among scientists for giving credit or recognition to the value of previous work [2]. There are commercial and free citation databases available via Web of Science, Scopus, Google Scholar and any subject-specific databases as such ADS: a Digital Library portal for researchers in astronomy and physics operated by the Smithsonian Astrophysical Observatory (SAO) under a NASA grant, now which also provide scientometrics of scientists. Web of Science is an online subscription-based scientific citation indexing service maintained by Thomson Reuters that provides a comprehensive citation search while Scopus is the largest abstracting and citation analysis database of peer-reviewed literature: scientific journals, books and conference proceedings. These citation databases are enriched with advanced features to enable detailed citation analysis along with h-index value. There is free citation databases such Google Scholar Citations, which provide a simple way for authors to keep track of citations to their articles and can check, who is citing his/her publications, graph citations over time, and compute several citation metrics and ADS is an online database of over twelve million astronomy and physics papers from both peer reviewed and non-peer reviewed sources. Astrophysics Data System Beta: A powerful new interface, which has chosen to explore out of the above said four databases being maximum used by IUCAA library users in their research, education requirement and citation analysis [3].

1.1 Need of Study

Publications are the main source of scientific communication. Research productivity in higher education and in research activities is gaining momentum and importance in the last ten years in India. It is the measurement of publication productivity that provides value to the research and denotes trends in that particular field. Citation analysis is used as an indicator of the impact valuing the quality and qualitative research. The work carried out by the Indian Scientists in the field of AA has international importance. Hence it is proper to measure/evaluate their publications to give the value at an international level. Researchers are always hungry for their new findings to give a new direction and contribute to their specific subject area, in forms of articles, patents, conference proceedings, books, chapter of books, etc. The result of such research has always been a source of further experiments and incalculable benefit to the society. Keeping this idea, the authors have made efforts to present this study with special reference to AA as a case study of Professor Jayant V. Narlikar, a globally well-known AA scientist, using of ADS beta version. There is no scientific research related to a study of the AA field in India. Hence, the present study is to find out some solutions for Library and Information Science (LISc) professionals as well as users. The major research questions are: Are we ready to explore citation analysis of scientists as individual and as group/collaborators via citation databases such as web of Science, SCOPUS, Google Scholar and ADS, etc.?, Are there problems by organizing/managing e-resources?, Are users of IUCAA aware of the capabilities (attitude, skill and knowledge) of accessing above citation databases?

1.2 Objectives

(i) To measure the publication productivity patterns of Prof. Jayant V. Narlikar in AA by using ADS beta: a powerful interface. (ii) To study and identify the contribution of Prof. Jayant V. Narlikar in the field of AA by measuring year wise, search result of refereed and non-refereed papers, citations, including self-citations, read and downloaded papers, author network, research collaborations, etc. (iii) To find out factors to motivate other scientists to write and publish their findings. (iv) To find out the authorship pattern of Prof. J. V. Narlikar. (v) To find out core collaborators of Prof. J.V. Narlikar in his research career.

1.3 Scope and Limitations

The IUCAA is a pioneer international institute for its teaching and research in the field of AA which includes main subjects: Astronomy, Astrophysics, Mathematics, Computer Science, Physics and its allied branches with more than 300 users comprising of senior faculty members, scientists, visiting associate, research scholars and non-teaching technical and supporting staff/trainees as well casual visitors and to facilitate them. Researchers have access to citation databases such as Web of Science and ADS database (free). Since it is a case study, scope and limitations are restricted to IUCAA and scientific publications of Prof. Jayant V Narlikar and scientometrics analysis as well as his aforesaid publications published during the period 1961-2015 [4].

2 Review of Literature

For the present study, many references were used such the ADS database and its related literature to explore citation analysis and get details of published papers/citations for the year. This is directly related to the present study covering AA subjects and its specialized users. The primary as well as the secondary sources of literature has been consulted. A number of studies have been conducted and no comprehensive and substantial studies have been carried out as a case study to understand and get citation analysis of an author via the ADS database. However, some study has been done by Christoph and Hans-Dietr, 2008, Stefanie, et al, 2014 and Helsinki University Main Library, 2016 but not directly related to a case study [5-9].

3 Research Methods

Scientific publications are the best source of scientific communications, while scientometrics is the best method to measure the contribution of individual scientists and/or growth of any discipline. To explore the details of any particular author's publications (Prof. Jayant V Narlikar) and citation patterns, the authors have used the scientometrics/citation analysis to evaluate and measure the research impact of astronomers/astrophysicist using an ADS Beta database. To explore the analysis, a case study of Prof. Jayant V. Narlikar has been taken and explored. For the purpose of the study, data was used from ADS database. A total of 472 publications have been analyzed. The bibliographic data of work carried out by Prof. J.V. Narlikar for the study were accessed from ADS. MS Office is used to organize tables and figures and analyses the data[4, 7, 9].

4 Result, Discussion/Analysis and Major Findings

4.1 Author-wise Productivity and Citations

Prof. J. V. Narlikar has published 472 papers during 1961-2015 (Table 1-2).The information about a data set was searched through ADS, for the papers published by Prof. Jayant V. Narlikar for the period of 54 years (From 1961 to 2015). The database was searched with necessary refinements to get information on published papers and citations in the ADS database which contains a maximum numbers of astronomy and astrophysics scholarly publications. ADS contain 472 papers published by Prof. Narlikar in scholarly international journals, books and conference proceedings with total numbers of 3516 citations [3].

SN	Five years duration	Numbers of papers	Citations
1	1961-1965	15	665
2	1966-1970	22	348
3	1971-1975	39	419
4	1976-1980	49	281
5	1981-1990	88	398
6	1986-1990	55	186
7	1991-1995	50	575
8	1996-2000	79	362
9	2001-2005	36	222
10	2006-2010	30	55
11	2011-2015	9	5
	Total	**472**	**3516**

Table-1 : Year wise (in Five Years Duration) Papers Published & Citations Received (ADS, 2016)

Name of Co-Authors	Nos of Publications
Hoyle, F.	87
Burbidge, G.	38
Wickramasinghe, N.	23
Padmanabhan, T.	21
Chitre, S.	17
Vishveshwara, C.	14
Kembhavi, A.	11
Packer, J.	11
Arp, H.	8
Vishwakarma, R.	8

Table-2 : Status of Research Collaborations (ADS, 2016)

Table-1 summarizes that Prof. Narlikar has published a total 472 papers, during 1961-2015 and an average of 8.7 papers per year with a total of 3516 citations with maximum papers published by him during the period 1981-1990. Also, a maximum numbers of citations (i.e. 665) for papers published during 1961 to 1965 for these papers. It can be concluded that period between 1981 - 1990 was the peak period for Prof. Narlikar from all points of view and maximum impact. Hence there is continued growth of publications. His total publications in the form of research output through different channels, includes journal articles (64.40%), books (07.62%), book chapters (03.38%) and in the conference proceedings/reprints (14.40%). His most preferred journals where Nature (35 papers), Bulletin of Astronomical Society of India (27 papers), Monthly Notices of the Royal Astronomical Society (21 papers) and so on.

After performing searches in the ADS modern form by Author surname "narlikar", the result is shown in table 1 that his total listed number of papers is 472 with the name of his research collaborators year wise during period 1961 – 2015. The maximum was during 1981-1985, while the minimum was during 2011-2015. It can be summarized that between 1961-1965 was beginning of his career while 2011-2015 has minimum research results. It also shows that there is a trend of growth during 1961 -1995, and it was increasing, while 1985 – 1990 and 2011-2015 was irregular rather than decreasing. Table 4 points out that Professor Jayant V. Narlkar has maximum research collaborations with Professor F. Hoyle (87 papers), Professor G. Burbidge (38 papers), Prof. N. Wickramasinghe (22 papers) and Prof. T. Padmanabhan (21 papers) (Table-2).

4.2 Citation Metrics of Research papers by Prof. Jayant V. Narlikar

Table – 1 & 4 shows that the author has total number of citations at 3516 as on September 15, 2016, a total number of self-citations at 641, which is increasing day by day. It's also shown that the author

https://doi.org/10.1051/epjconf/201818612012

has a total number of refereed citations - 3011 out of 3516. It can be concluded that the citations by others are much more than the self-citations.

Index	Total	Referred
h-index	33	30
m-index	0.6	0.5
g-index	50	45
i10-index	89	82
i100-index	4	3
tori-index	94.5	70.5
riq-index	176	152
Read10-index	73.3	52.7

Table- 3: Year-wis Status of Numbers of Indices of Author (ADS, 2016)

Publications	Total	Referred
Numbers of citing papers	2107	1724
Total citations	3616	2965
Numbers of self-citations	641	597
Average citations	7.4	12.1
Median citations	0	4
Normalized citations	1849.2	1462.6
Referred citations	3011	2616
Average refereed citations	6.4	10.7
Median refereed citations	0	3
Normalized refereed citations	1567.4	1293.1

Table-4: Status of Numbers of Citations, Self- citations, etc and its Count Year-wise (ADS, 2016)

The aim of the above indices (Table-3) is to capture productivity as well as impact by counting how many of an author's papers have been cited, and by how many times; to have a h-index of 5, five of a scholar's publications must have been cited by others at least five times each. While more sophisticated than plain citation counts, the h-index shares the limitations of in-comparability across fields and across career stages (Mcinearey Lab, 2016). It has been observed that there are a great variety of paper downloaded, cited and reads. The table 6 shows that the author hash -index count of 33, meaning his papers has been cited 33 times by others at least 33 times each.. It's also shows (table 8) the status of different indices such as m-index 0.6, g-index 50, i10-index 89, i100 index 4, tori index 176, risk index 176 and read 10-index is 67.1. It has been observed that via ADS Beta version database one can explore details of any particular author research impact such as its total citations, number of self-citations, average citations, median citations, normalized citations, refereed citations, average refereed citations, median refereed citations, normalized referred citations year wise, status of his research impact in types of indices such h-index, m-index, etc. year wise, status of numbers of reads, number of download with its count year wise, Status of groups of papers based on shared reference, status of frequently appearing and unique words in search result, status of result of read counts of author papers year wise. Prof. Narlikar, with his collaborators have published 472 papers during 1961-2015 in various areas of astronomy and astrophysics and allied subjects. Prof. Narlikar, with his collaborators have communicated their research findings and published their papers in mainly international journals of AA, which are indexed regularly in the ADS database and have a high impact factor. Prof. Narlikar has h-index count 33 that means his papers have been cited 33 times by others at least 33 times each.

5 Conclusions

It can be concluded that citation analysis of Prof. Narlikar has proved to be an important assessment tool for his research evaluation publication productivity, specifically the study of 54 years, during which time he had published 472 papers/8.7 per year. There are a number of co-authors with whom he did a major research collaboration during his scientific career. Many commercial and free citation databases are available such Web of Science, SCOPUS, Google Scholar, etc., But the ADS is free and powerful research tool and has had a significant impact on the efficiency of astronomical research since it was launched in 1992 as well as covering many old astronomical journals, proceedings, conferences. The use of ADS is almost universal among astronomers worldwide, and therefore ADS usage statistics as well as way to explore of citation analysis in unique, compare two available citation databases. The study suggests that citation-enhanced databases need to be examined carefully, with regard to both their potentialities and their limitations for citation analysis. Hence, on the above pattern, it is suggested to measure the research impact of other eminent scientists working on AA

References

[1] C. Neuhaus and D. Hans-Dieter, *Data source for performing citation analysis: an overview* in *Journal of Documentation* **64-2**, 193-210 (2008)

[2] R.K. Merton, *The Matthew Effect in Science, II: Cumulative Advantage and the Symbolism of Intellectual Property*. ISIS. **79**, 606-623 (1998). http://www.garfield.library.upenn.edu/merton/matthewii.pdf

[3] Astrophysical Data System. (2016). http://ads.harvard.edu

[4] Inter-University Centre for Astronomy and Astrophysics (IUCAA). 2016). http://www.iucaa.in

[5] Helsinki University Main Library. 2016. Measuring research impact: a citation analysis tools. http://libraryguides.helsinki.fi/c.php?g=212738&p=1403603

[6] Mach's principle. (2016). https://en.wikipedia.org/wiki/Mach%27s_principle

[7] L. Meho, K. Yang, *A new era in citation and bibliometric analysis: web of science, scopus and google scholar.* (2006) https://arxiv.org/abs/cs/0612132v1

[8] *Steady State Theoty.* (2016). https://en.wikipedia.org/wiki/Steady_State_theoty

[9] S. Haustein, T.D. Bowman, K. Holmberg et al., *Astrophysicists on Twitter: An in-depth analysis of tweeting and scientific publication behavior* in *Aslib Journal of Information Management* **66-3**, 279-296 (2014). https://doi.org/10.1108/AJIM-09-2013-0081

[10] Mcinearey Lab. (2016). http://mcinerneylab.com/research/h-index-m-index-and-google-citations

The CoBiS Linked Open Data Project and Portal

Luisa Schiavone[1,⋆], *Federico* Morando[2], and The CoBis Communication Working Group[3]

[1] *INAF Osservatorio astrofisico di Torino, Via Osservatorio 20, 10025 Pino Torinese, Italy*
[2] *Synapta Srl, Via S. Quintino 31, 10121 Torino, Italy*
[3] *www.cobis.to.it*

Abstract. The CoBiS is a network formed by 65 libraries. The project is a pilot for Piedmont that is aiming to provide the Committee with an infrastructure for LOD publishing, thus creating a triplification pipeline designed to be easy to automate and replicate. This is being realized with open source technologies, such as the RML mapping language or the JARQL tool that uses Linked Data to describe the conversion of XML, JSON or tabular data into RDF. The first challenge consisted in making possible the dialog of heterogeneous data sources, coming from four different library software (Clavis, Erasmo, SBNWeb and BIBLIOWin 5.0web) and different types of data (bibliographic, multimedia, and archival). The information contained in the catalogs is progressively interlinked with external data sources, such as Wikidata, VIAF, LoC and BNF authority files, Wikipedia and the Dizionario Biografico degli Italiani. Partners of the CoBiS LOD Project are: National Institute for Astrophysics (INAF), Turin Academy of Sciences, Olivetti Historical Archives Association, Alpine Club National Library, Deputazione Subalpina di Storia Patria, National Institute for Metrological Research (INRIM). The technical realization of the project is entrusted to Synapta, and it is partially sponsored by Piedmont Region.

1 Introduction

The **CoBiS** ("Coordinamento delle Biblioteche Speciali e Specialistiche di Torino" i.e. Coordination of Special and Specialized Libraries of Turin) is an informal network of 65 libraries, collaborating to provide continuing professional development and to offer a better service to their users.

CoBiS libraries are heterogeneous from many points of view: holdings, cataloguing softwares and OPACs. The **LOD project** started in 2015 as a training program, in collaboration with Prof. Vivarelli from the University of Turin. The program was divided in various topics: copyright, collaboration between libraries and Wikipedia, and Open Data.

2 Purpose

Turin Astrophysical Observatory Library and five other libraries from the CoBiS decided to participate in a **pilot project** with the purpose **to provide a unique access point to the collections of CoBiS libraries.**

⋆e-mail: luisa.schiavone@inaf.it ORCID: 0000-0003-2929-3727

CoBiS bibliographic data are not only becoming more interoperable among them, but the data are also opened in order to facilitate the collaboration with online communities and to connect them to the Linked Open Data Cloud.

During the project, **Synapta** made some improvements to the RML Processor to speed-up (by about two orders of magnitude) the triplication process.

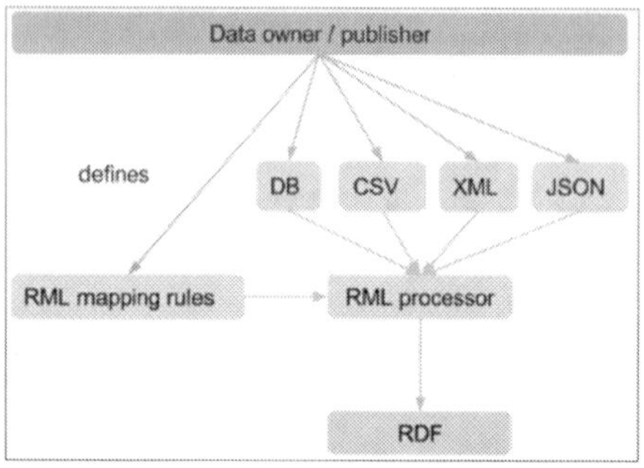

In order to further optimize the process, Synapta – in a joint effort with other open source developers – also realized and published JARQL (http://jarql.linked.solutions/), a new open source software tool for converting JSON data to RDF, using SPARQL 1.1 syntax, and using construct queries (see the project website or GitHub page for more details).

3 The Project

We used two main **ontologies**: Bibframe and Schema.org. We also used selected properties from RDFS, OWL, DCTerms, FAOF, and Culturalis (http://culturalis.org/ – used to describe CoBiS libraries as cultural heritage institutions).

With respect to the **Linked Data stack**, this is an overall picture of the **project**

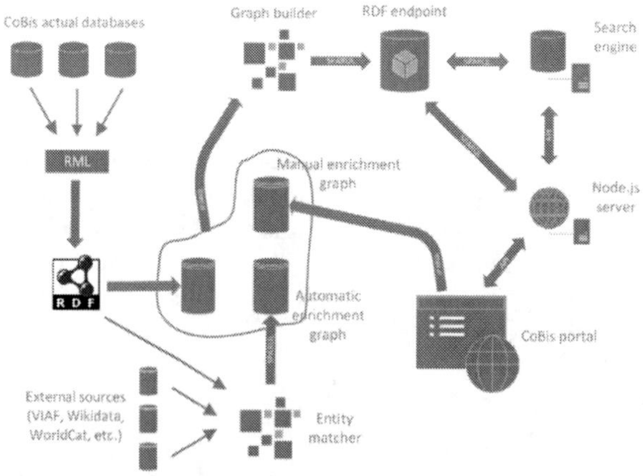

We started from heterogeneous sources, with the aim of combining them into RDF data. To reach this purpose, we used both **RML** (in particular in a first phase of the project) and our newly developed

JARQL library. Using a SPARQL query, we describe the transformation, obtaining an RDF file as result.

During the graph building, we also enriched data with external sources. We used both automatic algorithms and manual approaches.

For the latter, we exploited **OLAF (Open Linked Authority File), our crowd-sourcing interface for creating an authority file**. For instance, we want to match the correct Wikidata entity to a person named "Galilei", who wrote the books mentioned on the right side. Calling on the Wikidata SPARQL API, we obtain different entities that you can choose to identify the exact matching. In this case Galileo Galilei.

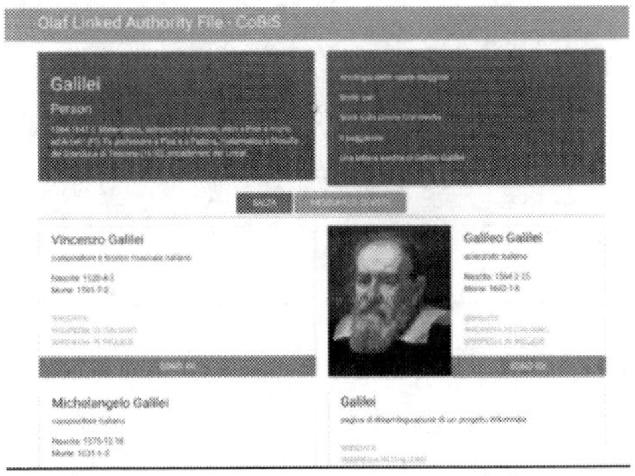

You may try OLAF live at https://olaf.synapta.io/ selecting "CONNETTI" near "Autori dei libri del CoBiS" and register, in order to annotate the suggested triples with your identity. All the interlinks, discovered via OLAF and the other enriching techniques, allow the portal to access external resources: VIAF, Wikidata, the Italian National Catalogue SBN, Enciclopedia Treccani.

4 The Portal

Finally, as result, we have exposed data through a **SPARQL endpoint** that feeds the CoBiS portal http://dati.cobis.to.it (will be online by Fall 2017).

The CoBiS LOD Project portal is online with its full Linked Data stack, including a public SPARQL end-point (configured to support federated queries), a full dump of the RDF data, and LodView to dereference URIs, etc.

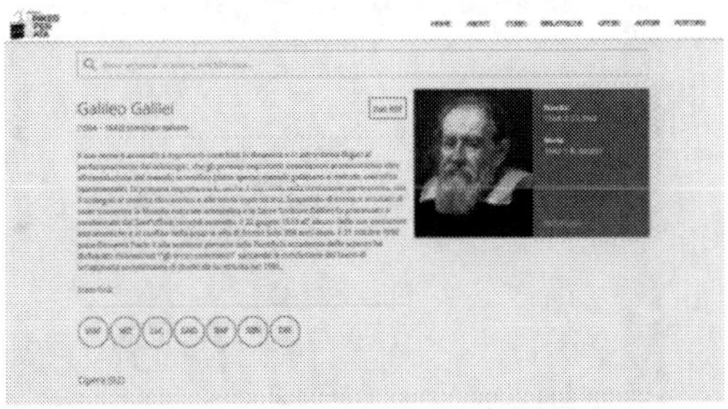

In the **author's page - dynamically generated through SPARQL queries** – you have biographical information and a list of interlinked resources coming from Wikidata and other bibliographic repositories (VIAF, Wikidata, LoC, Deutsche National Bibliothek GND, Bibliothèque Nationale de France BNF, Servizio Bibliotecario Nazionale SBN, Dizionario Biografico degli Italiani DBI).

On the right, an infobox with the author's data and an image, both coming from Wikipedia, thanks to the leveraging of Linked Data. Clicking on the **RDF button**, all the triples of the resource can be directly fetched.

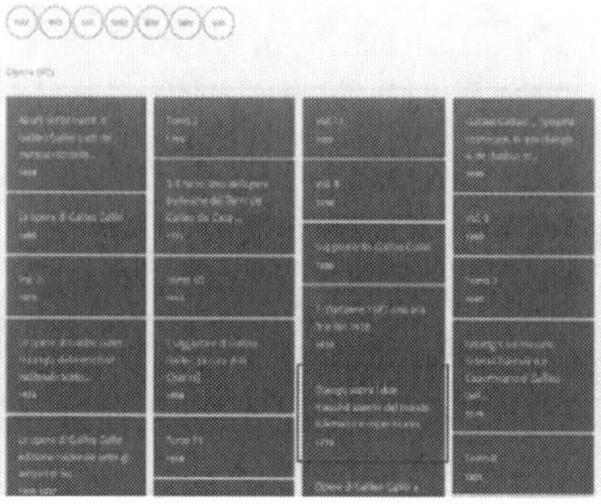

At the bottom of the page, all the author's **books inside the CoBiS database** are shown.

To explore information on such books, you can click one of the boxes or you can use the search bar and look for a title which is not listed.

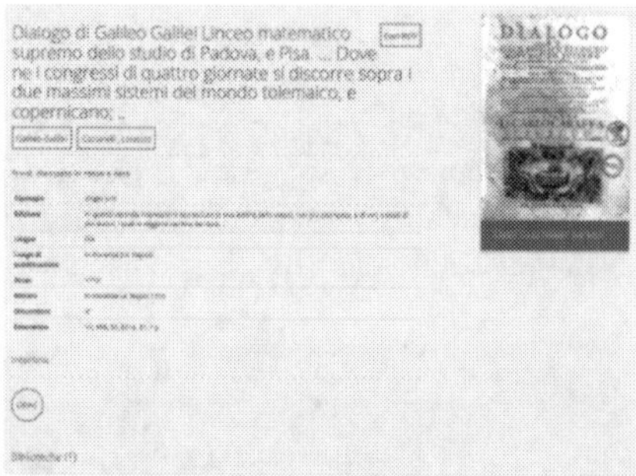

The image shows an **example search for the Dialogo**. On the left side of the page, you see bibliographic details with a collection of interlinked resources.

Exploiting the power of Linked Data, we are also able to read the **Internet Archive digital copy of the book**. A physical copy of the book is available in some CoBiS libraries. All details can be shown by clicking the **OPAC button.**

Acknowledgements

1. Regione Piemonte for the financial support
2. Synapta for the technical realization
3. Nexa Center for Internet & Society at Politecnico di Torino for the collaboration
4. All the participating Institutes: National Institute for Astrophysics (INAF), Turin Academy of Sciences, Olivetti Historical Archives Association, Alpine Club National Library, Deputazione Subalpina di Storia Patria, National Institute for Metrological Research (INRIM).

F. Woelfel and C. Halter (Photo: M. Brouty)

EPJ Web of Conferences **186**, 12014 (2018) https://doi.org/10.1051/epjconf/201818612014
LISA VIII

Cooperative Community Opportunities and the ADS

Donna Thompson[1,*] and *Carolyn* Grant[1]

[1] *Harvard-Smithsonian Center for Astrophysics*

Abstract. The ADS is used daily by the astronomy community and benefits from the cooperative nature of our contributors. In this poster we will outline some of the ways that the library community can continue to contribute to this resource. As online publishing trends continue to evolve, conference proceedings and presentations, PhD theses and other digital artifacts are being deposited to research data repositories such as Zenodo or Dataverse. We will describe the process that ADS follows to include relevant collections from such repositories when properly curated by librarians or researchers. In addition, while most of what we receive is vetted before it gets to us, we sometimes find ourselves in the position of having to determine whether or not an article, proceeding or other submission is refereed. ADS has guidelines on how we determine this but sometimes the lines are not so clear and sometimes the arguments on both sides are compelling. The process that we follow will be outlined and we would appreciate feedback on this topic. Finally, given the expertise of librarians in their field and within their organization, the ADS encourages librarians to consider submitting curated institutional bibliographies for inclusion in the ADS.

1 Introduction

The ADS seeks to provide comprehensive, but not all-inclusive, coverage of the fields of astronomy and astrophysics. To this end we welcome contributions to our collection that we may have missed and value added curation from the library community. In this paper we will describe some of the issues that we routinely face and the content which we would welcome.

2 Missing content

We receive regular feeds from both major and minor publishers, scientific societies, and other data providers such as CrossRef. We regularly monitor lists of conferences and check for their publication online or in print so that we can add this information to the ADS corpus. We encourage the ADS user community, particularly librarians, to contribute to our collection by communicating with us when gaps in our coverage are detected. One part of the collection where we know we are missing entries is the PhD theses. As many universities are no longer depositing theses into data depositories and opting instead to add them to their institutional archive, ADS no longer has an automated source for this valuable collection. We encourage submissions from both US and international institutions. We are happy

*e-mail: dthompson@cfa.harvard.edu ORCID: 0000-0001-6870-2365

to work with individuals to make this process streamlined on both sides. We download records from a select list of Zenodo collections on a weekly basis and welcome your suggestions for communities and topics to add to the content harvested by the ADS. These collections are typically contents of a conference proceeding which have been vetted by the organizer or other affiliated participant. Because Zenodo uses the Open Archive Initiative protocol (https://www.openarchives.org/pmh/), this allows additional records to be added after the initial batch has been submitted, which enables us to automate improving coverage of these proceedings. Additionally, we harvest the collection of astronomy PhD theses from Zenodo. Dataverse is a world-wide project designed to allow researchers to deposit data into open source archives on their own computer or computers at a publisher or institution. Dataverse links referenced in the astronomy literature can be included in the ADS by sending us the appropriate linking data.

3 What we include

We frequently receive requests from journal editors or users to add a certain journal title to the ADS. We verify that the publication is appropriate for inclusion by examining the journal for its relevance to ADS, including relevance to the scientific study of astronomy; content in any of the subfields of astronomy or physics; non-astronomy content of interest to astronomers in different subfields; and content in related fields that has astronomical relevance.

In addition, we expect journals to have:

- A quality and scope of interest to an international audience
- An ISSN
- A CrossRef registration
- Standard conventions for scholarly journals
- An informative title and descriptive article titles
- Full contact information for authors and editors
- Usable metadata, including abstracts, titles, and key words
- References that are current, from quality journals, and of an appropriate number for the article in a standard format
- All or mostly all original articles
- Articles that are written by professionals in their field.
- Publication on a regular schedule appropriate for the established frequency of the journal

Electronic-only journals must:

- Provide evidence of archiving arrangements, online or otherwise
- Include a posting date for articles and revisions
- Have published at least three issues

Features that support acceptance:

- Sponsorship by an astronomy or physics association or other society
- Astronomers or physicists among editorial board members and authors
- Indexed in other scholarly databases
- Online access

- English table of contents

For e-journals:

- Presence of version number
- Indication of where previous versions can be obtained (Links)

For a full list of our evaluation criteria see the FAQ section linked from the ADS search page.

We have created these metrics for evaluation in an effort to weed out journals which are predatory by nature. Journals which will publish any article for a fee, or reprint existing articles for additional exposure compromise the integrity of the data content in the ADS.

4 Refereed or not

Occasionally we receive an email from a user stating that his or her paper is refereed and why is it marked as non-refereed in the ADS? ADS considers articles refereed when they appear in journals that participate in peer review. Peer review is the process of having articles reviewed by experts in the field before publication. ADS staff check journal websites and author instructions to verify the peer review status. Sometimes periodical directories are consulted, such as Ulrich's International Periodicals Directory.

Non-refereed materials, such as conference proceedings, circulars and bulletin entries, are screened only by an editor and not sent out for peer review. The ADS does not consider materials that are verified only by an editor as refereed.

We actively collaborate with other information providers to improve the rubric for determining what qualifies as "refereed". We closely follow industry standards as they evolve through projects such as CrossMark (https://www.crossref.org/services/crossmark/), which can allow specification of metadata about the conference such as acceptance rate of papers and type of contributions which can help in such decisions.

5 Institutional bibliographies

ADS is used by many librarians to curate their institutional bibliographies. The API available through ADS-beta version can help to streamline the process (see the poster Managing Institutional Bibliographies using the ADS API: A new workflow using Google Sheets? from this conference for an example.) ADS can incorporate curated bibliographies into our holdings and add a filter for individual institutions. This will make it easier for ADS users to correctly identify researchers at each institution. By providing the ADS with an institutional bibliography we can create a "bibgroup" for that institution which will be entered into our list of filters and can be used easily by users to filter papers. This is one way to make author disambiguation easier.

Another effort to streamline author disambiguation is allowing users to search by affiliation, which has been implemented in its basic form in ADS-beta. Improvements to the basic affiliation search are underway, including normalized organization names and the ability to filter on the results.

ADS has partnered with ORCID to allow users to claim papers and associate the papers with an author's ORCID through the ADS. Users have the capability to claim papers in the ADS, have those claims uploaded to ORCID, and to search by ORCID id in ADS-beta. With the proper permission settings, users can grant ADS permission to track claims and enable other users to find one's articles more easily. More details about this can be found at http://adsabs.github.io/help/orcid/claiming-papers.

ADS is planning to integrate ORCID mappings curated by librarians into its database to complement data provided by publishers and claimed by users. A trial using data from the Harvard-Smithsonian Center for Astrophysics is planned for later this year.

6 Conclusion

The ADS is used by members of the astronomy community and the general public on a daily basis. We are continuously updating the data and metadata and will gladly accept verified input from community members. If you have or know of any relevant conference proceedings, journals or PhD theses to add to our collection, please contact us to see how we can incorporate the data. We can be reached by email at adscfa.harvard.edu or through our Feedback forms available from any search page on ads.harvard.edu.

Diane Bouquin (Photo: M. Wenger)

AUTHOR INDEX

AUTHOR INDEX